T0310070

Design, Development, and Applications of Engineering Ceramics and Composites

Design, Development, and Applications of Engineering Ceramics and Composites

Ceramic Transactions, Volume 215

A Collection of Papers Presented at the 8th Pacific Rim Conference on Ceramic and Glass Technology
May 31–June 5, 2009
Vancouver, British Columbia

Edited by
Dileep Singh
Dongming Zhu
Yanchun Zhou

Volume Editor
Mrityunjay Singh

The
American
Ceramic
Society

A John Wiley & Sons, Inc., Publication

Published by John Wiley & Sons, Inc., Hoboken, New Jersey.
Published simultaneously in Canada.

For general information on our other products and services or for technical support, please contact our Customer Care Department within the United States at (800) 762-2974, outside the United States at (317) 572-3993 or fax (317) 572-4002.

Wiley also publishes its books in a variety of electronic formats. Some content that appears in print may not be available in electronic format. For information about Wiley products, visit our web site at www.wiley.com.

Library of Congress Cataloging-in-Publication Data is available.

ISBN 978-0-470-88936-7

Printed in the United States of America.

10 9 8 7 6 5 4 3 2 1

Contents

ADVANCED CERAMIC COATINGS

GEOPOLYMERS

Preface

This proceedings contains 32 peer-reviewed papers (invited and contributed) from the following five symposia held during the 8th Pacific Rim Conference on Ceramic and Glass Technology (PACRIM 8), May 31–June 5, 2009 in Vancouver, Canada:

- Engineering Ceramics and Ceramic Matrix Composites: Design, Development, and Applications;
- Advanced Ceramic Coatings: Processing, Properties and Applications;
- Computational Approaches in Materials Research and Design;
- Geopolymers; and
- Advanced Thermal Management Materials and Technologies

We would like to thank all of the symposium organizers for their tireless work in the planning and execution of the excellent technical program. The editors also wish to extend their gratitude and appreciation to all the authors for their contributions, to all the participants and session chairs for their time and effort, and to all the reviewers for their valuable comments and suggestions. Finally, we thank the dedicated and courteous staff at The American Ceramic Society for organizing and sponsoring this meeting and making this proceedings volume possible.

DILEEP SINGH, *Argonne National Laboratory, USA*
DONGMING ZHU, *NASA Glenn Research Center, USA*
YANCHUN ZHOU, *Shenyang National Laboratory for Materials Science, China*

Introduction

The 8th Pacific Rim Conference on Ceramic and Glass Technology (PACRIM 8), was the eighth in a series of international conferences that provided a forum for presentations and information exchange on the latest emerging ceramic and glass technologies. The conference series began in 1993 and has been organized in USA, Korea, Japan, China, and Canada. PACRIM 8 was held in Vancouver, British Columbia, Canada, May 31–June 5, 2009 and was organized and sponsored by The American Ceramic Society. Over the years, PACRIM conferences have established a strong reputation for the state-of-the-art presentations and information exchange on the latest emerging ceramic and glass technologies. They have facilitated global dialogue and discussion with leading world experts.

The technical program of PACRIM 8 covered wide ranging topics and identified global challenges and opportunities for various ceramic technologies. The goal of the program was also to generate important discussion on where the particular field is heading on a global scale. It provided a forum for knowledge sharing and to make new contacts with peers from different continents.

The program also consisted of meetings of the International Commission on Glass (ICG), and the Glass and Optical Materials and Basic Science divisions of The American Ceramic Society. In addition, the International Fulrath Symposium on the role of new ceramic technologies for sustainable society was also held. The technical program consisted of more than 900 presentations from 41 different countries. A selected group of peer reviewed papers have been compiled into seven volumes of The American Ceramic Society's Ceramic Transactions series (Volumes 212-218) as outlined below:

- **Innovative Processing and Manufacturing of Advanced Ceramics and Composites, Ceramic Transactions, Vol. 212,** Zuhair Munir, Tatsuki Ohji, and Yuji Hotta, Editors; Mrityunjay Singh, Volume Editor
  Topics in this volume include Synthesis and Processing by the Spark Plasma

Method; Novel, Green, and Strategic Processing; and Advanced Powder Processing

- **Advances in Polymer Derived Ceramics and Composites, Ceramic Transactions, Vol. 213,** Paolo Colombo and Rishi Raj, Editors; Mrityunjay Singh, Volume Editor
  This volume includes papers on polymer derived fibers, composites, functionally graded materials, coatings, nanowires, porous components, membranes, and more.

- **Nanostructured Materials and Systems, Ceramic Transactions, Vol. 214,** Sanjay Mathur and Hao Shen, Editors; Mrityunjay Singh, Volume Editor
  Includes papers on the latest developments related to synthesis, processing and manufacturing technologies of nanoscale materials and systems including one-dimensional nanostructures, nanoparticle-based composites, electrospinning of nanofibers, functional thin films, ceramic membranes, bioactive materials and self-assembled functional nanostructures and nanodevices.

- **Design, Development, and Applications of Engineering Ceramics and Composite Systems, Ceramic Transactions, Vol. 215,** Dileep Singh, Dongming Zhu, and Yanchun Zhou; Mrityunjay Singh, Volume Editor
  Includes papers on design, processing and application of a wide variety of materials ranging from SiC SiAlON, ZrO$_2$, fiber reinforced composites; thermal/environmental barrier coatings; functionally gradient materials; and geopolymers.

- **Advances in Multifunctional Materials and Systems, Ceramic Transactions, Vol. 216,** Jun Akedo, Hitoshi Ohsato, and Takeshi Shimada, Editors; Mrityunjay Singh, Volume Editor
  Topics dealing with advanced electroceramics including multilayer capacitors; ferroelectric memory devices; ferrite circulators and isolators; varistors; piezoelectrics; and microwave dielectrics are included.

- **Ceramics for Environmental and Energy Systems, Ceramic Transactions, Vol. 217,** Aldo Boccaccini, James Marra, Fatih Dogan, Hua-Tay Lin, and Toshiya Watanabe, Editors; Mrityunjay Singh, Volume Editor
  This volume includes selected papers from four symposia: Glasses and Ceramics for Nuclear and Hazardous Waste Treatment; Solid Oxide Fuel Cells and Hydrogen Technology; Ceramics for Electric Energy Generation, Storage, and Distribution; and Photocatalytic Materials.

- **Advances in Bioceramics and Biotechnologies, Ceramic Transactions, Vol. 218;** Roger Narayan and Joanna McKittrick, Editors; Mrityunjay Singh, Volume Editor
  Includes selected papers from two cutting edge symposia: Nano-Biotechnology and Ceramics in Biomedical Applications and Advances in Biomineralized Ceramics, Bioceramics, and Bioinspiried Designs.

I would like to express my sincere thanks to Greg Geiger, Technical Content Manager of The American Ceramic Society for his hard work and tireless efforts in the publication of this series. I would also like to thank all the contributors, editors, and reviewers for their efforts.

MRITYUNJAY SINGH
Volume Editor and Chairman, PACRIM-8
Ohio Aerospace Institute
Cleveland, OH (USA)

Developments in
Engineering Ceramics

ELASTIC AND VIBRATION PROPERTIES OF DIAMOND-LIKE B-C MATERIALS

J E Lowther
DST - NRF Centre of Excellence in Strong Materials and School of Physics,
University of the Witwatersrand, Johannesburg, South Africa.

ABSTRACT

Various Boron Carbon diamond like phases are examined using ab-initio computational modeling. Trends show that compressibility of B-C diamond like structures remains good but Shear and Young moduli decrease with increasing B concentration. Calculated vibration spectra are also reported and in light of these implications for recently reported experimental synthesis of the materials discussed.

1 INTRODUCTION

Boron, carbon and nitrogen form the atomic basis for the hardest materials known to date. Diamond is the hardest known material with cubic boron nitride (c-BN) second - both materials having a simple cubic structure [1,2]. Recently reports of other hard materials [3-7] with an apparent diamond like structure have appeared and these materials present enormous potential for industrial application. Such materials also have fundamental significance as they contain substantial concentrations of boron thus making very lightweight superhard materials that are lighter than even diamond.

As point defect boron is a well know impurity in diamond that makes the material semiconducting. It gives rise to a defect level lying at 0.37eV above the valence band edge and is responsible for making the material p-type[8]. Infra red optical spectroscopy has now revealed that the ground state of the bound hole of the boron acceptor is pinned to the valence band[9,10]. Under high doping concentrations of boron in diamond, various complexes exist and some of these also form shallow acceptor states but with energy levels far different to that of the simple substitutional center[11] These centers give rise to a complicated electrical conductivity[12,13] and, if diamond is subject to extreme conditions such as pressure, possibly even to superconductivity[14]. The additional states of boron also give rise to defect energy levels that lie above the valence band and some of these have been observed to give a hopping mechanism for the electrical conductivity[13,11,12] and thought to be associated with different forms of B-related centers[12]. At high concentrations of boron in diamond a metal insulation transition occurs[13] and then with even higher boron concentrations superconductivity appears[14-16] that is also tentatively associated with Boron pairs[13]. The vibration properties of boron in diamond have revealed that a 500cm^{-1} band clearly emerges in the Raman spectra that is thought to originate from local vibrations of boron pairs[17]

For even higher concentrations of boron relative to carbon, ultra hard structures emerge of which a B-C compound has now been synthesized [5,6]. This material had been investigated theoretically[18] where it was found that several of these structures are metastable. The stoichiometry of the synthesized metastable B-C phase has now been suggested to be in the stoichiometry BC_5 with an apparent diamond-like structure[6]. Calculations[18] however had indicated that an exact cubic structure is not probable with heavy B concentrations, and at least in the case of a BC_3 stoichiometry, slight tetragonal or trigonal distortions are expected. More recently a hexagonal structure has been considered as being a possibility for BC_5[19]

There have been at least two reports of the Raman spectra of synthesized B-C structures. Zinin et. al [20] reported a spectrum of a nominally BC_3 phase that had been synthesized at 50GPa and temperatures of 2033K. They observed a rather broad spectrum consisting of two rather broad bands, the around peaks $496cm^{-1}$ to $676cm^{-1}$ and the second band extending from $997cm^{-1}$ to $1413cm^{-1}$. They concluded that the various peaks relate to the structure of diamond BC_3 and that the low energy peaks are similar to boron-doped diamond. More recently Solozhenko et. al.[6] reported Raman spectra also consisting of two broadened bands centered at just below $600cm^{-1}$ and the other just above $1200cm^{-1}$. They suggested that the lower energy peaks were associated with B-B and B-C bonds and assigned the structure to a BC_5 diamond like phase. In this work there was no report of a vibration structure above 1200 cm^{-1}. In both reports the origin of the various vibration structures was somewhat tenuous and not really understood.

There have also been extensive studies of graphitic phases of boron containing graphitic phases[20; 21] especially in the stoichiometry BC_3. It is likely that such graphitic materials would play an important role as precursors to a diamond - like B-C phase in a similar way that graphite is the precursor to diamond. There is also need to be precise on the exact stoichiometry of the diamond like B-C phases and how this will influence the overall physical properties of the material. In this regard a comparison with theory is needed.

In the present work we examine potential models for diamond-like B-C phases and deduce various elastic constants. In addition we also report calculations of the vibration spectra associated with the phases. Comparison is made with similar calculations on Boron pairs in diamond. Suggestions are made concerning recent claims regarding the synthesis of BC_3 and BC_5 superhard phases.

2 COMPUTATIONAL

The first-principles calculations have been performed using the VASP (Vienna Ab -Initio Simulation Package) code[22], which performs plane - wave pseudopotential total energy calculations. The total energy and the electronic structure of the materials are calculated using density functional theory (DFT) with the local density approximation (LDA) of Ceperley and Alder[23] used to treat the exchange and correlation functional. The k-points are generated according to the Monkhorst–Pack[24] scheme usually on a 8x8x8 grid with PAW pseudopotentials[25] used throughout. All the calculations carried out are fully relaxed and optimized with respect to volume with an energy criterion of 10^{-5} eV/atom. The equilibrium volume and the Bulk modulus were calculated using the Birch–Murnaghan equation of state[26] and elastic constants using the Energy - displacement method through a least squares fit to a second order polynomial. From these values an effective isotropic Bulk, Shear and Young modulus as well as Poisson ratio can be obtained using the following general expressions:

$$B = \frac{1}{9}(c_{11} + c_{22} + c_{33} + 2(c_{12} + c_{13} + c_{23}))$$

$$G = \frac{1}{15}(c_{11} + c_{22} + c_{33} + 3(c_{44} + c_{55} + c_{66}) - c_{12} - c_{13} - c_{23})$$

$$E = \frac{9BG}{3B+G}; \quad v = \frac{(3B-2G)}{2(3B+G)}$$

Starting with the optimized structures we also calculated that vibration spectra at the Gamma point of the Brillouin Zone which would be appropriate to Raman spectra. In the present work we do not report on symmetry of the Raman modes but only on the overall vibration spectra as obtained from relative intensities of the vibrations as deduced from the eigenvectors of the dynamical matrix. For example considering three eigenvectors for a phonon of frequency ω directed along the cubic directions x, y, and z for a specific atom as being $\alpha_x, \alpha_y, \alpha_z$ we can define a probability of localization for that atom at the phonon frequency as:

$$L(\omega) = \alpha_x^2 + \alpha_y^2 + \alpha_z^2$$

Thus we can consider displacements of specific atoms and in this way we are able to associate the localization of the various vibration bands with either B or C in a typical B-C structure.

3. BORON IN DIAMOND

A 64-atom unit cell was used to study properties of boron in diamond with boron being placed as a substitutional atom and then as boron pairs either on adjacent lattice locations or next nearest neighbor positions. In this unit cell a Monkhorst-Pack grid of size 4x4x4 was used.

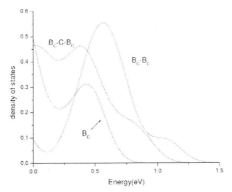

Figure 1 Electronic density of electron states (arb. Units/B atom) as measured from the top of the valence band for adjacent substitutional B dimers, next neighbor B pairs and single substitutional B.

The density of states is shown in Figure(1). In the case of substitutional boron a defect level that follows quite closely - but not exactly - the bulk valence band edge highlights the shallow nature of the B defect and is consistent with the experimental observations discussed earlier that the defect is an acceptor in diamond. In the case of B pairs the electrical energy level structure is different as seen from Figure(1). When the B dimer is formed from B ions on nearest neighbor locations additional levels arise around the valence band edge. The location of the Fermi level also shows that the defect levels will be partially empty and will thus facilitate a conducting process. As is evident from Figure(1) the nature of the energy levels depends on the precise location of the two B atoms.

Using the same supercell we evaluated vibration properties of the various boron related structures and using the same 4x4x4 Monkhorst Pack grid. Without symmetry restrictions, this now lead to a considerably larger number of k-points in the calculation with the energy convergence mentioned earlier. As described above using eigenvectors of the dynamical matrix we can precisely determine atoms amplitudes of the various vibration energies and thus associate the vibration with different atoms. For the various boron atoms in the configurations discussed above this is now shown in Figure(2).

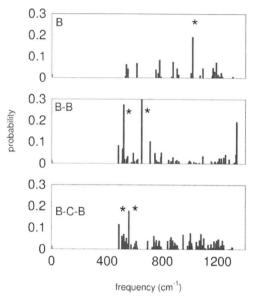

Figure 2. Vibration spectra of various B defects in diamond as represented by a 64-atom unit cell. Vibrations strongly associated with boron atoms are indicated. The proability is obtained from $L(\omega)$ defeined in Section(2).

The calculation has clearly indicated that lower energy vibration modes emerge for boron pairs in the range of about 600cm^{-1} and that these move to lower energy as the boron are further separated. This feature we suggest, as we shall see, will have implications for boron carbon based new ultra hard materials.

4. BORON - CARBON DIAMOND RELATED STRUCTURES

Initially to simulate the BC structures we follow the earlier approach[18] of considering an 8-atom unit cell filled with B or C atoms leading to possible structure representing stoichiometries BC, BC_3, BC_7 and diamond as shown in Figure(3).

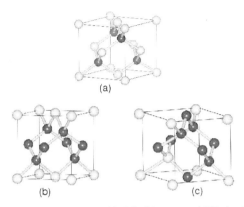

Figure 3 Structure of BC and BC₃ phases. (a) cubic BC, (b) tetragonal BC_3 (no B-B bonds), (c) trigonal BC_3 (one B-B bond). Dark spheres represent C atoms.

The results of the calculations are given in Table (2) with the various elastic constants in Table (3). There is a clear decrease in all elastic moduli with increasing B concentration. We note that in the case of the BC_3 structure elastic moduli have been calculated in the cubic approximation and also for tetragonal or trigonal symmetry of the distorted 8-atom cell. There are only small differences in the isotropic moduli B, G and E as obtained from expressions above.

Table 1 Number of atoms used in unit cell (Z), cellular structure, Birch equation of state and total energy of BC, BC_3, BC_7 and hexagonal BC_3 and BC_5. *crystal structure taken from [19]

	Z	a (A)	c (A)	B (GPa)	B'	E_{tot} (eV/atom)
BC	8	3.745	3.745	301	3.57	-8.346
BC_3(a)	8	3.509	3.876	379	3.59	-9.265
BC_3(b)	8	3.642	3.611	378	2.94	-9.239
BC_3(c) hexagonal	16	5.190	6.782	209	2.27	-9.005
BC_3(d) hexagonal	16	5.150 (5.200)	4.272	345	3.55	-9.109
BC_5* hexagonal	5	2.564	6.326	404	3.6	-9.530
BC_7	8	3.581	3.581	418	3.62	-9.642
diamond	8					

Table 2: Elastic constants of diamond like BC_3, BC_5 and BC_7 structures. *crystal structure of BC_5 taken from [19]

	c_{11}	c_{12}	c_{13}	c_{33}	c_{44}	c_{66}	B	G	E	v
BC	307	299	-	-	38	-	302	24	71	.46
BC_3(a)	749	195	-	-	178	-	380	217	548	.26
	720	206	220	788	464	268	391	344	798	.16
average							385	280	673	.21
BC_3(b)	650	220	-	-	382	-	363	315	733	.16
	628	230	239	380	380	-	367	270	651	.20
average							364	292	692	.18
BC_3(c) (hexagonal)	687	314	0.0	14	0.0	-	223	109	281	.29
BC_3(d) (hexagonal)	844	227	70	694	140	-	346	252	608	.21
BC_5* (hexagonal)	872	205	113	1042	371	-	405	372	855	.15
BC_7	789	233	-	-	502	-	418	412	931	.13
Diamond	1045	99	-	-	584	-	430	539	1125	.05

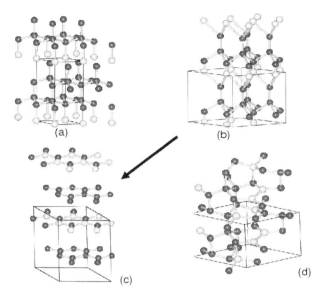

Figure 4: Structure of various B-C phases. (a) final relaxed BC₅ phase [after [19]] (b) starting BC₃ phase after hexagonal diamond (c) final relaxed phase starting from phase (b) and (d) final relaxed phase with a B atoms from the B plane shown in phase(b) .

We have also considered a hexagonal diamond BC_3 phase. This structure is initially derived from the hexagonal diamond structure and in Figure(4b). A recently suggested structure[19] for hexagonal BC_5 is shown in Figure(4a). While we find that the hexagonal BC_5 structure is quite stable, this is not so for hexagonal BC_3 at least when all the boron atoms are in a single 111 plane as indicated in the starting structure of Figure(4b). In fact the relaxed structure is quite graphitic as shown in Figure(4c) where the emergence of two graphitic sheets – one consisting of BC and the other a graphene related sheet is more energetically favorable. Overall the structure still remains hexagonal.

We continued to investigate the hexagonal BC_3 structure but now interchanging a boron and carbon atom. The final structure is shown in Figure(4d) – a very near hexagonal structure is still evident. Thus we point out that boron located in the structure of hexagonal BC_3 is quite critical in avoiding a breakdown of the structure to graphite like phase.

Results for all crystal structures are contained in Table (1) – in relation to the graphitic BC_3 structure we note that the phase with B incorporated into the structure (Fig (4d)) is energetically favored over the graphitic structure (Fig (4c)). But even so the nether of the hexagonal BC_3 phases is energetically favored over the diamond like BC_3 phases – i.e. $BC_3(a)$ or $BC_3(b)$.

Figure (5) summarizes calculations of the isotropic moduli for the various phases as considered in Table (2). It is noted from Figure (5) that whereas boron carbon structures generally are incompressible there is a sharp drop in the shear and Young moduli with increasing boron concentration. This implies that the hardness of heavily doped boron carbon diamond like structures will rapidly decline.

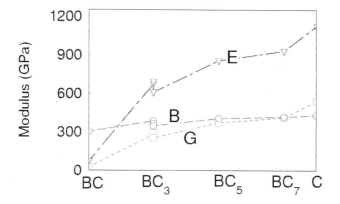

Figure 5: Variation of the elastic moduli of various B-C phases as obtained from the data contained in Table(2). C denotes diamond. As sen from values in Table(2), the Poisson ratio also shows similar trend – decreasing in value with increasing C concentration.

Next we come to the calculations of the vibration structure. Starting from the optimized crystal structures that were given in Table (1) we obtained the vibration frequencies and relative localization of the phonons for either boron or carbon atoms using the method described earlier. For each of the diamond structures of BC, BC_3 and BC_7 the results are now shown in Figure (6). We note the vibration peaks of the boron atoms. At a stoichiometry of BC_7 there is no vibration structure below about 800 cm^{-1}. This is consistent with our earlier results on the substitutional B defects in diamond – as shown in Figure (2). However for higher stoichiometries of boron relative to carbon, lower energy vibration structures do appear.

For BC_3 there is a difference in the two possible structures of $BC_3(a)$ or $BC_3(b)$ corresponding to the presence of boron pairs. As seen from the Figure, boron vibrations lie higher when there are no boron pairs ($BC_3(a)$) as opposed to when there is boron pairs ($BC_3(b)$). Indeed the presence of boron pairs in the structure does see the emergence of a boron related band at about 600cm^{-1}. For the very highly metastable structure BC the boron related band has moved to far lower energy around 400cm^{-1}.

Now we consider hexagonal structures that were shown in Figure (4). We recall there were two stable hexagonal forms of BC$_3$ – but in the case of the structure with no boron bonds (BC$_3$(c)) there was the emergence of two graphitic sheets. It is not unexpected the vibrations of this structure are rather higher in as they are associated with a carbon graphite structure but because the unit cell is different from graphite they occur at slightly lower energies than for (say) graphene. We find a high energy mode to be present around 1400cm^{-1} and there are vibrations associated with the related B-C graphite structure around 800cm^{-1}. In fact this trend is consistent with results reported earlier for such isolated graphite structures[27]. However the situation when a single boron atom is incorporated into the hexagonal (rather than all being in a layer) leads to a rather complicated system associated with varying carbon bonds. This vibration structure is in no way as intense as the graphitic structure as seen for BC$_3$(d). Here there is a rather large unresolved vibration structure emerging - other locations of B in the hexagonal BC$_3$ structure therefore cannot be excluded.

Thus we suggest that the BC$_3$ material that has reportedly been synthesized by Zinin et. al [20] where a rather broad vibration spectrum consisting of two rather broad bands, first between 496cm^{-1} and 676cm^{-1} and the second extending from 997cm^{-1} to 1413cm^{-1}. is indeed associated with a phase of BC$_3$. The higher energy band suggests graphitization is present. The origin of the lower energy peak that these authors observe is far more uncertain but is consistent with either the diamond like BC$_3$(b) system – the vibration spectrum shown in Figure(6) or the hexagonal BC$_3$(d) system the spectrum shown in Figure(7). Clearly more work is encouraged here.

Finally we considered the vibration structure of BC$_5$ and this is also shown in Figure (7). There is a strong boron related peak at about 800cm^{-1} and a lower energy structure at about 400cm^{-1} but mainly associated with carbon. A carbon related peak is at about 1200cm^{-1}.

We consider implications for the vibration spectrum recently reported by Solozhenko et. al.[6] . This consisted of two broadened bands centered at just below 600cm^{-1} and the other just above 1200cm^{-1} . The energy of the higher is consistent with the calculated structure shown in Figure (7) as is the lower energy band around 400cm^{-1}. But the peak we calculate to lie around 700cm^{-1} and associated with boron does not seem to be present in the experimentally reported spectrum[6] . Again some further work is encouraged to elucidate this structure.

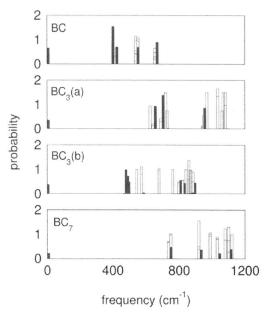

Figure 6 Vibration spectra of B-C phases shown in Figure(3). Dark bars show boron - related vibrations, open bars carbon related vibrations. The essential difference between BC3(a) and BC3(b) is presence of a boron-boron bond.

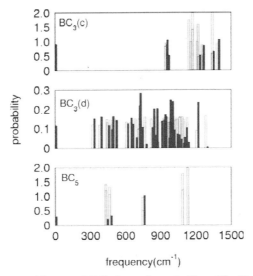

Figure 7 Vibration spectra of hexagonal B-C phases shown in Figure(5). Dark bars show boron - related vibrations, open bars carbon related vibrations. The essential difference between BC₃(c) (a graphitic structure) and BC₃(d) is presence of a boron-carbon bond that stabilizes the hexagonal structure.

5 CONCLUSION

It is anticipated that many boron – carbon complexes will be metastable relative to diamond at least under ambient conditions. As such these materials will require extreme conditions to be synthesized although it may be possible to stabilize these materials at ambient conditions. In the present work we have considered various possible structures for some boron – carbon structures that have some form of diamond related crystal structure. It was found that diamond-like structures have quite large bulk modulus suggesting such structures would be incompressible new materials. However the calculations also indicate that with very high boron concentrations the other moduli, such as the shear and Young values, decrease rapidly. This would imply that heavy boron concentrations in diamond related structure whilst being incompressible may not exhibit extreme hardness. We also examined the role played by boron in such structures and this is shown in the observed vibration spectra. Comparison is made with point defect boron structures in diamond. We see that the low energy modes – around $600cm^{-1}$ – 700 cm^{-1} are primarily associated with boron dimer structures. This is not surprising as boron – carbon bonds are far weaker than carbon-carbon bonds and thus vibrations will be at lower energy. The influence of boron bonds continues to be evident in the various boron-carbon potentials ultra hard structures with strong bands again emerging at the $600cm^{-1}$ region of the vibration spectra. This region of the vibration spectra is an important indicator of the potential structure and as boron bonds will essentially weaken the structure of the overall of the hardness of the material.

We have seen that our results have consistencies with recently reported Raman spectra[6; 20] and the suggestions of BC_3 or BC_5 stoichiometries as reported in both works. In the case of the BC_3 spectra however we suggest that there is some graphitization present but the presence of an ultra hard phase cannot be ruled out. At the same time the observed BC_5 spectra does appear consistent with our calculations albeit boron associated structure around 700cm^{-1} should be further examined.

ACKNOWLEDGEMENTS

This work has been supported by the National Research Foundation (South Africa).

REFERENCES

[1]A. Badzian and T. Badzian, "Recent developments in hard materials," *International Journal of Refractory Metals and Hard Materials*, **15** 3-12 (1997).

[2]J. Haines, J. M. Leger and G. Bocquillon, "Synthesis and design of superhard materials," *Ann. Rev. Mater. Res.*, **31** 1-9 (2001).

[3]H. W. Hubble, I. Kudryashov, V. L. Solozhenko, P. V. Zinin, S. K. Sharma and L. C. Ming, "Raman studies of cubic BC2N, a new superhard phase," *Journal Of Raman Spectroscopy*, **35** 822-25 (2004).

[4]V. L. Solozhenko, D. Andrault, G. Fiquet, M. Mezouar and D. C. Rubie, "Synthesis of superhard cubic BC2N," *Appl. Phys. Lett.*, **78** 1385-88 (2001).

[5]V. L. Solozhenko, N. A. Dubrovinskaia and L. S. Dubrovinsky, "Synthesis of bulk superhard semiconducting B-C material," *Applied Physics Letters*, **85** 1508 (2004).

[6]V. L. Solozhenko, O. O. Kurakevych, D. Andrault, Y. Le Godec and M. Mezouar, "Ultimate Metastable Solubility of Boron in Diamond: Synthesis of Superhard Diamondlike BC5," *Physical Review Letters*, **102** **015506-10** (2009).

[7]S. N. Tkachev, V. L. Solozhenko, P. V. Zinin, M. H. Manghnani and L. C. Ming, "Elastic moduli of the superhard cubic BC2N phase by Brillouin scattering," *Phys Rev B*, **68** 052104 (2003).

[8]J. F. Custers, "Boron in Diamond" *Nature*, **176** 173-8 (1955).

[9]A. T. Collins and E. C. Lightowlers, "The boron defect in Diamond," *Physical Review*, **171** 843-50 (1968).

[10]P. A. Crowther, P. J. Dean and W. F. Sherman, "Photoconductivity of Boron in Diamond," *Physical Review*, **154** 772-81 (1967).

[11]R. F. Mamin and T. Inushima, "Conductivity in Boron-Doped Diamond," *Phys. Rev. B*, **63** 033201-08 (2001).

[12]W. B. Wilson, "Several Boron Types of Conductivity in Diamond," *Physical Review*, **127** 1549-57 (1962).

[13]M. J. R. Hoch, J. E. Lowther and T. Tshepe, "Boron in diamond at high concentrations," *Proceedings of the 25th International Conference on the Physics of Semiconductors, Pts I and Ii*, **87** 154-55 (2001).

[14]E. A. Ekimov, V. A. Sidorov, E. D. Bauer, N. N. Mel'nik, N. J. Curro, J. D. Thompson and S. M. Stishov, "Superconductivity in Diamond," *Nature*, **428** 542-6 (2004).

[15]T. Yokoya, T. Nakamura, T. Matsushita, T. Muro, Y. Takano, M. Nagao, T. Takenouchi, H. Kawarada and T. Oguchi, "Origin of the metallic properties of heavily boron-doped superconducting diamond," *Nature*, **438** 647-50 (2005).

[16]N. Dubrovinskaia, G. Eska, G. A. Sheshin and H. Braun, "Superconductivity in polycrystalline boron-doped diamond synthesized at 20 GPa and 2700 K," *Journal of Applied Physics*, **99** 033903-6 (2006).

[17]M.Bernar, C. Bar and A. Deneuville, "About the origin of the low wave number structures of the Raman spectra of heavily boron doped diamond films," *Diamond & Related Materials*, **13** 896-95 (2004).

[18]J. E. Lowther, "Potential super-hard phases and the stability of diamond-like boron-carbon structures," *J. Phys.-Condes. Matter,* **17** 3221-29 (2005).

[19]M. Calandra and F. Mauri, "High-Tc superconductivity in superhard diamondlike BC5," *Phys Rev Lett,* **101**[1] 016401-8 (2008).

[20]P. V. Zinin, L. C. Ming, I. Kudryashov, N. Konishi and S. K. Sharma, "Raman spectroscopy of the BC3 phase obtained under high pressure and high temperature," *Journal of Raman Spectroscopy,* **38** 1362-67 (2007).

[21]Vladimir L. Solozhenko, Oleksandr O. Kurakevych, Elena G. Solozhenko, Jiuhua Chen and J. B. Parise, "Equation of state of graphite-like BC," *Solid State Communications* **137** 268-71 (2006).

[22]G. Kresse and J. Hafner, "Ab initio molecular dynamics for liquid metals " *Phys. Rev. B,* **47** 558-65 (1993).

[23]D. M. Ceperley and B. J. Alder, "Ground State of the Electron Gas by a Stochastic Method " *Phys. Rev. Lett.,* **45** 566-70 (1980).

[24]H. J. Monkhorst and J. D. Pack, "Special points for Brillouin-zone integrations " *Phys. Rev. B,* **13** 5188-96 (1976).

[25]G. Kresse and D. P. Joubert, "From Ultrasoft Pseudopotentials to the Projector Augmented Wave Method," *Phys. Rev. B,* **59** 1758-64 (1999).

[26]F. Birch, "Finite elastic strain of cubic crystals," *Phys. Rev.,* **71** 809-14 (1947).

[27]J. E. Lowther, P. V. Zinin and L. C. Ming, "Vibrational energies of graphene and hexagonal structured planar B-C complexes," *Physical Review B,* **79** 033401-5 (2009).

TOWARDS SIMULATION-BASED PREDICTIVE DESIGN OF GLASSES

Liping Huang[1,*] and John Kieffer[2]
[1]Department of Materials Science and Engineering, Rensselaer Polytechnic Institute, Troy, NY 12180, USA
[2]Department of Materials Science and Engineering, University of Michigan, Ann Arbor, MI 48109-2136, USA

ABSTRACT
    A normal solid becomes stiffer when squeezed and softer when heated. In contrast, silica glass behaves the opposite way: its elastic moduli decrease upon compression and increase upon heating. Silica glass is also known to permanently densify under compression. These have been long-standing mysteries in glass science. Using computer simulation, we uncovered the structural origins of the anomalous thermo-mechanical behaviors in silica glass. Accordingly, these anomalies can be attributed to localized structural transitions, analogous to those that occur in the crystalline counterparts. Moreover, we showed that these anomalies are universal in network glasses, and can be explained by similar mechanisms, whether the network structures are formed from tetrahedral or trigonal building blocks (as demonstrated for silica and boron oxide). Our simulations further revealed that the permanent densification in silica glass is achieved through irreversible structural transition involving bond breaking and re-formation under a combination of high pressure and temperature. The inherent instability of network rings in normally quenched glass, as evident by a progressive weakening of the structure upon initial compaction, is a key to facilitating permanent densification. Thus, our simulations predicted that by processing in ways that eliminates anomalous thermo-mechanical behaviors, the propensity of the glass to undergo irreversible densification can be eradicated. This provides the conceptual foundation for the bottom-up design of new glasses with novel properties.

INTRODUCTION
    Discoveries in the past three decades strongly suggest that two or more distinct amorphous states may exist for the same material. This phenomenon, termed 'polyamorphism,' is one of the most intriguing and puzzling topics of condensed matter physics. It has been observed in various classes of materials, such as amorphous ice, silica, boron oxide, silicon, and chalcogenide glasses[1-15]. These materials tend to have open and less dense network structures. As one manifestation of their unique topologies, these network structures exhibit a number of anomalous behaviors, some of which are shared among different systems. For instance, both water and silica shrinks when heated in certain temperature regimes[16]. Among these materials, silica glass is one of the most widely and well studied, not only as an archetypical amorphous material, but also because of its anomalous thermo-mechanical properties. The elastic moduli of silica glass increase with increasing temperature,[17-20] and the bulk modulus passes through a minimum upon compression at ~2-3 GPa[21-27]. Furthermore, this material can undergo irreversible densification under pressure.[2, 3, 14, 28-32] Models proposed to explain these phenomena are in one way or another based on the assumption that two or more energetically distinct amorphous states coexist in proportion that vary with pressure and temperature,[25, 33] but for the most part little detail is provided as to the structural entities that constitute these states and the atomic-scale mechanisms that underlie the structural transitions between the states. Experiments,[9-11, 13, 34] to some extent led by computer simulations,[35-57] are only recently starting to provide interpretations of these behaviors based on glass structure and elementary atomistic processes. A basic understanding of polyamorphism from computer simulation, complementary to experiment, will not only offer a new

_____

*Electronic mail: huangl5@rpi.edu

perspective on the amorphous state of matter in general, but also provide explanations on the structural origins of the anomalous behaviors of network-forming systems. This will further accelerate the design of novel materials with desired macroscopic properties from the microscopic level.

COMPUTATIONAL DETAILS

To simulate systems like SiO_2 and B_2O_3, in which the structural building blocks can exhibit multiple coordination states, it is essential to use a potential model that allows for coordination changes during simulations. We have designed such a coordination-dependent charge-transfer three-body potential model for oxides with mixed covalent-ionic bonding characteristics.[52, 58] There are three main features of our potential model: (1) *dynamic charge redistribution* — a charge transfer term controls the extent of charge polarization in a covalent bond, as well as the amount of charge transferred between atoms upon rupture or formation of such a bond; (2) *conditional three-body interactions* — the directional character of the covalent bonding is coupled to the degree of covalency in atomic interactions and vice versa; (3) *variable coordination number* — both two- and three-body interactions depend on the effective number of nearest neighbors of an atom, which is evaluated dynamically based on the local environment of this atom.

Our molecular dynamics (MD) simulations of SiO_2 and B_2O_3 glass were carried out for 3000-particle (1000 Si and 2000 O) and 8232-particle (2744 Si and 5488 O), respectively for 640-particle (256 B and 384 O) and 2560-atom (1024 B and 1536 O) systems, with periodic boundary conditions. No system size effects in excess of statistical errors could be detected for either system. The room temperature glasses were obtained by heating and melting crystalline silica and boron oxide and subsequently quenching the liquids at different cooling rates. Temperature ramping was achieved by velocity scaling and the density adjusts according to the Anderson constant-pressure algorithm[59]. The hydrostatic compression-decompression was carried out at different temperatures, between –20 GPa and 20 GPa, at 1 GPa intervals. The rate of compression was 0.5 GPa/ps. The bulk modulus of glasses was calculated directly from the equation of state according to $B = \rho(dP/d\rho)$. Further computational details can be found in our previous publications.[50, 52, 53]

RESULTS AND DISCUSSION

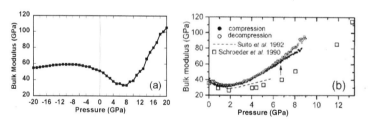

Figure 1. Bulk modulus of silica glass versus pressure from (a) MD simulations and (b) experiments.[26]

We have studied amorphous-amorphous transitions under various thermo-mechanical conditions using MD simulations[50, 51, 53, 60] and reproduced all commonly observed anomalous temperature or pressure dependence of elastic properties in silica glass (Figure 1). Simulated silica glass exhibits an initial decrease in bulk modulus with pressure until it reaches a minimum at approximately 6-8 GPa. By applying a hydrostatic tensile stress (negative pressure), the bulk modulus of silica glass continues to increase until pressure reaches –8 GPa. Below –8 GPa and above 8 GPa, silica glass behaves like a normal solid, but in between these pressures, it behaves anomalously. This anomalous behavior of silica glass has long been known from experiments, except for the return to normalcy at large tensile stresses. Our MD simulations reveal that thermo-mechanical anomalies of silica glass are due to localized reversible structural transitions similar to that underlying the α–β cristobalite silica phase transformation. These 'transitions' involve an abrupt rotation of Si–O–Si bridges and thus take place without any change in the local order (bond length and angles, ring size distribution) but affect the ring geometry so as to become more symmetric under mechanical or thermal expansion (Figure 2). Since the α-cristobalite phase is characterized by a lower modulus than the β-phase,[50] when a portion of Si–O–Si bridges undergo a rotation into the β configuration, the elastic modulus increases. Silica glass is made of a mixture of local structural motifs with α- and β-like rings, i.e., close to those in high-density low-modulus α-cristobalite and low-density high-modulus β-cristobalite, respectively. Accordingly, compressive volume changes of silica glass lead to the collapse of the network rings, which assume α-like geometries, and hence the structure softens. Conversely, upon expansion the local network character changes from α-like to β-like, and the bulk modulus increases.

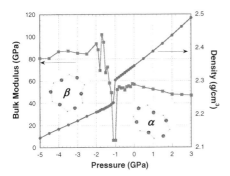

Figure 2. Bulk modulus (squares) and density (circles) of cristobalite silica through the α- to β- transformation under pressure. Insets show the comparison of the geometries of a 6-membered ring in α- and β-cristobalite. Note: α-cristobalite has a higher density, lower symmetry rings, and lower bulk modulus; β-cristobalite has a lower density, higher symmetry rings, and higher bulk modulus.

Early studies showed that B_2O_3, a simple glass former consisting of a highly connected planar triangle network, only exhibits normal elastic properties, and the anomalous elastic moduli in silica seem to be strongly correlated to the highly connected tetrahedral network and to the nature of the link between atoms.[61, 62] However, more recent Brillouin light scattering experiments showed that the anomalous increase of the elastic modulus with temperature occurs for B_2O_3 in the liquid state at high temperatures.[63] Thus, it appears that this anomaly is ubiquitous to network glass formers with tetrahedral or trigonal building blocks. In SiO_2 the modulus increases continuously, and the transition between glass and liquid is at best apparent through a minute change in slope of the modulus vs. temperature data. In contrast, for B_2O_3 the elastic modulus exhibits very little temperature dependence in the glassy state, drops precipitously above T_g but then reaches a persistent positive slope at high temperatures. Hence, at first glance B_2O_3 behaves more like a normal material, at least under ambient conditions. Conversely, the steady increase of the elastic modulus upon heating at temperatures 1000 degrees above the melting point of crystalline B_2O_3 is quite unusual and indicative of a strong networking tendency in this material. In our previous simulations, we have successfully reproduced the anomalous temperature-dependent of elastic moduli in silica[50] and boron oxide glass[52]. To identify what causes the anomalies in B_2O_3, and in fact to establish that the underlying mechanisms are in essence ubiquitous to all strongly networked glasses, we used simulations to explore the behavior of this material under conditions that are difficult, if not impossible to realize in experiments. In our previous studies we demonstrated that the effects of thermal and mechanical influences are equivalent, i.e., silica glass becomes stiffer when its structure expands, regardless of whether the expansion is achieved via heating or by applying a tensile stress. Note that the anomalous behavior for B_2O_3 only sets in at high temperatures, i.e., when the structure is expanded to a significant degree. Hence, if indeed the emergence of the anomalous behavior in elastic properties is uniquely related to the specific volume of the material, we should be able to incite this anomaly in B_2O_3 at room temperature by subjecting the material to isotropic tensile deformation.

In Figure 3, we plotted the bulk modulus vs. pressure for boron oxide glass on top of that for silica glass. In the compressive stress regime, boron oxide behaves normally, i.e., it becomes stiffer upon compression. While in the tensile stress regime, after reaching a minimum at about −1 GPa, the modulus increases again, i.e., the pressure dependence of the elastic modulus becomes negative. Our data show that there appear to be some definite parallels in the modulus vs. pressure behaviors of three- and four-coordinated network glasses. While the bulk modulus minimum in silica glass is at ~2-3 GPa, which can be easily measured by experiments,[21-23, 26] in B_2O_3 glass the minimum occurs at a negative pressure, and has so far eluded experiments. Our simulations also show that the minimum for B_2O_3 shifts toward the positive pressure range at higher temperatures, which is consistent with the notion that the occurrence and extent to which the network structures behave anomalously can be sensibly mapped onto a density change, whether this is brought on by thermal expansion or tensile stress. For each glass-forming system, the anomalous negative modulus vs. density dependence intercepts normal behavior and consequently extends between a maximum and a minimum. Silica glass behaves anomalously at ambient conditions. At room temperature, the maximum therefore occurs at negative and the minimum at positive pressures, and because of the small expansion coefficient of this substance, the anomalous behavior persists over a wide temperature range at ambient pressure. In vitreous B_2O_3 the anomalous regime is shifted towards negative pressures at room temperature, but due to a larger thermal expansion coefficient, the anomalous behavior re-surfaces at high temperature under ambient pressure. Our results show that the anomalous increase of the elastic moduli upon expansion are one and the same whether brought upon by tensile deformation or thermal expansion, and that it is a universal trait of network-forming glasses, tetrahedral and trigonal. Moreover, the phenomenon appears to be unaffected by the glass transition, i.e., it persists from the glassy into the liquid state.[52, 63]

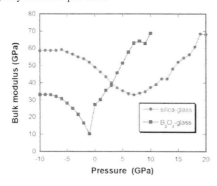

Figure 3. Bulk modulus vs. pressure in silica glass at 500 K and boron oxide glass at 300 K.

To argue that anomalous thermo-mechanical properties are universal for all network glasses, a similar transition-based mechanism must be responsible for the behavior of vitreous B_2O_3. The question then arises as to what might be the local structural motifs, and whether corresponding crystalline counterparts exist. According to the literature, two crystalline forms are known for boron oxide, i.e., B_2O_3-I, which is stable at ambient pressures and has boron coordinated by three oxygen atoms, and B_2O_3-II, which is stable at higher pressure and has tetrahedrally coordinated boron. B_2O_3-I falls within the trigonal space group of $P3_1$,[64] and has a density of 2.56 g/cm^3, much higher than 1.80 g/cm^3 of B_2O_3 glass.[65] In our simulations, subject to tensile isotropic stress our simulated B_2O_3-I expands and at –6 GPa transforms into a low-density phase with a very open structure. We tentatively call this phase B_2O_3-0 in light of the succession it assumes in terms of density relative to B_2O_3-I and B_2O_3-II. Figure 4 shows views of the low- and high-density modifications of B_2O_3-0 along the [010] direction. The partial collapse of network rings is particularly obvious for the smaller ring seen from this perspective. In fact, the resemblance to the ring geometries in α- and β-cristobalite is remarkable, and we therefore refer to the low- and high-density modifications of B_2O_3-0 as β-like and α-like, respectively. Similar to β-cristobalite, β-like B_2O_3-0 not only has more symmetric rings, but also the higher bulk modulus of the two modifications[52]. Amorphous B_2O_3 can be thought of as a mixture of "puffed" β-like and "puckered" α-like rings. Accordingly, compression converts "puffed" rings into "puckered" rings, causing the structure to compact and soften. Conversely, upon expansion the network rings puff up and the structure stiffens. The bulk modulus of B_2O_3 glass increases gradually as it expands under tension due to the increasing population of β-like rings in the system.

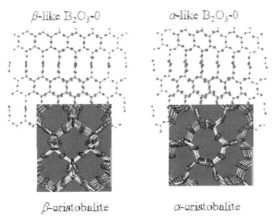

Figure 4. Structure of α- and β-modification of B_2O_3-0, comparing to that of α- and β-cristobalite SiO_2. Note: large spheres are O atoms, small ones are boron in B_2O_3-0 and silicon in SiO_2.

Our simulations also successfully reproduced the permanent densification in recovered silica glass after isothermal compression,[2, 66] and further revealed that this process in silica glass is achieved through an irreversible structural transition involving bond breaking and re-formation under a combination of high pressure and temperature. More importantly, our simulations demonstrated that the inherent instability of network rings in normally quenched glass, as evidenced by a progressive weakening of the structure upon initial compaction (moduli decrease with pressure), is a key to facilitating permanent densification. If we can process silica glass in ways that eliminate the inherent instability of network rings, we should be able to eradicate both the anomalous behaviors and the permanent densification in this material.

Pressure and temperature are important basic thermodynamic variables that determine the structure, dynamics, and macroscopic properties of a material.[67] Varying temperature has been used for eons to transform matter from one state to another. On the other hand, varying pressure has only infrequently been applied probably due to procedural difficulties, even though available literature showed that pressure could be very effective in synthesizing novel glasses with superior properties[68] and in providing a better understanding of the glass transition.[69, 70] By quenching silica glass under high pressures followed by pressure releasing at room temperature (Figure 5), the bulk modulus becomes larger, but the negative pressure dependence of bulk modulus becomes smaller, and eventually disappears in the system quenched while subject to 6 GPa or larger pressure (Figure 6a). The bulk modulus of this system is almost independent of pressure in the range from −10 to 10 GPa. It is important to note that the propensity of the silica glass to undergo irreversible densification under pressure is almost eradicated when the anomalous thermo-mechanical behaviors are eliminated (Figure 6b). In silica glass quenched at 6 GPa, the amount of permanent densification after a compression and decompression cycle is much less than that in a normally processed glass, namely quenched without any pressure applied to it. Thus, by processing in a way that eliminates anomalous thermo-mechanical behaviors, the degree to which the glass will undergo irreversible densification can be predicted by modifying the microscopic structure in a controllable way.

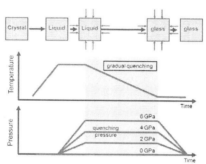

Figure 5. Pressure-quenching of silica glass under various pressures.

It should be pointed out that the degree of the anomalous thermo-mechanical behaviors and the propensity of the glass to undergo irreversible densification can be also controlled through isothermal compression without going through the liquid state and glass transition. Under isothermal compression, saturation densities are reached in the released specimens. While these maximum densities are temperature dependent, at any given temperature this maximum cannot be exceeded, even after repeated compression-decompression cycles, i.e., any hydrostatic compaction beyond this point is elastic and reversible, as shown in Brillouin light scattering[2], Raman light scattering[9], x-ray diffraction[11] experiments, and in computer simulations[51]. Once this density is achieved, anomalous decrease in bulk modulus with pressure is absent.[2, 51] However, it should be noted that the structure of glass obtained from isothermal compression is not identical to that prepared by quenching from the liquid under high pressure. The former does not undergo glass transition at high pressure and represents the densified glasses, and the latter represents the structure of supercooled liquids at high pressure at T_g. Previous studies showed that densified boron oxide glass from isothermal compression remains trigonally coordinated,[7, 12, 52, 56, 71] while that from pressure-quenching of the liquid has a mixture of tetrahedrally and trigonally coordinated B atoms.[72]

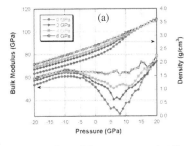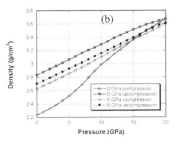

Figure 6. (a) Bulk modulus vs. pressure in silica glass quenched at various pressures from 0–6 GPa. (b) Density vs. pressure during a compression and decompression cycle in silica glass quenched at pressures of 0 and 6 GPa.

Even though our MD simulations are preliminary, the atomistic understanding of the glass structure we ascertained not only helps explain long-standing mysteries (e.g. the anomalous thermo-mechanical behaviors in silica glass) in glass science, but also opens the door to controlling the glass structure and properties through processing in a predictable way. Future improvement in the understanding of the link between the structural modifications and glass properties is expected with the involvement of large-scale simulation in the description of glass behavior, by taking advantage of the many multi-core, multi-processor machines now available[73]. Although further extensive complementary experiments and simulations are necessary to describe the exact structure of the polyamorphic states, we predict that it is possible to stabilize new polyamorphs at ambient conditions. As these new polyamorphs are expected to have different mechanical, optical, electrical or biological properties, it will open new perspectives in materials design from the atomic level.

CONCLUSION

MD simulations, based on a coordination-dependent charge transfer potential, were used to study the behavior of SiO_2 and B_2O_3 glass in response to various thermal and mechanical constraints. This interaction potential allows for the charges on atoms to re-distribute upon the formation and rupture of chemical bonds, and dynamically adjusts to multiple coordination states for a given species. Our simulations reveal the structural origin of the anomalous thermo-mechanical behaviors of network forming glasses, such as the increase of mechanical moduli upon expansion of the structure. The mechanism we found for B_2O_3 is analogous to the one we identified earlier as underlying the anomalous behaviors of SiO_2, and appears to be universal for network-forming glasses. Thus, these anomalous behaviors can be explained as the result of localized structural transformations between two motifs of different stiffness that are similar to those found in the material's crystalline counterparts. Furthermore, our simulations predicted that by processing in ways that eliminates anomalous thermo-mechanical behaviors, the propensity of the glass to undergo irreversible densification can be eradicated. This opens the door for the bottom-up design of new glasses with novel properties.

REFERENCES

[1]O. Mishima, L. D. Calvert, and E. Whalley, An apparently first-order transition between two amorphous phases of ice induced by pressure, *Nature* **314**, 76 (1985).
[2]M. Grimsditch, Polymorphism in Amorphous SiO_2, *Phys. Rev. Lett.* **52**, 2379 (1984).
[3]M. Grimsditch, R. Bhadra, and Y. Meng, Brillouin scattering from amorphous materials at high pressures, *Phys. Rev. B* **38**, 7836 (1988).

[4]O. Ohtaka, H. Arima, H. Fukui, W. Utsumi, Y. Katayama, and A. Yoshiasa, Pressure-induced sharp coordination change in liquid germanate, *Physical Review Letters* **92** (2004).

[5]M. Guthrie, C. A. Tulk, C. J. Benmore, J. Xu, J. L. Yarger, D. D. Klug, J. S. Tse, H. K. Mao, and R. J. Hemley, Formation and structure of a dense octahedral glass, *Physical Review Letters* **93** (2004).

[6]S. Sastry and C. A. Angell, Liquid-liquid phase transition in supercooled silicon, *Nature Materials* **2**, 739 (2003).

[7]J. D. Nicholas, S. V. Sinogeikin, J. Kieffer, and J. D. Bass, Spectroscopic evidence of polymorphism in vitreous B_2O_3, *Phys. Rev. Lett.* **92**, 215701 (2004).

[8]S. N. Tkachev, M. H. Manghnani, and Q. Williams, In situ brillouin spectroscopy of a pressure-induced apparent second-order transition in a silicate glass, *Physical Review Letters* **95** (2005).

[9]B. Champagnon, C. Martinet, C. Coussa, and T. Deschamps, Polyamorphism: Path to new high density glasses at ambient conditions, *Journal of Non-Crystalline Solids* **353**, 4208 (2007).

[10]B. Champagnon, C. Martinet, M. Boudeulle, D. Vouagner, C. Coussa, T. Deschamps, and L. Grosvalet, High pressure elastic and plastic deformations of silica: In situ diamond anvil cell Raman experiments, *Journal of Non-Crystalline Solids* **354**, 569 (2008).

[11]Y. Inamura, Y. Katayama, and W. Utsumi, Transformation in intermediate-range structure of vitreous silica under high pressure and temperature, *Journal of Physics-Condensed Matter* **19** (2007).

[12]S. K. Lee, P. J. Eng, H. K. Mao, Y. Meng, M. Newville, M. Y. Hu, and J. F. Shu, Probing of bonding changes in B_2O_3 glasses at high pressure with inelastic X-ray scattering, *Nat. Mater.* **4**, 851 (2005).

[13]V. V. Brazhkin, Y. Katayama, K. Trachenko, O. B. Tsiok, A. G. Lyapin, E. Artacho, M. Dove, G. Ferlat, Y. Inamura, and H. Saitoh, Nature of the structural transformations in B_2O_3 glass under high pressure, *Physical Review Letters* **101** (2008).

[14]C. Meade, R. J. Hemley, and H. K. Mao, High-pressure x-ray diffraction of SiO_2 glass, *Phys. Rev. Lett.* **69**, 1387 (1992).

[15]G. D. Mukherjee, S. N. Vaidya, and V. Sugandhi, Direct observation of amorphous to amorphous apparently first-order phase transition in fused quartz, *Physical Review Letters* **87** (2001).

[16]C. A. Angell and H. Kanno, Density maxima in high-pressure supercooled water and liquid silicon dioxide, *Science* **193**, 1122 (1976).

[17]M. Fukuhara and A. Sanpei, High Temperature-Elastic Moduli and Internal Dilational and Shear Frictions of Fused Quartz, *Jpn. J. Appl. Phys.* **33**, 2890 (1994).

[18]R. E. Youngman, J. Kieffer, J. D. Bass, and L. Duffrène, Extended Structural Integrity in Network Glasses and Melts, *J. Non-Cryst. Solids* **222**, 190 (1997).

[19]A. Polian, D. Vo-Thanh, and P. Richet, Elastic properties of a-SiO_2 up to 2300 K from Brillouin scattering measurements, *Europhys.Lett.* **57**, 375 (2002).

[20]H. K. Mao, P. M. Bell, J. W. Shaner, and D. J. Steinberg, Specific Volume Measurements of Cu, Mo, Pd, and Ag and Calibration of Ruby R1 Fluorescence Pressure Gauge from 0.06 to 1 Mbar, *Journal of Applied Physics* **49**, 3276 (1978).

[21]P. W. Bridgman, *Am. J. Sci.* **10**, 359 (1925).

[22]P. W. Bridgman, *Proc. Am. Acad. Arts Sci* **76**, 9 (1945).

[23]P. W. Bridgman, *Proc. Am. Acad. Arts Sci* **76**, 71 (1948).

[24]C. Meade and R. Jeanloz, Frequency-dependent equation of state of fused silica to 10 GPa, *Phys. Rev. B* **35**, 236 (1987).

[25]M. R. Vukevich, *J. Non-Cryst. Solids* **11**, 25 (1972).

[26]O. B. Tsiok, V. V. Brazhkin, A. G. Lyapin, and L. G. Khvostantsev, Logarithmic Kinetics of the Amorphous-Amorphous Transformations in SiO_2 and GeO_2 Glasses under High Pressure, *Phys. Rev. Lett.* **80**, 999 (1998).

[27]F. S. El'kin, V. V. Brazhkin, L. G. Khvostantsev, O. B. Tsiok, and A. G. Lyapin, In situ study of the mechanism of formation of pressure-densified SiO2 glasses, *Jetp Letters* **75**, 342 (2002).

[28]H. M. Cohen and R. Roy, Densification of glass at vey high pressure, *Phys. Chem. Glasses* **6**, 149 (1965).

[29]M. Grimsditch, Annealing and relaxation in the high-pressure phase of amorphous SiO_2, *Phys. Rev. B* **34**, 4372 (1986).

[30]R. J. Hemley, H. K. Mao, P. M. Bell, and B. O. Mysen, Raman Spectroscopy of SiO_2 Glass at High Pressure, *Phys. Rev. Lett.* **57**, 747 (1986).

[31]S. Susman, K. J. Volin, D. L. Price, and M. Grimsditch, Intermediate-range order in permanently densified vitreous SiO_2: A neutron-diffraction and molecular-dynamics study, *Phys. Rev. B* **43**, 1194 (1991).

[32]G. D. Mukherjee, S. N. Vaidya, and V. Sugandhi, Direct Observation of Amorphous to Amorphous Apparently First-Order Phase Transition in Fused Quartz., *Phys. Rev. Lett.* **87**, 195501 (2001).

[33]C. L. Babcock, S. W. Barder, and K. Fajans, *Ind. Eng. Chem* **46**, 161 (1954).

[34]R. Le Parc, C. Levelut, J. Pelous, V. Martinez, and B. Champagnon, Influence of fictive temperature and composition of silica glass on anomalous elastic behaviour, *Journal of Physics-Condensed Matter* **18**, 7507 (2006).

[35]W. Jin, R. K. Kalia, and P. Vashishta, Structural Transformation, Intermediate-Range Order, and Dynamical Behavior of SiO_2 Glass at High Pressures, *Phys. Rev. Lett.* **71**, 3146 (1993).

[36]K. Trachenko and M. T. Dove, Compressibility, kinetics, and phase transition in pressurized amorphous silica, *Phys. Rev. B* **67**, 064107 (2003).

[37]R. J. Della Valle and E. Venuti, High-pressure densification of silica glass: A molecular-dynamics simulation, *Phys. Rev. B* **54**, 3809 (1996).

[38]J. S. Tse, D. D. Klug, and P. L. Page, High-pressure densification of amorphous silica, *Phys. Rev. B* **46**, 5933 (1992).

[39]Y. Liang, C. R. Miranda, and S. Scandolo, Temperature-induced densification of compressed SiO_2 glass: A molecular dynamics study, *High Pressure Research* **28**, 35 (2008).

[40]Y. F. Liang, C. R. Miranda, and S. Scandolo, Mechanical strength and coordination defects in compressed silica glass: Molecular dynamics simulations, *Physical Review B* **75** (2007).

[41]B. Liu, J. Y. Wang, Y. C. Zhou, and F. Z. Li, Temperature dependence of elastic properties for amorphous SiO_2 by molecular dynamics simulation, *Chinese Physics Letters* **25**, 2747 (2008).

[42]A. Takada, Molecular dynamics study of pressure induced structural changes in B_2O_3, *Phys. Chem. Glasses* **45**, 156 (2004).

[43]A. Takada, Molecular dynamics simulation of deformation in SiO_2 and Na_2O-SiO_2 glasses, *Journal of the Ceramic Society of Japan* **116**, 880 (2008).

[44]A. Takada, C. R. A. Catlow, and G. D. Price, Computer modeling of B_2O_3: part I. New interatomic potentials, crystalline phases and predicted polymorphs, *J. Phys.: Condens. Matter* **7**, 8659 (1995).

[45]A. Takada, C. R. A. Catlow, and G. D. Price, Computer modeling of B_2O_3: part II. Molecular dynamics simulations of vitreous structures, *J. Phys.: Condens. Matter* **7**, 8693 (1995).

[46]A. Takada, P. Richet, C. R. A. Catlow, and G. D. Price, Molecular dynamics simulation of polymorphic and polyamorphic transitions in tetrahedral network glasses: BeF_2 and GeO_2, *Journal of Non-Crystalline Solids* **353**, 1892 (2007).

[47]K. Trachenko, V. V. Brazhkin, G. Ferlat, M. T. Dove, and E. Artacho, First-principles calculations of structural changes in B_2O_3 glass under pressure, *Physical Review B* **78** (2008).

[48]K. Trachenko and M. T. Dove, Densification of silica glass under pressure, *J. Phys.: Condens. Matter* **14**, 7449 (2002).

[49]J. S. Tse, D. D. Klug, and Y. Le Page, High-pressure densification of amorphous silica, *Phys. Rev. B* **46**, 5933 (1992).

[50]L. P. Huang and J. Kieffer, Amorphous-amorphous transitions in silica glass. I. Reversible transitions and thermomechanical anomalies, *Phys. Rev. B* **69**, 224203 (2004).

[51]L. P. Huang and J. Kieffer, Amorphous-amorphous transitions in silica glass. II. Irreversible transitions and densification limit, *Phys. Rev. B* **69**, 224204 (2004).

[52]L. P. Huang and J. Kieffer, Thermomechanical anomalies and polyamorphism in B_2O_3 glass: A molecular dynamics simulation study, *Phys. Rev. B* **74** (2006).

[53]L. P. Huang and J. Kieffer, Anomalous thermomechanical properties and laser-induced densification of vitreous silica, *Appl. Phys. Lett.* **89** (2006).

[54]L. P. Huang and J. Kieffer, Polyamorphic transitions and thermo-mechanical anomalies in network glasses, *Glass Sci. Technol.* **77**, 124 (2004).

[55]L. P. Huang, L. Duffrene, and J. Kieffer, Structural transitions in silica glass: thermo-mechanical anomalies and polyamorphism, *J. of Non-Cryst. Solids* **349**, 1 (2004).

[56]L. Huang, J. Nicholas, J. Kieffer, and J. Bass, Polyamorphic transitions in vitreous B_2O_3 under pressure, *Journal of Physics-Condensed Matter* **20** (2008).

[57]G. Ferlat, T. Charpentier, A. P. Seitsonen, A. Takada, M. Lazzeri, L. Cormier, G. Calas, and F. Mauri, Boroxol rings in liquid and vitreous B_2O_3 from first principles, *Physical Review Letters* **101** (2008).

[58]L. P. Huang and J. Kieffer, Molecular dynamics study of cristobalite silica using a charge transfer three-body potential: phase transformation and structural disorder, *J. Chem. Phys.* **118**, 1487 (2003).

[59]H. C. Andersen, Molecular dynamics simulations at constant pressure and/or temperature, *J. Chem. Phys.* **72**, 2384 (1980).

[60]L. P. Huang, L. Duffrene, and J. Kieffer, Structural transitions in silica glass: thermo-mechanical anomalies and polyamorphism, *J. Non-Cryst. Solids* **349**, 1 (2004).

[61]M. N. Kul'bitskaya, S. V. Nemilov, and V. A. Shutilov, *Sov. Phys.: Solid State* **16**, 2319 (1975).

[62]M. A. Ramos, J. A. Moreno, S. Vieira, C. Prieto, and J. F. Fernandez, Correlation of elastic, acoustic and thermodynamic properties in B_2O_3 glasses, *Journal of Non-Crystalline Solids* **221**, 170 (1997).

[63]R. E. Youngman, J. Kieffer, J. D. Bass, and L. Duffrène, Extended structural integrity in network glasses and liquids, *J. Non-Cryst. Solids* **222**, 190 (1997).

[64]G. E. Gurr, P. W. Montgomery, C. D. Knutson, and B. T. Gorres, The Crystal Structure of Trigonal Diboron Trioxide, *Acta. Cryst.* **B26**, 906 (1970).

[65]P. A. V. Johnson, A. C. Wright, and R. N. Singlair, A neutron diffraction investigation of the structure of vitreous boron trioxide, *J. Non-Cryst. Solids* **50**, 281 (1982).

[66]C. S. Zha, R. J. Hemley, H. K. Mao, T. S. Duffy, and C. Meade, Acoustic Velocities and Refractive-Index of SiO_2 Glass to 57.5-Gpa by Brillouin-Scattering, *Physical Review B* **50**, 13105 (1994).

[67]J. S. Schilling, The use of high pressure in basic and materials science, *Journal of Physics and Chemistry of Solids* **59**, 553 (1998).

[68]T. Grande, J. R. Holloway, P. F. McMillan, and C. A. Angell, Nitride Glasses Obtained by High-Pressure Synthesis, *Nature* **369**, 43 (1994).

[69]K. L. Ngai and S. Capaccioli, Impact of the application of pressure on the fundamental understanding of glass transition, *Journal of Physics-Condensed Matter* **20**, 244101 (2008).

[70]L. Wondraczek, S. Sen, H. Behrens, and R. E. Youngman, Structure-energy map of alkali borosilicate glasses: Effects of pressure and temperature, *Physical Review B* **76** (2007).

[71]A. C. Wright, C. E. Stone, R. N. Sinclair, N. Umesaki, N. Kitamura, K. Ura, N. Ohtori, and A. C. Hannon, Structure of pressure compacted vitreous boron oxide, *Phys. Chem. Glasses* **41**, 296 (2000).

[72]S. K. Lee, K. Mibe, Y. W. Fei, G. D. Cody, and B. O. Mysen, Structure of B_2O_3 glass at high pressure: A B-11 solid-state NMR study, *Phys. Rev. Lett.* **94**, 16557 (2005).

[73]ORNL: http://www.nccs.gov/jaguar/. The Cray XT Jaguar, has more than 180,000 processors, each with 2 gigabyte of local memory. Rated at 1.64 petaflops, it is currently the fastest general purpose supercomputer in the world.

DYNAMIC BEHAVIOR OF THICK ALUMINA PLATES WITH TUNNELED INTERFACES

F. Orgaz (*), T. Gómez del Rio (**), A.Varela (***) and J.F. Fernández (*).
(*) Instituto de Cerámica y Vidrio. Campus de Cantoblanco. Madrid. Spain
(**) LICAM. Universidad Rey Juan Carlos. Campus de Mostotes. Madrid.
(***) Instituto Nacional de Técnica Aeroespacial. Torrejón de Ardoz. Madrid. Spain

ABSTRACT
        The dynamic behaviour of alumina plates thermally joined with different tunneled interfaces has been studied and compared with monolithic alumina. The tunneled interface was designed to consist of three tape casted layers of which the inside is the weakest. The structure has been built by sandwiching the tunneled interface between thick sintered alumina uni-axially pressed plates. The mechanical behaviour of different tunneled interfaces under impact loads have been studied by means of dynamic compression tests performed in a split Hopkinson pressure bar (SHPB). In these experiments, the stress-strain curves of the laminates at high strain rates and the capacity of transmitting and reflecting the impact energy have been determined. The influence of the nature of the tunneled interface on the fragmentation of these materials has also been considered. The experimental results show a larger capacity of these laminated materials to reflect energy in comparison with monolithic alumina. This high capacity to reflect energy produces larger fragments. A direct relationship between reflected energy (%) and average fragment size is observed. The analysis of the SHPB recovered fragments also show that the prediction of the Mott and Linfoot´s statistical function provides an adequate fit to the experimental data.

INTRODUCTION

        Ceramic materials have been extensively used in armour applications for both personnel and vehicle protection. Ceramics such as Al2O3, SiC, B2Ti and B4C have been used in integrated armour for several decades. The next generation of armour systems require integration of several attributes with hybrid structures and configurations which can be accomplished though introduction of new concepts in the material architecture design. New materials /structures must be created in such a manner that they are light weight, impact resistive, have structural integrity and at the same time can have signature management and controlled communication capabilities. Layered armour concept has been used in many vehicle and personnel armour designs.[1, 2] It is necessary incorporate novel ideas in ceramic armour design so as to develop improved armour with minimal added mass.

        Adding a ceramic plate on top of a metal plate has been known to significantly enhance the ballistic protection over the monolithic metal armour since the 1960s and 1970s. Providing a fundamental understanding of the penetration and the dynamic fracture and damage processes produced by the projectile impact are primary objectives for the design of new armour systems. When a protective armour system is subjected to a high energy impact process, the incident energy of the penetrator is reflected, absorbed and transmitted through the armour system (residual energy). This residual energy must be zero to avoid the human and material catastrophic effects of not defeating the threats. Traditionally the strategy used to design armour systems has been focused to develop systems having a high capacity to absorb the energy transferred by the penetrator and a deep understanding of the penetration process has been developed in the scientific literature in the last 40 years.[3,4] Various techniques have been used to increase the ballistic performance to areal density ratio in ceramic armour materials. Extrinsic and intrinsic means have been used to damage mitigation in ceramics. Extrinsic processes such as prestressing, confinement, encapsulation [5-8] and surface copper buffers [9] have been used to enhance the ballistic performance. Improved ceramic performance is obtained by improved back plate support and lateral confinement in addition to improved toughness. Front confinement of ceramics results in a greater overall fragmentation and less amount of very fine ceramic powder formed. Intrinsic approaches are more directly related to the knowledge of the impact damage, creative material design and low -defect materials processing. More recent advances have been addressed to the development of improved ceramic armour materials capable of mitigating significant damage using high toughness cermets [10] and membrane wrapping processes[11]. This front face confinement produced by membrane wrapping alters the flow and the velocity of the pulverized ceramic that is ejected out and produces a greater erosion and reduced velocity of the projectile. This improvement is mainly a result of impact face constraint that the tape provides. However, this desired understanding on the impact response of materials is far from comprehensively developed.

The fundamental understanding of the mechanisms by which ceramic materials deform and fracture during dynamic impact loading has been advanced tremendously by the development, modification and augmentation of such techniques as Taylor impact, Kolski and split-Hopkinson bar (SHPB), plate impact, explosive cylinder and spherical cavity expansion. Even when higher-rate experiments with sufficient time resolution are desired to reveal the dynamic fracture and failure processes in ceramic targets because the local strain rate in a ceramic target just ahead of the penetrator is higher than those achievable on a Hopkinson bar, predictive ballistic performances have been inferred using these techniques. A review has been recently published [12] on the dynamic fracture of ceramics in the framework of the ceramic armour penetration processes.

The split Hopkinson bar is being widely used to determine the dynamic compressive strength of ceramics and ceramic composites. Pulse shaping techniques to obtain compressive strain-stress data for testing brittle materials with SHPB were developed by Frew et al [13] Stachler et al [14] used SHPB tests to study the failure of a high strength of alumina at strain rates of the order of 1000 s^{-1}. There appears to be a critical strain rate above which the traditional method of using the transmitter bar signal to calculate the stress in the specimen is no longer valid. Ravichandran and Subash [15] have shown that the highest strain rate at which ceramics can be tested using SHPB without violating the underlying assumptions is found to be in the range of 2500-3000s^{-1}. Dynamic fracture under multi-axial stresses have been studied by Nie and Chen [16] using cuboid borosilicate glass specimens with the material axis inclining to the loading direction at different angles. Recent investigations have also been addressed to visualize the origin, growth, coalescence and propagation of cracks produced by SHPB experiments in confined and unconfined materials. High speed photographs were correlated in time with measurements of the stresses in the sample[17-18,12] Dynamic behaviour of damage/ comminuted ceramics is also been investigated.[12]

In this paper, the dynamic behaviour of alumina plates thermally joined with different tunneled interfaces is analysed by performing dynamic compression tests in Split Hopkinson pressure bar. This technique is used to obtain the stress strain curves at high strain rates as well as energy reflection and energy transmission coefficients. The research is focused in the framework of ceramic-armour penetration processes. The ceramic system of alumina and tunneled interface was designed to stop the penetration of KE projectiles. The concept of tunneled interface was performed to introduce ductility thought the thickness of a ceramic material by layered methods. We study the dynamic fracture behaviour of these ceramic laminates with special interfaces under various conditions in loading rates, stress states, and loading histories. Our interest have also been to analyse the post mortem fragmentation effects produced by the tests and to correlate the fragmentation results with the reflected, transmitted and absorbed energies involved in the rupture produced by these dynamic compressive tests. Because the transmitted KE may have severe consequences on many light structures or personnel behind the armours, it is thus desired to develop novel concepts for armour designs that will not only stop penetrators but also dissipate KE effectively by reflection or other mitigation mechanisms. When ceramic tiles are impacted by bars or projectiles fail through a complex combination of processes resulting from the shock wave propagation and reflection. These processes include fragmentation and formation of radial and circumferential macro-cracks, and pulverization of the ceramic into fine powder. Because the fracture and pulverization of the ceramic material are effective ways to dissipate part of the kinetic energy (KE) generated by the impact, it is expected that from the statistical distribution function and from the morphology of the pulverized ceramic fragments and their causing energies, relevant information can be drawn on the factors governing the impact and penetration resistance of these materials.

PREPARATION OF MATERIALS

A typical layered system consisting of two alumina plates and a triple tunneled interface is shown schematically in Fig. 1. The tunneled interface was designed to consist of three tape casted layers of which the inside is the weakest and tougher to dissipate energy. The internal layers studied were constituted either by steel fibres (sample E8C1) or by a mixture of large grain size silicon carbide and aluminium metal (samples E8C4, E8C9 and E8C10). The external layers were also formed by a silicon-sodium-calcium glass in order to join the internal layers to the alumina plates. The structure has been built by sandwiching the tunneled interface between sintered alumina uni-axially pressed plates.

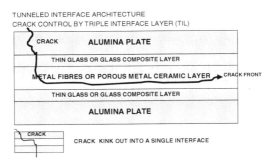

Figure1. A typical layered system consisting of two alumina plates and a triple tunneled interface

Commercially available alumina powders (Grade sg3000. Alcoa) was used as the starting material for the preparation of thick alumina plates. The powder was shaped by uniaxial pressing and sintered at 1575 °C for two hours at a heating rate of 5°C/minute. The microstructure of the sintered alumina is shown in figure 1. Several types of specimens were used: rectangular specimens of dimensions 38x 29x 5 mm for determination of Young´ modulus and for estimation of the interfacial fracture resistance. Small discs of diameter 6,5 mm and thickness 7 mm were used for SHPB tests

For the preparation of triple tunneled interfaces a tape casting process was used. In table I the composition of the central and external layers of the tunneled interface is shown. For each layer the tapes were prepared from water slips formed by an ammonium salt of a polyacrylic acid as dispersant agent to produce low-viscosity stable suspensions, a copolymeric styrene-acrylic latex (Mowilith DM 765S, Hoechst Perstorp AB) as binder and polyethylene glycol as plasticizer. The solid content of the slurries was 50 wt%. The polymer layer formed on drying has a glass transition temperature which makes it flexible and elastic at room temperature. Batches of 500 g of powder were dispersed in water by ball milling for 2 h in 4-litre HDPE plastic jars with 500 g of alumina balls and the optimal amount of dispersant. After ball-milling, the latex binder was added and gently mixed for approximately 1 h. The slip was then sieved through a cloth to remove agglomerates and entrapped air. Maximisation of solid loading of the starting suspension is important to avoid particle segregation effects, to minimise the shrinkage during drying and to achieve a high green density.

Table I. Composition of the tunneled interfaces

Sample	Triple tunneled interface	
	Two external layers	Central layer (weight ratio)
E8C1	Silicon sodium calcium commercial glass powder (<100 microns)	Metallic fibres (RIMSA) + glass powder (40%)
E8C4	Silicon sodium calcium commercial glass powder(<100 microns)	SiC + Aluminium metal (40%)
E8C9	Silicon sodium calcium commercial glass powder (<100 microns)	SiC + Aluminium metal (60%)

In the tape casting experiments, a continuous tape caster with a stationary casting head was used. The machine is specially designed for making tapes with a wide range of thicknesses. Most of the tapes formed had a thickness of approximately 1,5- 2 mm. The drying section was designed in such a way as to allow rapid drying and, at the same time, reduce the risk of skin formation and cracking. Heat is supplied from underneath the tape to improve the diffusion of solvent to the surface. Tape casting experiments were made at speeds from 1 to 5 cm/s with doctor blades. The drying conditions were the same in all experiments, the temperature of the heated sections was 50°C and the exhaust air 30°C.The carrier film was a commercial Mylar plastic film.

For all the single tapes, a simultaneous thermal analysis was performed (Netzsch STA 409 or Setaram Setsys instrument). Figure 2 a and 2.b show the results obtained by steel fibre tapes (E8C1) and for a silicon carbide- aluminium metal system (sample E8C4)

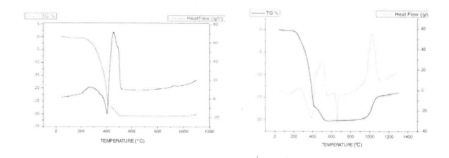

Figure 2a and 2b.-Thermal analysis of tapes E8C1 (left side) and E8C4 (right side) respectively

As shown, the most part of the weight loss began at about 200°C and was completed at around 500°C. According to these results, the adopted debinding heating cycle was 2 °C/min up to 500° C, and a dwell of 1 h at 500°C to allow complete binder removal. No oxidation of the metallic fibres was observed up to 1100°C. Glass powder was mixed with the metallic fibres to facilitate the dispersion into the tape casting slip and to avoid fibre oxidation by coating. Figure 3 shows the DTA- TGA curves for the sample E4 formed by a mixture of silicon carbide and aluminium metal powder. As shown in figure 3, the aluminium powder melts at 700°C and starts oxidizing at approximately 950°C

In the lamination step, disc-shaped specimens of 100 mm diameter were punched from the green tapes and stacked to form a triple tunneled interface. These stacks were then hot pressed for 5 min at pressures of 70 MPa and 80°C. Disc-shaped specimens of 6 mm diameter were punched from these stacks for SHPB tests and samples of 60x20 mm to perform interfacial fracture-resistance measurements. The tunneled interfaces were glued to the alumina plates by a commercial adhesive. Laminated plates and discs with tunneled interfaces were joined at temperatures of 1050°C for 1/4 h and a heating rate of 5°C/min. with furnace cooling, in order to increase the cohesion between alumina and the tunneled interface Apparent density of the laminated and green tapes was calculated by measuring the thickness and weight of the disc-shaped specimens. Densities of the sintered bodies were measured by Archimedes' technique in distilled water.

RESULTS

Microstructure of the tunneled interfaces.
Figures 3 and 4 show the microstructure of the interfaces corresponding respectively to the samples E8C1 and E8C4. As observed, the interface of the sample E8CI is formed by metallic fibres coated by glass with a very porous and interlocking interface. A strong bond to the alumina plates is also formed through the external glass layer. A porous interface is also formed between the silicon carbide grain and the aluminium phase. For the interface corresponding to the sample E8C4 an inhomogeneous and porous interface was formed. Large grains of silicon carbide are observed surrounded by molten aluminium.

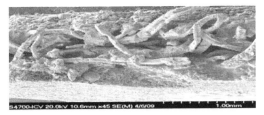

Figure 3. SEM of the interface corresponding to the sample E8C1.

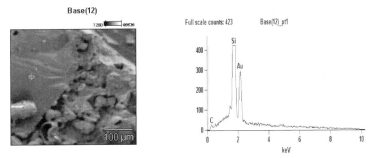

Figure 4. SEM of the interface corresponding to the sample E8C4

Modulus of Young

The Young's modulus of the specimens was measured using an impulse-excitation technique (Grindo-Sonic MK4x), according to ASTM Method C 1259-94. The Young's modulus (E) results are shown in Table II. The values of E for the laminate are lower than the alumina value and seems to increase as the aluminium content in the interface decreases following the Voigt rule of mixtures. The density values measured are also shown in Table II

Table II. Young's modulus and volumetric density of the laminates

Sample	Young's modulus(E) GPa	Volumetric density (g/cm3)
Alumina	312	3,70
E8C1	223	3,49
E8C4	96,3	3,25
E8C9	73,3	3,23
E8C10	117	3,25

DYNAMIC FRACTURE UNDER UNIAXIAL COMPRESSION.SPLIT HOPKINSON PRESSURE BAR (SHPB) RESULTS.

The dynamic compressive stress–strain responses of the materials was studied using a modified Split Hopkinson pressure bar (SHPB) with ramp loading pulses to generate families of stress–strain curves at controlled strain rates. This experimental apparatus permits time resolved analysis of material response to transient compressive loading SHPB, originally developed by Kolsky [19] , which has been modified to determine the dynamic constitutive behaviours of a variety of brittle materials including concrete and ceramics. The focus of these experimental investigations is typically on the dynamic stress– strain response, which is another important aspect of ceramic impact response, rather than on dynamic fracture behaviour. The details of SHPB and its working principle are well described by Follanbee [20]

Experimental set-up

A sketch of the SHPB device used is described by T. Gómez del Rio [21]. The device consists of a gas gun, an input bar and an output bar, the supports, and a data acquisition system. Both bars are of high-yield silver-steel, 22 mm in diameter and 1 m long, which can move horizontally without any restriction. The experimental set-up works in the following way: an air gun impels a projectile against one end of the incident bar connector and generates a tensile pulse which travels along the incident bar. When this pulse reaches the specimen, it is reflected partially and partially transmitted to the specimen and to the output bar. The stress pulses in the bars are measured by strain gauges (VISHAY CEA-06-12UN-350) attached to the bars at the central point. The strain gauge signals are recorded using a VISHAY 2200 signal conditioner and a TEKTRONIX TDS 420A digital oscilloscope for data point acquisition. The projectile is launched at a few metres per second. In all the tests of this work the projectile was impelled at 10 m/s. To prevent the ceramic specimen from indenting into bar-end faces and thus causing stress concentrations along specimen edges, a pair of high-stiffness tungsten carbide (CW 3%Co) plates with mechanical impedance matching with the bars were placed between the specimen and the bars. These plates have a diameter of 12 mm and a thickness of 8 mm. The plates remain fully elastic and sustain no damage during the experiments. A pulse shaper was used at the impact end of the incident bar to generate incident pulses of linear ramps, which was necessary to deform the specimen

with a nearly linear response at constant strain rates. The loading was provided by the striker impacting on the incident bar through a composite pulse shaper consisting of a 10 mm diameter and 4 mm thickness annealed copper disk. The dimensions of the cylindrical specimens used in this research were 6.35 mm in diameter and 6.35 mm in length. The specimens were uniaxially pressed within 0.005 mm cylindricity. A cylindrical specimen is placed between two high strength steel bars and a stress pulse is applied to the system by means of the impact of a projectile. The stress wave propagates along the bar up to the bar/specimen interface where it is partially reflected and partially transmitted to the second bar. Assuming that both bars behave elastically, measurements of strains in the bars provide the dynamic stress-strain curve of the specimen material or the energy reflected and transmitted in the bar/specimen interface, according to the following expressions

$$\sigma(t) = (S_b / S_s) E \varepsilon_T(t), \qquad \varepsilon(t) = \int_0^t (2c/l) \varepsilon_R(t) \qquad (1)$$

$$E_R = (1/2) E S_b c \int_0^t \varepsilon^2_R(t)\, dt \qquad (2)$$

$$E_T = (1/2) E S_b c \int_0^t \varepsilon^2_T(t)\, dt \qquad (3)$$

where $\sigma(t)$ and $\varepsilon(t)$ are the stress and strain in the specimen; ε_R and ε_T the strain measured in the bars corresponding to the reflected and transmitted wave, S_s and l the section and length of the specimen; S_b, E and c the section, young modulus and wave velocity of the bar; and E_R and E_T the energy reflected and transmitted by the specimen, respectively. After impacting, the specimen fragments were collected for post-mortem analysis and characterization of microscopic failure. Small plastic bags partially wrapping the incident and transmission were placed below the test samples to collect all fragments produced.

Results and discussion

Figure 5 shows the original oscilloscope wave records of the sample E8C1 heat treated at 1050°C for 15 minutes subjected to a dynamic compressive test with the Hopkinson bar. A high amplitude of the reflected wave and a low amplitude of the transmitted wave is observed.

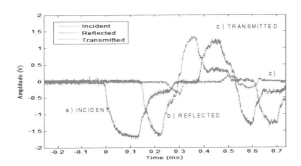

Figure 5   Original oscilloscope records for the sample E8C1 heat treated at 1005°C for 15 minutes subjected to a dynamic compressive test using the Hopkinson bar

Figures 6 show the dynamic stress-strain curves for Alumina and E8C4 laminates heat treated at 1050°C for 15 minutes. The strain stress curves using the Hopkinson bar exhibit three characteristic stages: an initial nearly linear elastic deformation region, a yield point, a plateau and a deep (Alumina) or smooth decrease to crushing (laminates). The alumina samples showed a higher yield stress and a lower strain than the laminated samples. The dynamic stress-strain curves are similar for all the laminates tested. For the layered systems it is noted that the higher the porosity, the longer the plateau region is, but higher porosity implies lower yield stress. These phenomena indicate that the system could deform and yield more easily as the porosity and the plastic deformation increases. It can be seen from the stress-strain curves that there exits only one crushing stress at low strain for monolithic alumina, which is different from that of the layered system. However the strength (yield stress of near 3000 MPa) is higher for monolithic alumina that for laminates (yield stresses between 500-700 MPa) due to its high density. It is to be considered that the stresses are always obtained directly from the transmitted wave to the second instrumental bar, because this is the usual way to determine the stress in the specimen in the Hopkinson bar experiments. It should also be noted that the maximum value reached by the stress in the dynamic tests does not represent any material failure, but only the maximum stress applied to the specimen during the test before the unloading. Therefore, the yield stress is not the better way to analyse the dynamic fracture of materials but its capacity to reflect, transmit or absorb energy at different strain rates. Another important aspect to be considered is the relation between the energies involved during the dynamic impact tests and the level of fragmentation and pulverization produced, as analysed later on.

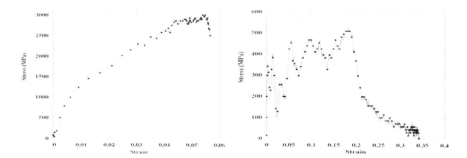

Figures 6 Dynamic stress- strain curves for Alumina (left side) and for E8C4 laminates (right side) heat treated at 1050°C for 15 min

The capacity of the laminates studied to reflect or transmit the impact energy in relation to the alumina monolith is analysed in Table III. Two conclusions can be directly derived: first, the laminated samples reflect much more energy that monolithic alumina; and second, the capacity reflected depends on the nature of each laminate and on the heat treatment performed to joint tunneled interface to alumina plates. As observed, the monolithic alumina reflects an energy of 39,2%, a transmitted energy of 34,6 and an absorbed energy of 25,9% for a strain rate of 3480 s^{-1}. The laminated specimens show higher reflected energies ranging from 80.54% for E8C9-1050°C to 88.79% for the sample E8C9. Only 1-2% of the incident energy is transmitted by the alumina laminates. The absorbed energies range from 10.35 to 18.63%. It is also observed that the samples heat treated at lower temperatures of 900°C show a slight increase at the transmitted energy. This can be explained as due to a larger porosity of the samples heat treated at lower temperatures as a consequence of a lower glass viscous flow and therefore a lower interpenetration of the external layer into the internal interface of the tunneled interface.

Table III. Distribution of energies in dynamic compression experiments with a Hopkinson bar

Sample	Incident energy (J)	Reflected energy (%)	Transmitted energy (%)	Absorbed energy (%)	Strain rate s^{-1}
Alumina	63,11	39,20	34,86	25,94	3480
E8C1-1050°C	58,81	88,79	1,46	10,35	2620
E8C4-1050°C	71,05	81,52	1,18	17,65	2430
C8C9-1050°C	63,96	80,54	1,30	18,63	2470
E8C1-900°C	39,65	89,24	0,97	9,8	1790
E8C4-900°C	41,67	88,51	1,96	9,52	1570
C8C9-900°C	40,43	89,04	2,66	8,30	1570

DYNAMIC FRAGMENTATION. PATTERNS AND MODELLING.

Experimental procedure
        Post-mortem analysis of the fractured samples after being tested with the Hopkinson bar was performed by analysing the fragmentation size and by fractography of the fractured samples using optical and Scanning electron microscopy (SEM). Distribution of fragment sizes was characterized through sieving techniques where the particles are classified in terms of their ability or inability to pass through an aperture of controlled size. SHPB recovered fragments were introduced onto a stack of sieves with successively finer apertures below and agitated by mechanical pulses to induce translation. Sieves with apertures of 50,100, 315, 1000 and 2000 microns were used. After meshing, the direct measurement of the mass of each fraction of fragments was weighted.. For the specimens analysed, the total weight of the fragments was around M ≈ 0.60 g. All the fragments were considered spherical because the precision of sieve analysis depends somewhat on the aspect ratio of the particle.

Results and discussion
        The measured distributions of recovered fragments are shown in Fig. 7 which represents the cumulative weight of fragments with size > n as a function of the fragment size, n. As observed, larger fragments were recovered from laminated in comparison to monolithic alumina, which can be explained as due to a lower rupture stress and a higher toughness of the laminates.

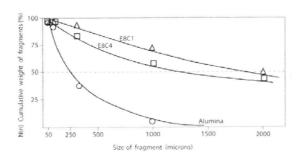

Figure 7 Cumulative weight of SHPB recovered fragments as a function of the fragment size

A preliminary relationship between average fragment size and SHPB reflected energy (%) is observed in Figure 8. The average fragment size was calculated from the value at which the cumulative weight of fragments is 50%.

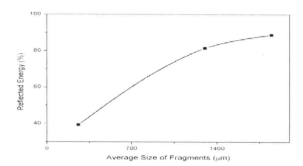

Figure 8. Relationship between SHPB reflected energy and average size of fragments

Statistics, energy, cohesive and computational models have been used to analyse the dynamic fragmentation process. Substantial progress has been made in the theoretical and numerical treatment of fragmentation, and especially in the cohesive treatment of the initiation, propagation, and coalescence of microcracks. Subhash et al [22] have reviewed recent advances in fragmentation modelling. The fragmentation under dynamic conditions involves the nucleation, growth, and coalescence of a network of cracks [23] which depend on the loading rates [24] and significantly influenced by the microscopic heterogeneities, microstructure[25] and on the nature and configuration of the materials and systems. Lienau [26], Mott and Linfoot [27] and Grady and Kipp [28] and Grady [29] 'have proposed models based on statistical assumptions

Mott and Linfoot proposed the seminal work on fragmentation in the classified literature and later in the open literature. They forwarded a geometric statistics-based theory of fragmentation based on the random creation of cracks and their interaction through unloading waves to explain the fragment size distribution observed in dynamically expanding rings. One of the central assumptions of this work is that energy requirement in the actual fracture process is negligible and that fracture is instantaneous while Griffith assumed that all available energy is used up in creating free surfaces. According to Mott and Linfoot, the cumulative number of fragments N (n) is described by

$$N(m) = N_0 \ \exp. (- \ (3N_0 m)^{1/3})$$
(4)

or
$$N(n) \ = \ K \ \exp. (- \ (2 \ \lambda \ n)^{1/2})$$
(5)

where K is the total number of fragments, n is the fragment area, and λ is a fitting parameter given by Ck=A with A denoting the total area of fragments and C a constant.

Grady and Kipp [29] have further improved the models. They assume that the probability of fracture is spatially uniform and that all points in a body are accessible to fracture, and adjacent fracture sites can be arbitrarily close to each other. In application, an event can be regarded as continuous if the average fragment size is large relative to the minimum fragment size. For the specimens analysed here, the minimum fragment size is typically less than one-fourth of the average fragment size. Thus, the assumption of continuous fracture is valid. Under such conditions, they have obtained the following distribution function for dynamic fragmentation:

$$N(m) = N_0 \exp. \{ (M/\mu - 1) \ln(1 - m/M) \} \qquad (6)$$

where M is the total mass of the body and μ the average fragment mass.

Grady and Kipp [29] have also proposed energy-based models in terms of the propagation of the microcracks and their interaction and coalescence during the fragmentation event. Their models rely on an energy balance between the surface energy released due to fracture and the kinetic energy of the generated fragments, and provides a simple relation between the average fragment size L and the strain rate $\acute{\epsilon}$ as

$$L = 24^{1/3} (K_{IC}/\rho c \acute{\epsilon})^{2/3} \qquad (7)$$

where K_{IC}, r and c respectively denote the fracture toughness of the material, density and dilatational wave speed of the material

Figure 9 shows the above results with the predictions of Equation 5 where Neperian logarithm of the cumulative weight of fragments is plotted against the square root of the fragment size.

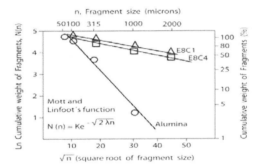

Figure 9. Neperian logarithm of the cumulative weight of fragments against the square root of the fragment size

As observed the prediction of Mott and Linfoot (equation 5) provides an adequate fit to the experimental data. Both plots clearly show the samples configured with tunneled interfaces give rise to higher average fragment sizes than those obtained with monolithic alumina when they are subjected to dynamic fracture using SHPB experiments .Small differences are observed between samples E8C1 and E8C4. According to the predictions of Gray and Kipp given by equation 7, a larger L means a higher dynamic toughness of the laminated samples in relation to monolithic alumina at the same strain rate. The fitting parameter λ from equation 5 can then be associated to the dynamic toughness of the sample. Therefore, the analysis of the dynamic fragmentation process using SHBP experiments can be used for comparing the dynamic toughness of different materials.

Research is in progress to establish relationships between the slopes of the Mott and Linfoot distribution function of the recovered fragments and the toughness of different materials subjected to SHPB experiments Figure 10 shows the morphology of the fracture surface (fractography) of SHPB recovered fragments of the sample E8C4. As seen, the failure process bears resemblance to the morphology of the alumina microstructure. The alumina fragments are smooth indicating trans and intra granular cleavage of the Al2O3 grains. .The cleavage planes span over multiple grains.

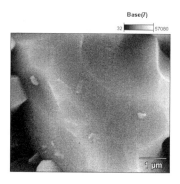

Figure.10   SEM of SHPB fragments. Sample E8C4 heat treated at 1050°C

CONCLUSIONS
        Several conclusions can be drawn from the dynamic behaviour of laminated ceramics formed by alumina plate heat joined by different tunneled interfaces.
        1.- The mechanical response under impact loads by means of dynamic compression tests performed in a split Hopkinson pressure bar (SHPB) shows a different behaviour between alumina monoliths and laminates with tunneled interfaces
        2.-The stress-strain curves of the laminates at high strain rates always show a lower strain and a higher yield stress for monolithic alumina.
        3.-The experimental results show a larger capacity of these laminated materials to reflect energy during the SHPB tests. Values of 80-85 % of reflected energy were observed
        4.- A direct relationship between reflected energy (%) and average fragment size is observed.
        5.- The analysis of the SHPB recovered fragments also show that the prediction of the Mott and Linfoot´ statistical function provides an adequate fit to the experimental data.

Further studies are necessary to study the dynamic behaviour under impact loads and the capacity to reflect energy of triple tunneled interfaces located on the front face of the alumina plates, to analyse more deeply the relation between reflected energy and microstructure of the interfaces and to evaluate the relationships between these SHPB dynamic results and the ballistic performances.

REFERENCES

[1] W.A.Gooch. Overview of ceramic armour applications. *Ceramic transactions,Vol 134. 3-21. Ceramic Armour materials by design. Edited by M.McCauley et al. American Ceramic Society,Westerville.Ohio,2001*

[2] W.A.Gooch. Overview of the development of ceramic armour technology-Past, present and the future. Presentation at the 30th international conference on advanced ceramics and composites. American Ceramic Society Cocoa Beach, FL, January 22-27,2006

[3] M.L.Wilkins R.L.Landingham and C.A.Honodel. Fifth progress report of light armour program. UCRL-50694.Lawrance Radiation Laboratory,1971

[4] D.M.Steep. Damage mitigation in ceramics: Historical developments and future directions in army research. *Ceramic transactions, Vol 134, 421-428. Ceramic Armour materials by design. Edited by M.McCauley et al. American Ceramic Society, Westerville.Ohio,2001*

[5] J.M.Wells. On the role of impact damage in armour ceramic performance. *Advances in ceramic armour II, 227- 235. 30th international conference on advanced ceramics and composites. John Wiley and Sons.2007*

[6] G.E. Hauver, P.H. Netherwood, R.F. Benck and L.J. Kecskes. Ballistic performance of ceramic targets. *Army Symposium on solids mechanics, USA (1993)*

[7] N.L. Rupert, W.H.Green, K.J. Doherty and J.M.Wells. Damage assessment in TiB2 ceramic armour targets, Part II. Radial cracking. *Proc MSMS2001,AU 137-143 (2001)*

[8] W.J. Bruchey and E.J. Horwath. System consideration concerning the development of high efficiency ceramic armours. www.dtic.mil/matris/sbir/sbir012/a01-039a pdf

[9] T.J. Holmquist, C.E. Anderson and T. Behner. Design, analysis and testing of an unconfined ceramic target to induce dwell .*Proceeding of 22 nd IBS,APA ,v2,860-868 (2005)*

[10] M. Wilkins and R.L. Landingham. Prologue. *International Journal of Applied Ceramic Technology. 1,203-204 (2004)*

[11] S. Nemat-Nasser, S. Sarva, J.B. Isaacs and W. Lischer. Novel ideas in multifunctional ceramic armour design *Ceramic transactions, Vol 134, 421-428. Ceramic Armour materials by design. Edited by M. McCauley et al. American Ceramic Society, Westerville.Ohio,2001*

[12] W.W. Chen, A.M. Rajendran, B. Song and X. Nie. Dynamic fracture of ceramics in armour applications. *J.Am. Ceram. Soc 90, 1005-18 (2007)*

[13] D.J. Frew, M.J. Forrester, and W. Chen. Pulse shaping techniques for testing brittle materials with a split Hopkinson pressure bar. *Experimental Mechanics., 42,93-106 (2002)*

[14] J.M. Staechler, W.W. Predebon, B.J. Pletka and J. Landford. Testing of high-strength ceramics with the split Hopkinson pressure bar. *J.Am.Ceram. Soc 76, 536-538 (1993).*

[15] G. Ravichandran and G. Subash. Critical appraisal of limiting strain rates for compression testing of ceramics in a split Hopkinson pressure bar. *J.Am.Ceram. Soc 77, 263-67 (1994)*

[16] X. Nie and W.W. Chen. Dynamic failure of borosilicate glass under compressive/shear loading experiments. *J.Am.Ceram. Soc 90, 2556-62 (2007)*

[17] T. Jiao,Y. Li, K.T. Ramesh and A.A.Wereszczak. *Int.J.Appl. Ceram. Technol. 1,243-53 (2004)*

[18] B. Paliwal, K.T. Ramesh and J.W. McCauley. Direct observation of the dynamic compressive failure of a transparent polycrystalline ceramic (ALON) *J.Am.Ceram. Soc 89, 2128-33, (2007)*

[19] H. Kolsky. An investigation of the mechanical properties of materials at very high rates of loading. Proc.Phys. Soc B62,676-700 (1949)

[20] P.S. Follansbee. The Hopkinson bar. *Metals handbook, Vol 8, 198-203(1985). Ed. American Society of metals. Ohio*

[21] T. Gomez-del Rio, E. Barbero, R. Zaera and C. Navarro. Dynamic tensile behaviour at low temperature of CFRP using a split Hopkinson pressure bar. Composites Science and Technology 65, 61-71 (2005)

[22] G. Subhash, S. Maiti, P.Geubelle, and D.Ghosh. Recent advances in dynamic indentation fracture, impact damage and fragmentation of ceramics. *J. Am. Ceram. Soc.,* 91, 2777-91 (2008)

[23] D.E. Grady and M.E. Kipp. Mechanisms of dynamic fragmentation. Factors governing fragment size. *Mech. Mater.,* 4, 311-20 (1985)

[24] D.A. Shockey. Discussion of mechanisms of dynamic fragmentation. Factors governing fragment size. *Mech. Mater.,* 4, 321-24 (1985).

[25] A.R. Keller and M. Zhou. Effect of microstructure on dynamic failure resistance of titanium diboride / alumina ceramics. *J. Am. Ceram. Soc.,* 86, 449-57 (2003)

[26] C.C. Lienau. Random fracture of a brittle solid. J. Franklin Inst., 221, 485-94 (1936)

[27] N.F. Mott and E.H. Linfoot. Ministry of supply, Report AC 3348 (1943)

[28] D.E. Grady. Particle size statistics in dynamic fragmentation. J. Appl. Phys. 68,6099-105 (1990)

[29] D.E Grady and M.E. Kipp. Geometric statistics and dynamic fragmentation. J.Appl. Phys. 58, 1210-22 (1985)

DEPOSITION PHASE DIAGRAMS FOR CHEMICAL VAPOR DEPOSITION OF BCl$_3$-CH$_4$-H$_2$ SYSTEM

Shanhua Liu[*], Litong Zhang, Yongsheng Liu, Laifei Cheng, Xiaowei Yin, Qingfeng Zeng

National Key Laboratory of Thermostructure Composite Materials, Northwestern Polytechnical University, Xi'an 710072, China

ABSTRACT

Based on the minimization of the Gibbs free energy of the reaction system, applying the FactSage Thermodynamic Software, the phase diagrams of BCl$_3$-CH$_4$-H$_2$ system were calculated. The effect of partial pressure of reactants, total pressures, and deposition temperatures on the phase diagrams was well explained by corresponding law. Optimization of process parameters for a desired deposition could be established from the phase diagrams.

INTRODUCTION

Carbon fiber-reinforced silicon carbide ceramic matrix composites (C/SiC) with outstanding mechanical properties especially at high temperatures are considered as a potential high-temperature thermalstructure materials. However, the gasification of carbon fibers limits their long-term applications in high-temperature oxidizing environments [1,2].

B$_4$C is a candidate to protect C/SiC by forming a B$_2$O$_3$ due to the oxidation reaction which protects carbon fibers from the oxidation at interfacial zones[3]. Chemical vapor deposition (CVD) is a general method to fabricate B$_4$C, and BCl$_3$-CH$_4$-H$_2$ system is seemed as the most commonly adapted mixtures for obtaining the B$_4$C[4-6]. Thermodynamic evaluation is regarded as the first step to grasp the CVD process which is difficult to be controlled and optimized[7,8].

2. PROCEDURES

The thermodynamic data for all chemical species is from the Factsage database (Thermfact/CRCT and GTT-Technologies, Canada and Germany). The temperature of preparation for B$_4$C for self-healing materials of C/SiC is on the order of 1000°C with low pressure[9-11]. Thermodynamic calculations in present work were performed at varying partial pressures and temperatures of reactants. The total pressures range from 1 to 100 kPa and temperatures from 800°C to 1200°C. The phase diagrams were obtained by fitting the equilibrium species as a function of BCl$_3$ and CH$_4$ partial pressures.

3. RESULTS AND DISCUSSION

3.1 The effect of reaction gas partial pressure

The phase diagram at 900°C and a total pressure of 100 kPa is shown in fig.1.

As a whole, besides gaseous phase region, the phase diagram included five condensed phase regions such as single B region, B+B_4C region, B_4C region, B_4C+C region, and free C region. The change in the diagrams can be well explained by affinity of chemical reaction and Gas reaction balance theory [12]

$$A = -RTln(J_P / K^\Phi) \tag{1}$$

$$J_P = \prod(P_B / P^\Phi)^{VB} \tag{2}$$

$$mA_{(g)}+nB_{(g)}=xC_{(s)}+yD_{(g)} \tag{3}$$

$$K^\Phi=P^y_D / (P^m_A * P^n_B) \tag{4}$$

Where A is chemical affinity, related to the negative partial derivative of Gibbs energy with respect to extent of reaction at constant pressure and temperature. It is positive for spontaneous reactions. J_P is the product of the partial pressure of the reactants. V_B is negative when corresponding to reactants and positive value when V_B is corresponding to the species. K^Φ is equilibrium constant. P_B is the partial pressure of each gas component. P^Φ is standard gas pressure.

When $J_p < K^\Phi$, $A > 0$, reaction can be carried out spontaneously.

$J_p = K^\Phi$, $A=0$, reaction achieves balance.

$J_p > K^\Phi$, $A < 0$, reaction could not be carried out (reverse reaction itself possible to carry out).

The effect of H_2 in the reaction system is reduction. Three kinds of condensed species exist in BCl_3-CH_4-H_2 reaction system, such as B, B_4C and C. The related chemical reactions as follows:

$$2BCl_{3(g)}+3H_{2(g)}=B_{(s)}+6HCl_{(g)} \tag{5}$$

$$4BCl_{3(g)}+CH_{4(g)}+4H_{2(g)}=B_4C_{(s)}+12HCl_{(g)} \tag{6}$$

$$CH_{4(g)}= C_{(s)}+ H_{2(g)} \tag{7}$$

When $P_{BCl3} > 0.1Pa$, $P_{CH4} < 4$ Pa, H_2 is excess, the reaction affinity of Eq. (5) increases with the increasing of P_{BCl3}, which makes just the Eq. (5) carry out, However, Eq. (6), (7) could not be carried out because of the low reaction affinity. So there is only B region appears.

When $P_{BCl3} > 0.1$ Pa, $P_{CH4} > 4$ Pa, the partial pressure of H_2 decreases along with the increasing partial pressure of P_{CH4}, which makes the J_P value of Eq. (6) increase. Then Eq. (6) carries out. The phase region changes from B to B+B_4C region. B+B_4C region is seemed as the transition region to the pure B_4C region. From B_4C region, it is noticed that the ratio of P_{CH4}/ P_{BCl3} is greater than 1.

When $P_{BCl3} > 0.1$ Pa and $P_{CH4} > 2$ kPa, free C appears because of the Eq. (7)

carrying out. The condensed species are B_4C+C. When $P_{BCl3} < 0.01$ Pa and $P_{CH4} > 2$ kPa, the condensed specie is only free C.

3.2 The effect of the system total pressure

Figure 2(a, b, c, d) give different thermodynamic phase diagrams at 1000°C and different total pressures. When the system total pressures are increased from1 kPa to 100 kPa, B+B$_4$C, B$_4$C and B$_4$C regions reduce. However, single-B, gas-phase and free C regions expand with increasing total pressure.

The Eq. (5-7) all result in the volume expansion. The percent of expansion are 20%, 33.3% and 100%, respectively. As we know Le-Chatelier law, which states that any change in a system at equilibrium results in a shift of the equilibrium in the direction which minimizes the change. Volume expansion may have a tendency to inhibit Eq. (5-7) to carry out. For each reaction, the less percent of volume expansion, the larger percent of condensed phase region expansion.

Single-B region expands with the total pressure increases significantly because of the least percent of volume expansion of Eq. (5). However, the free C region also expands because of the pyrolysis of excess CH$_4$. The effect to Eq. (7) can be seen from the B$_4$C+C region, which reduces significantly. B+B$_4$C and B$_4$C regions reduce compared with B region. Consequently, from Fig. 2(a, b, c, d), the boundaries among the regions shift up and right with the increasing total pressure, which indicates that it needs more partial pressures of BCl$_3$ and CH$_4$ to obtain the corresponding species.

3.3 Effect of the temperature

The phase diagrams at different temperatures and the same total pressure of 1 kPa are shown in Fig. 3(a, b, c, d). The phase diagrams include three regions such as B+B$_4$C, B$_4$C, B$_4$C+C region. The phase diagrams exhibit a reducing B4C phase region with increasing temperature. As the temperature decreases, the pyrolysis of CH$_4$ becomes slow[13]. Hence, CH$_4$ becomes more stable at lower temperatures, and the deposited carbon has to be obtained at higher partial pressure. Compared with the phase diagrams, the carbon-rich region at 800°C is less than those at high-temperatures. The B$_4$C region reduces with the increasing temperature.

4. CONCLUSION

Thermodynamic phase diagrams of BCl$_3$-CH$_4$-H$_2$ system were researched for understanding the deposition process by corresponding law. Valuable information can be summarized to determine the parameters for fabricating desired deposition from the phase diagrams. The effect of H$_2$ is reduction. The phase diagrams are affected mainly by the ratio of P$_{CH4}$/ P$_{BCl3}$.

The ratio of P$_{CH4}$/ P$_{BCl3}$ is greater than 1 is seemed as the necessary condition to deposit pure B$_4$C. With the increasing total pressure it is more difficult to obtain B$_4$C phase, and the same situation also exists with the increasing temperature.

However, thermodynamics calculation is only conducted on the ideal conditions; there may be some differences from the results on the factual conditions. Complementary experiments are still needed to obtain the desired phases in practice.

REFERENCES
[1]R. Naslain, "Design preparation and properties of non-oxide CMCs for application in engines and nuclear reactors: an overview," *Compos. Sci. Technol.*, 64:155-170(2004).
[2]Y.S. Liu, "Preparation and application bases of B-C ceramic by CVD/CVI," *PhD thesis.*, North Western Polytechnical University, Xi'an, China (2008).

[3]Y. Liu, L, Zhang, L Cheng, W. Yang, W. Zhang, Y. Xu, "Preparation and oxidation protection of CVD SiC/a-BC/SiC coatings for 3D C/SiC composites," *Corrosion Science.*, doi:10.1016/j.corsci. 01.026 (2009).

[4]K.W. Lee, S.J. Harris, "Boron carbide films grown from microwave plasma chemical vapor deposition," *Diamond Relat. Mater.*, 7 [10] 1539–1543(1998).

[5]U. Jansson, J.O. Carlsson, B. Stridh, S. Soederberg, M. Olsson, "Chemical vapour deposition of boron carbides. I. Phase and chemical composition," *Thin Solid Films.*, 172 [1] 81–93 (1989).

[6]J. C. Oliveira, O. Conde, "Deposition of boron carbide by laser CVD: a comparison with thermodynamic predictions," *Thin Solid Films.*, 307 [1–2] 29–33(1997).

[7]M. Ducarrior, C. Bernard, "Thermodynamic domains of the various solid deposits in the B-C-H-Cl vapor system," *Journal of the Electrochemical Socirty: Solid-state Science and Technology.*, 123[1]:136-140(1976).

[8]L. Vandenbulcke, "Theoretical and experimental studies on the chemical vapor deposition of boron carbide," *Industrial and Engineering Chemistry Prodceeding Research.*, 24: 568-575(1985).

[9]H. Hannache, F. Langlais, R. Naslain, "Kinetics of Boron Carbide Chemical Vapor Deposition and Infiltration," pp. 219–33 in Proceedings of the Fifth European Conference on *Chemical Vapour Deposition.*, Uppsala University, Department of Chemistry, Uppsala, Sweden, 1985.

[10]J. Berjonneau, F. Langlais, G. Chollon, "Understanding the CVD process of (Si)–B–C ceramics through FTIR spectroscopy gas phase analysis," *Surface & Coatings technology.*, 201, 7273–7285 (2007).

[11]J. Berjonneau, G. Chollon, F. Langlais, "Deposition process of Si–B–Cceramics from CH$_3$SiCl$_3$/BCl$_3$/H$_2$ precursor," *Thin Solid Films.*, TSF-23-290; No of Pages 10(2007).

[12]S. Song, G. Zhuang, Z. Wang, "Physical Chemistry (the first volume)," *Higher Education Press.*, 271-275(1995).

[13] M. Fitzsimmons. V. K. Sarin, "Comparison of WCl$_6$-CH$_4$-H$_2$ and WF$_6$-CH$_4$-H$_2$ Systems for Growth of WC Coatings," *J. Surf. Coating Tech.*, 76-77:250-255(1995).

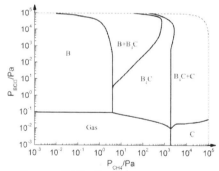

Fig. 1, Phase diagram of BCl$_3$-CH$_4$-H$_2$ system, P$_{total}$/100 kPa; T/900°C

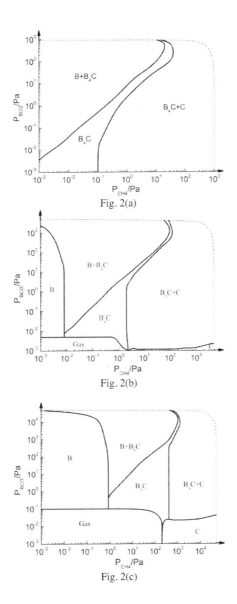

Fig. 2(a)

Fig. 2(b)

Fig. 2(c)

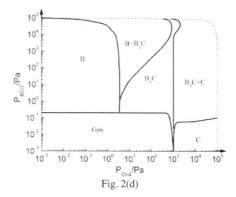

Fig. 2(d)

Fig. 2, Phase diagrams of BCl$_3$-CH$_4$-H$_2$ system at 1000 °C and different total pressures
(a) P$_{total}$/1 kPa; (b) P$_{total}$/5 kPa; (c) P$_{total}$/50 kPa; (d) P$_{total}$/100 kPa;

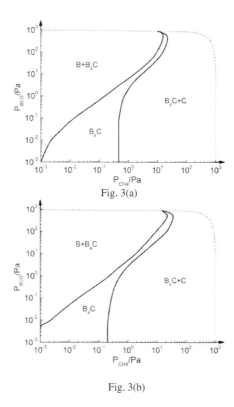

Fig. 3(a)

Fig. 3(b)

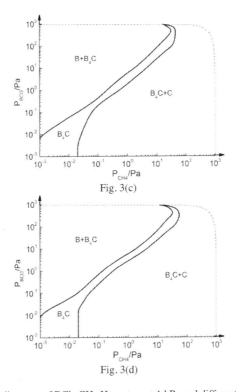

Fig. 3(c)

Fig. 3(d)

Fig. 3, Phase diagrams of BCl$_3$-CH$_4$-H$_2$ system at 1 kPa and different temperatures
(a) T/800°C;   (b) T/900°C;   (c) T/1100°C;   (d) T/1200°C;

THE RELATION BETWEEN OPTICAL PROPERTIES AND LATTICE DEFECTS OF TRANSLUCENT ALUMINUM NITRIDE CERAMICS FABRICATED BY THE NOVEL ANNEALING PROCESS

Yukihiro Kanechika[1,2], Yuriko Kaito[1], Masanobu Azuma[1], Hiroshi Fukushima[2]

[1] Tokuyama Corp., Shunan, Yamaguchi, 745-8648 Japan

[2] Graduated School of Engineering, Hiroshima University, Higashi-Hiroshima, 739-8527 Japan

Yukihiro Kanechika: y-kanechika@tokuyama.co.jp

ABSTRACT

Tube-shaped translucent aluminum nitride (AlN) ceramics of 98 % total transmittance have been prepared by the novel annealing process. Positron Annihilation Lifetime (PAL) spectroscopy, Electron Spin Resonance (ESR) and Cathode Luminescence (CL) analyses have been carried out to investigate the relation between transmittance and lattice defects in AlN. After UV irradiation, mean positron lifetime became shorter, and the ESR signal of $g = 2.005$ attributable to trapped holes by O_N (substituted oxygen at nitrogen site) - V_{Al} (aluminum-site vacancy) complexes became higher. The ESR signal intensity of the annealed AlN was lower than that of the pre-annealed AlN. We concluded that shorter mean positron lifetime suggested the change of electronic charge state of V_{Al} by trapped holes. ESR analyses revealed decreasing amount of O_N-V_{Al} complexes in the annealed AlN with increasing oxygen concentration, and also the intensity of CL around 360 nm assigned to O_N-V_{Al} became lower after the annealing. These results suggest the possibility that the novel annealing process reconstructs existing oxygen related lattice defects in the conventional translucent AlN ceramics.

INTRODUCTION

Since the band gap energy of wurtzite type AlN was determined to be 6.2 eV at room temperature by Yim et al [1], translucency has been one of the attractive properties of AlN such as high thermal conductivity [2], excellent electrical insulation [3-5] and strong corrosion resistance [6]. There have been many researches on sintering of highly translucent AlN ceramics and their evaluation [3,4,7-9]. In the early 1980's, Kuramoto et al [4] reported translucent and high thermal conductive AlN ceramics produced by pressureless sintering using high purity and good formability AlN powder and sintering aid of $3CaO\text{-}Al_2O_3$. They were able to produce excellent AlN ceramics by applying the following roles of the sintering aids, that is, promotion of particle rearrangement and grain growth via liquid phase, elimination from sintering body after full densification, and prevention of oxygen diffusion into AlN grains. Furthermore, Harris and Youngman [10] have reported the character of photo-induced defects in AlN ceramics. In their work, it was shown that UV-light irradiation induced defects were associated with oxygen impurities dissolved in the AlN lattice and the defect sites were the O_N-V_{Al} complexes. The defect center responsible for the photo-induced absorption process was an aluminum vacancy - oxygen impurity complex which resided in the AlN lattice [10]. Trinkler and Berzina [11] have reported the study of AlN ceramics using analyses of luminescence process after exposure to UV (λ = 240 nm) irradiation. They have found a peak between 360 nm and 385 nm after UV irradiation, and suggested recombination between donor $(O_N)^-$ centers and acceptor $(O_N$-$V_{Al})$ centers within the band gap of 6.2eV in AlN. Honma et al [9] have reported that the increase in the transmittance of AlN ceramics was related to the decrease in the oxygen induced defect density based on Cathode Luminescence (CL) measurement.

As it is described above, to examine the behavior of defects created by dissolved oxygen in an AlN crystal is very important to understand the properties of AlN ceramics. To investigate lattice defects in AlN ceramic, Electron Spin Resonance (ESR) is one of the effective methods. Kai et al [12] have studied the relation between the defects and thermal conductivity in AlN using ESR, and found signals of g = 1.999 and g = 2.006. They concluded that defects responsible for the g = 2.006 signal was assigned as a hole trapped in an O_N-V_{Al} complex and the g = 1.999 signal was possibly due to an electron-trapped center. Positron annihilation lifetime (PAL) spectroscopy is also effective to investigate vacancy type defects in materials, since positrons tend to be trapped in negatively charged vacancy-type defects [13,14,15]. Therefore, PAL spectroscopy can detect aluminum vacancies in AlN ceramics without sample destruction.

Recently we have reported highly translucent AlN ceramics fabricated by the novel annealing process [15]. Total transmittance of tube-shaped AlN ceramics achieved 98 % in visible light range, which was almost equivalent value to that of translucent Al_2O_3. No machining and polishing treatments of the surface have been performed for tube-shaped AlN. In the previous work [15], we have discussed impurity analysis and aluminum vacancies in AlN ceramics using PAL analysis and TEM observation, and we have found increase in the concentration of oxygen and aluminum-site vacancies in the annealed AlN. Slack et al [16] proposed that oxygen was incorporated with substitution on nitrogen sites with subsequent formation of aluminum-site vacancies for charge compensation. Indeed, thermal conductivity of AlN in our previous work was decreased with increasing oxygen concentration by the novel annealing process [15].

We investigated the optical properties and defects in AlN ceramics fabricated by the novel annealing process using PAL spectroscopy, CL and ESR analyses before and after UV irradiation. The relation between the optical properties and defects in AlN ceramics is discussed in the present paper, considering how to develop highly translucent AlN ceramic tubes.

EXPERIMENT

Preparation of AlN specimen

High purity AlN powder (Tokuyama Corp. F-grade) was used as a starting material. The specific surface area and grain size of AlN powder was 3.3 m^2 / g and 1.25 micrometer, respectively. The impurities of AlN powder were oxygen 0.85 mass %, carbon 300 ppm, Ca 10 ppm, Si 9 ppm, Fe 5 ppm. $Ca_3Al_2O_6$ powder was used as a sintering aid. The mixture of AlN powder and sintering aid were mixed with plastic binder. Ceramic green body of tube-shaped AlN (15 mm in diameter, 20 mm in length and 0.8 mm in thickness) was made by injection molding. And disk shaped green body (30 mm in diameter and 5 mm in thickness) was made by uniaxial pressing. The green body was sintered at 1880 degrees C in reduced nitrogen atmosphere after de-waxing in air. The as-sintered AlN ceramics were named sample-A. After sintering, the as-sintered body was annealed with metal oxide (Al_2O_3 powder, Showa Denko Corp. UA grade) at 1880 degrees C in nitrogen atmosphere. The resultant annealed AlN ceramics were named sample-B. Total transmittance of tube-shaped sample-A and sample-B were 65% and 98 %, respectively [15].

Method of UV irradiation

UV irradiation system (As-One Corp.) was used with the wavelength of 253.7 nm. The distance between Hg lamp and samples was 10 cm. The intensity of UV illumination was about 0.9 mW / cm^2.

Analysis of AlN samples

The total transmittance of tube-shaped AlN was measured by the total transmittance measurement system for both sample-A and sample-B of 0.6 mm in thickness. And the total transmittance of disk shaped AlN was measured by UV-VIS (Ultraviolet-Visible light) spectrometer (Shimazu Corp., Type UV-2100, integral sphere, Type ISR-260, reference air). Size of AlN disk was 30 mm in diameter, and 0.3 mm in thickness, and surface roughness was less or equal to Ra 50 nm by mirror polishing. Thermal conductivity was measured by laser flash method using thermal diffusivity measurement system (Kyoto Densi Corp., LFA-502). In order to analyze the oxygen concentration in AlN sintered body, SIMS (Secondary Ion Mass Spectrometry, PHI Corp., ADEPT-1010) was used. First ion species was Cs^+ and acceleration voltage was 5.0 kV. SIMS analyzed area was square shape of 0.06×0.06 mm. ESR was carried out using ESP350E (BRUKER Corp.,) to investigate the unpaired electron in AlN. Measurement temperature: 20 K and R.T., magnetic field center: 3370 G, frequency of micro wave: 9.45 GHz (4 mW). The shape of ESR specimen was 18 mm in length, 2 mm in width, and 1 mm in thickness. And its weight was 0.116 g / sample. Furthermore, CL at 29 K with acceleration voltage of 10 kV and PAL spectroscopy were used, to evaluate lattice defects in AlN. A shape of samples for PAL spectroscopy was rectangular parallelepiped of $5 \times 5 \times 0.3$ mm.

RESULTS AND DISCUSSION

Morphology and impurity analysis of AlN

Figure 1 shows backscattered electron micrographs of the fractured surface of AlN ceramics. Both samples are in good shape and there is no appearance of secondary phase. $3CaO\text{-}Al_2O_3$ added as sintering aid was evaporated after densification in reduced nitrogen atmosphere during sintering.

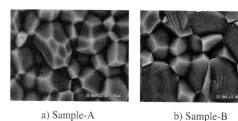

a) Sample-A   b) Sample-B

Figure 1. Backscattered electron micrographs of the fractured surface of AlN ceramics. Sample-A: as-sintered AlN body, Sample-B: sintered and annealed AlN body

Table I shows the results of SIMS analysis of AlN samples of the present work. Oxygen concentration of sample-B ($1 \times 10^{+20}$ atoms / cm^3) which shows higher total transmittance in visible light range is three times larger than that of sample-A ($3 \times 10^{+19}$ atoms / cm^3). Therefore, it can be concluded that oxygen concentration in AlN was increased by the annealing with Al_2O_3 powder. These results agree well with the previously reported result of oxygen concentration by the combustion method [15].

Table I. Oxygen concentration in AlN samples by SIMS

	Sample-A	Sample-B
Oxygen concentration in AlN ceramic	$3\times10^{+19}$	$1\times10^{+20}$

*Unit: atoms / cm^3

Optical properties of translucent AlN ceramics after UV irradiation

Figure 2 shows the change of total transmittance of sample-B with UV irradiation time. Total transmittance within 260 and 600 nm range decreases from 70 % to 48 % after UV irradiation. The absorption edge of short wavelength region should be 200 nm, because the band gap energy of AlN is 6.2 eV at room temperature [1]. However the absorption edge of sample-B is 260 nm as shown in figure 2, and the result suggests that sample-B contains many defects. No shift of the absorption edge of 260 nm can be observed during UV irradiation.

Figure 3 shows the change of difference spectra of the total absorption for sample-B with UV irradiation time and that for sample-A of 10 minutes UV irradiation, which are calculated using total transmittance and reflectance spectra of the initial states and each UV irradiation state. Total absorption of AlN within 260 and 600 nm range increases with UV irradiation for both samples A and B, however the intensity of absorption for sample-B is lower than that for sample-A after 10 minutes UV irradiation. Comparing absorption spectra of sample-A and sample-B, the change of absorption around 360 nm and 440 nm in sample-B is 10 % lower than that in sample-A. These results suggest the reduction of photo sensitive defects (color centers) in AlN by the novel annealing procedure. Trinkler et al [11] reported the formation of a peak at 380 nm in the UV irradiated AlN ceramics by photoluminescence analysis. And they attributed the peak signal at 380 nm to oxygen related defects dissolved in AlN crystal. Youngman et al [17] reported cathode luminescence spectrum from a typical oxygen doped AlN sample and concluded that the peak signal around 375 nm was due to oxygen related defects. The peak at 360 nm found in the present sample is consistent with these results and is attributable to the defect produced by oxygen. Slack et al [18] reported that the optical absorption at 433 nm was due to oxygen impurities and suggested that the yellow coloration was probably caused by a nitrogen deficiency in the AlN lattice.

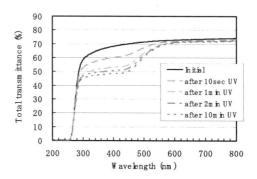

Figure 2. The change of total transmittance of sample-B during UV irradiation

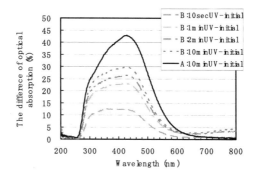

Figure 3. The change of difference spectra of the total absorption during UV irradiation

Moreover, from the peak deconvolution analysis of spectra in figure 3, it is found that absorption spectra are composed of five separated peaks at 290 nm, 315 nm, 360 nm, 440 nm, and 550 nm. To investigate the intrinsic absorption peak in AlN, figure 4 shows the difference spectrum between the spectrum for AlN of 0.3 mm in thickness and that of 0.1 mm in thickness. From figure 4, six peaks at 260 nm, 290 nm, 315 nm, 360 nm, 420 nm, and 485 nm exist in AlN before UV irradiation. Table II indicates absorption peaks in AlN before and after UV irradiation. As shown in table II, three absorption groups are observed. Four peaks at 290 nm, 315 nm, 360 nm, and 440 nm already exist in AlN and their intensity increase with UV irradiation. The absorption peak of 260 nm exists before UV irradiation but shows no change of intensity after UV irradiation. The peak at 550 nm is newly created after UV irradiation.

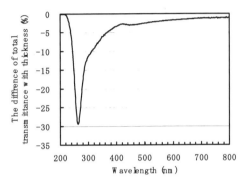

Figure 4. The difference spectrum of total transmittance for sample-A 0.3 mm in thickness and that for sample-A 0.1 mm in thickness

Table II. Classification of absorption peak observed before and after UV irradiation by UV-VIS spectra

Group	Wavelength (nm)
1	290, 315, 360, 420(440)
2	260, 480
3	550

Feature of absorption peak: Group 1: Peaks already exist in AlN and their intensities increase after UV irradiation, Group 2: Peaks alredy exist in AlN and show no change after UV irradiation, Group 3: Peak is newly created after UV irradiation

Cathode luminescence analysis of AlN

Comparison of CL analysis for the samples A and B is shown in figure 5. Three peaks of luminescence, around 360 nm, 550 nm, and 700 nm, can be observed in both samples. The peak intensities around 550 nm and 700 nm are nearly the same in both samples, however the intensity around 360 nm of sample-B was lower than that of sample-A. Therefore, it can be concluded that color centers related to luminescence of 360 nm decrease by the novel annealing procedure, and also concluded that defects related to luminescence of 360 nm affect the transmittance of AlN ceramics. In the previous study, CL intensity of 360 nm is directly correlated with oxygen concentration in AlN ceramics as reported by Honma et al [9]. CL spectra around 360 nm are attributable to the defect related to oxygen in AlN. However, in the present work, despite oxygen concentration in AlN grain increases, CL intensity at 360 nm decreases. Therefore, the lattice defects species created by oxygen dissolved in AlN crystal or oxygen aggregate structure in the present study should be different to that of the previous study. We must consider not only oxygen concentration of AlN ceramic but also defect species and defect cluster structure in order to understand the relationship between oxygen dissolved in AlN crystal and transmittance of AlN.

In the previous work, it was investigated that aluminum vacancy in AlN was increased by the novel annealing process by PAL measurement [15]. Color center creation of sample-B was lower than that of sample-A, despite oxygen concentration of sample-B was higher than that of sample-A. There is a possibility that oxygen introduced to AlN lattice exist as another defect cluster structure which does not contribute to luminescence under UV irradiation. The possible structure is aluminum octahedrally coordinated to oxygen proposed by Youngman et al [17] and/or oxygen compensation of V_N intrinsically existed in AlN lattice. Nepal et al reported the study of AlN epilayers grown by metalorganic chemical vapor deposition by deep UV photoluminescence and energy level of the nitrogen vacancy is about 260 m eV below the conduction band from the band to impurity transition [19]. Also, ionization levels for the aluminum vacancies are in the range of 0.5-1.7 eV above the valence band maximum, which shows these defects act as deep acceptors as mentioned by Mattila et al [20]. Furthermore, their calculations predict that O_N-V_{Al} complexes a deep acceptor level about 1 eV above the valence band maximum. On the other hand, oxygen donors were found to have a deep character with first ionization level about 3 eV above the level related to the O_N-V_{Al} complex [20]. This value correlates 3.2-3.3 eV of the violet luminescence in AlN [20]. We supposed the optical absorption after UV irradiation observed in our samples of 360 nm (3.5 eV) and 440 nm (2.8 eV) would be absorption between V_{Al}-O_N complex and valence-band, and the optical absorption of 290 nm (4.3 eV) and 315 nm (3.9 eV) would be between donor and accepter levels like V_{Al} and O_N^- or V_N.

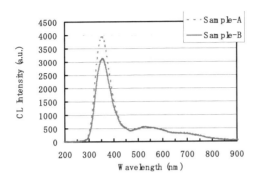

Figure 5. Comparison of CL spectra for AlN ceramics before and after the annealing

Positron annihilation lifetime spectroscopy of AlN

Figure 6 shows the results of PAL analysis of AlN ceramics under UV irradiation. Mean positron lifetime for both irradiated surface and opposite (rear) surface of sample-A were 138 ps before UV irradiation. On the other hand, mean positron lifetime of sample-B was 168 ps which is longer than that of sample-A. The first lifetime component annihilated in pure AlN crystal is about 132 ps, and second lifetime component attributable to aluminum vacancy is from 230 to 240 ps [15]. An intensity of second lifetime component in sample-A is 2 %, on the other hand, that of sample-B is 23 %. These results indicate increased concentration of aluminum-site vacancies in AlN crystal due to doped oxygen by the novel annealing process. During UV irradiation, no change of mean positron lifetime of sample-A is observed, but the mean positron lifetime of sample-B drastically decreases from 168 ps to 159 ps after 30 min UV irradiation. Consequently, the electronic charge state of V_{Al} may be changed due to trapping of holes by aluminum vacancies which are excited by UV irradiation, and then conversion of an electrical charge state leads to shorter mean positron lifetime because of the Coulomb repulsion between a positron and a hole at an aluminum-site vacancy. Furthermore, we can conclude that no aluminum vacancy affect the optical properties according to the present PAL analysis. PAL measurement clearly points out that the quantity of aluminum-site vacancies in sample-B is increased with increasing oxygen concentration by the novel annealing process. Consequently, optical absorption from 280 nm to 800 nm (mainly around 300 nm) decreases after the annealing process [15]. There is also a possibility that oxygen atoms doped in AlN lattice make defect structures not to contribute to the luminescence under UV irradiation, like aluminum octahedrally coordinated to oxygen proposed by Youngman et al [17] and/or oxygen compensation of V_N intrinsically existed in AlN lattice.

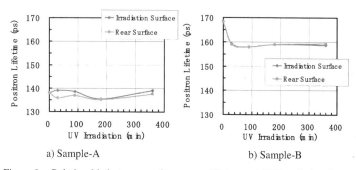

a) Sample-A   b) Sample-B

Figure 6.   Relationship between positron mean lifetime and UV irradiation time

Electron spin resonance analysis of AlN

As shown in figure 7, ESR measurement at 20 K was carried out to investigate the unpaired electron in AlN samples. A broad signal at g = 2.005 is detected in both samples of A and B. Spin density and g-value for both of AlN samples are summarized in table III. It is found that the electronic state of unpaired electrons in the defects of AlN is the same in both samples as a result of no difference in the signal at g = 2.005. Spin density of sample-B calculated from ESR signal intensity in figure 7 is a little lower than that of sample-A.

In general, g-value is represented as $g = g_e + \Delta g$, where g_e is g-value of a free electron (g_e = 2.0023) and Δg is a shift from g-value of a free electron. A signal with $\Delta g > 0$ corresponds to a hole-trapping center, and a signal with $\Delta g < 0$ corresponds to an electron-trapping center. Since g = 2.005 signal gives $\Delta g = 0.0027 > 0$, the signal can be attributed to a hole-trapping center. Shweizer et al [21] proposed that g = 2.004 in Y_2O_3 added AlN ceramics can be attributed to the model of a hole trapped by the O_N-V_{Al} complex. Also, Kai et al [12] concluded that g = 2.006 in AlN ceramics can be attributed to the O_N-V_{Al} complex, and they proposed that these defects exist at AlN grain boundaries. The g = 2.005

signal in the present study is probably due to O_N-V_{Al} complex, considering the results of oxygen impurity analysis and PAL spectroscopy. However, the ESR signal at g = 2.007 was found in the neutron irradiated AlN ceramics and assigned to F-centers trapping electrons in nitrogen-site vacancies by Atobe et al [22]. Nakahata [23] et al assigned the ESR signal at g = 2.007 in AlN ceramics to electrons trapped by V_N. They suggested that concentration of nitrogen vacancy is inversely proportionate to the concentration of O_N and pointed out that the thermal conductivity increased with the intensity of g = 2.007 ESR signal [23]. On the other hand, g = 2.005 ESR signal detected in our samples can be attributed to the O_N-V_{Al} complex as discussed above. However, the relation between ESR signal and thermal conductivity in our study is very similar to the results by Nakahata et al [23]. Therefore, we can conclude that the doping of oxygen atoms in AlN lattice by the novel annealing process introduce not only the creation O_N-V_{Al} complex but also annihilation of nitrogen-site vacancies intrinsically existed in AlN lattice. Consequently, it is possible to improve total transmittance of AlN ceramic by decreasing the concentration of nitrogen-site vacancies.

Table III.   The g-value and spin density of AlN ceramics at 20 K

	g-value	Spine density at 20 K (spins/g)
Sample-A	2.005	$6.3 \times 10^{+14}$
Sample-B	2.005	$4.7 \times 10^{+14}$

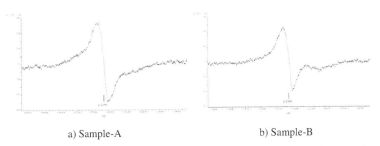

a) Sample-A   b) Sample-B

Figure 7.   ESR signals of AlN sintered body at 20 K. Sample weight: 0.116 g / sample

Figure 8 shows the change of ESR signal for AlN ceramics during UV irradiation at room temperature. The intensity of g = 2.005 signal having broad line width, increases with UV irradiation for both of samples A and B. The g = 2.004 excited by UV irradiation at 4.2 K is the acceptor due to hole trapped by the O_N-V_{Al} complex as reported by Schweizer et al [21]. The g = 2.005 ESR signal excited by UV irradiation in the present study can be assigned to hole trapped by the O_N-V_{Al} complex.

Figure 9 shows that the intensity of ESR signal of sample-B after 10 minutes UV irradiation is lower than that of sample-A. Therefore, it is suggested that the amount of hole trapped by O_N-V_{Al} complex in sample-B was lower than that of sample-A.

Furthermore, the correlation between spin density in AlN and total transmittance of 0.3 mm thick AlN is shown in figure 10. The total transmittance of AlN ceramics during UV irradiation decreases with increasing intensity of spin density in g = 2.005 ESR signal at room temperature. Considering the results of ESR analyses under UV irradiation, the amount of O_N-V_{Al} complex in sample-B is lower than that of sample-A, although the oxygen concentration of sample-B is higher than that of sample-A. Oxygen atoms in AlN lattice of sample-B probably form not only O_N-V_{Al} complex but also O_N, and/or make another defect clusters like aluminum octahedrally coordinated to oxygen. [17]

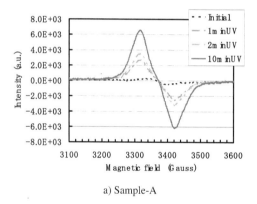

a) Sample-A

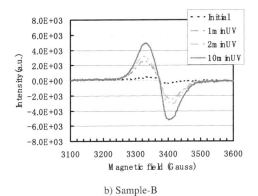

b) Sample-B

Figure 8. The time dependence of the ESR signal of AlN ceramics during UV irradiation

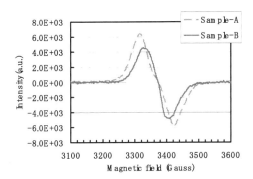

Figure 9. Comparison of ESR signals in AlN ceramics after 10minutes UV irradiation

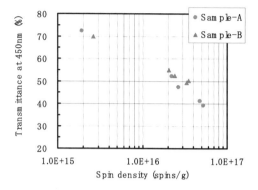

Figure 10. Correlation between spin density in AlN and total transmittance of AlN with 0.3 mm thickness at 450 nm by UV-VIS spectrometry

CONCLUSION

The relation between transmittance and defects in AlN ceramics have been investigated using UV-VIS spectrometry, positron annihilation lifetime, electron spin resonance and cathode luminescence analyses of the translucent specimens prepared by the novel annealing process. During UV irradiation, the mean positron lifetime has become shorter, and the intensity of the ESR signal of g = 2.005 attributable to a hole trapped by O_N-V_{Al} complex has become stronger, although the intensity of the ESR signal in the annealed AlN was lower than that of pre-annealed specimen. It can be concluded that the shorter mean positron lifetime indicates the change of electrical charge state of V_{Al} by a trapped hole. The ESR study revealed that quantity of O_N-V_{Al} complex in the annealed AlN was decreased despite increasing oxygen concentration. And the CL intensity of 360 nm assigned to O_N-V_{Al} was decreased after the novel annealing process in contrast to the increase in the concentration of oxygen. These results suggest the possibility that the novel annealing process reconstructs existing oxygen related lattice defects in the conventional translucent AlN ceramics. It can be concluded that the doping of oxygen atoms in AlN lattice by the novel annealing process shown in the present work may cause not only the creation of O_N-V_{Al} complex but also the annihilation of nitrogen vacancy intrinsically existed in AlN lattice and/or formation of another defects like aluminum octahedrally coordinated to oxygen.

ACKNOWLEDGEMENTS

The authors wish to acknowledge useful discussions on ESR with Dr. Ayako Kai of Yamaguchi University.

REFERENCES

[1] W.M.Yim, E.J. Stofko, P.J. Zanzucchi, J.I. Pankove, M. Ettenberg and S.L. Gilbert, Epitaxially grown AlN and its optical band gap, *J. Appl. Phys.*, **44**, 292-296, 1973

[2] G.A.Slack, Nonmetallic crystals with high thermal conductivity, *J. Phys. Chem. Solids*, **34**, 321-335, 1973

[3] N.Kuramoto, H.Taniguchi and I.Aso, Translucent AlN ceramic substrate, *IEEE Trans. Components, Hybrids and Manufacturing Technology*, **9**, 386-391, 1986

[4] N.Kuramoto, H.Taniguchi and I.Aso, Development of translucent aluminum nitride ceramics, *Ceramic bulletin*, **68**, No.4, 883-887, 1989

[5] W.A.Groen and P.F.van Hal, High temperature electrical conductivity of AlN ceramics, *British Ceramic Transactions*, **93**, No.5, 192-195, 1994

[6] S.Shimura, T.Ohashi and K.Watanabe, Fluorination of AlN ceramics, silicon and silica in an inductively coupled NF3 plasma, Finishing of advanced ceramics and glasses, *Ceramic transactions*, **102**, 341-349, 1999

[7] M.Hirano, K.Kato, T.Isobe, T.Hirano, Sintering and characterization of fully dense aluminum nitride ceramics, *J.Materials Science*, **28**, 4725-4730, 1993

[8] Y.Xiong, Z.Y.Fu, H.Wang, Microstructural effects on the transmittance of translucent AlN ceramics by SPS, *Mater. Sci. Eng. B*, **128**, 7-10, 2006

[9] T.Honma, Y.Kuroki, T.Okamoto, M.Takata, Y.Kanechika, M.Azuma, H.Taniguchi, Transmittance and cathodoluminescence of AlN ceramics sintered with $Ca_3Al_2O_6$ as sintering additive, *Ceramics International*, **34**, 943-946, 2008

[10] J.H. Harris and R.A. Youngman, Light-induced defects in aluminum nitride ceramics, *J. Mater. Res.*, **8**, No.1, 154-162, 1993

[11] L. Trinkler and B. Berzina, Radiation induced recombination processes in AlN ceramics, *J. Phys. Condens. Matter*, **13**, 8931-8938, 2001

[12] A. Kai, D. Tomohiro, Y. Kanechika and T. Miki, Electron Spin Resonance of defects related to thermal conductivity in AlN ceramics, *J.J. Applied Physics,* **47**, No.8, 6394-6398, 2008

[13] J. Moxom, J. Xu, R. Suzuki, T. Ohdaira, G. Brandes and J.S. Flynn, Characterization of Mg doped GaN by positron annihilation spectroscopy, *J. Applied Physics,* **92**, No. 4, 1898-1901, 2002

[14] F. Tuomisto, J.M. Maki, T.Yu. Chemekova, Yu.N. Makarov, O.V. Avdeev, E.N. Mokhov, A.S. Segal, M.G. Ramm, S. Davis, G. Huminic, H. Helava, M. Bickermann, B.M. Epelbaum, Characterization of bulk AlN crystals with positron annihilation spectroscopy, *J. Crystal Growth,* **310**, 3998-4001, 2008

[15] Y. Kanechika, M. Azuma and H. Fukushima, Optimum sintering conditions for optical properties of translucent aluminum nitride ceramics, *Chinese Science Bulletin,* **54**, No. 5, 842-845, 2009

[16] G.A. Slack, R.A. Tanzilli, R.O. Pohl and J.W. Vandersande, The intrinsic thermal conductivity of AlN, *J. Phys. Chem. Solids,* **48**, No.7, 641-647, 1987

[17] R.A.Youngman and J.H.Harris, Luminescence Studies of Oxygen-Related Defects in Aluminum Nitride, *J. Am. Ceram. Soc.,* **73**[11], 3238-46, 1990

[18] G.A. Slack and T.F. Mcnelly, GROWTH OF HIGH PURITY AlN CRYSTALS, *J. of Crystal Growth,* **34**, 263-279, 1976

[19] N.Nepal, K.B.Nam, M.L. Nakarmi, J.Y. Lin, H.X. Jiang, J.M. Zavada and R.G. Wilson, Optical properties of the nitrogen vacancy in AlN epilayers, *Applied Physics Letters,* **84**, No.7, 1090-1092, 2004

[20] T.Mattila and R.M.Nieminen, Point-defect complexes and broadband luminescence in GaN and AlN, *Physical Review B,* **55**, No.15, 9571-9576, 1997

[21] S.Schweizer, U. Rogulis, J.M. Spaeth, L. Trinkler, and B. Berzina, Investigation of Oxygen-Related Luminescence Centres in AlN Ceramics, *phys. stat. sol. (b),* **219**, 171-180, 2000

[22] K. ATOBE, M. HONDA, N. FUKUOKA, M. OKADA and M. NAKAGAWA, F-Type Centers in Neutron-Irradiated AlN, *J. J. Applied Physics,* **29**, No.1, 150-152, 1990

[23] S. Nakahata, K. Sogabe, T. Matsuura and A. Yamakawa, Electron Spin Resonance Analysis of Lattice Defects in Polycrystalline Aluminum Nitride, *J. Am. Ceram. Soc.,* **80**[6], 1612-14, 1997

THERMODYNAMIC CALCULATIONS OF ZrC-SiC SYSTEM FOR CHEMICAL VAPOR DEPOSITION APPLICATIONS FROM SiCl$_4$-ZrCl$_4$-CH$_4$-H$_2$

Qiaomu Liu[*], Litong Zhang, Yiguang Wang, Laifei Cheng
National Key Laboratory of Thermostructure Composite Materials, Northwestern Polytechnical University, Xi'an, China 710072

ABSTRACT

Thermodynamic equilibrium condensed phases for chemical vapor deposition of ZrC-SiC ternary system from SiCl$_4$-ZrCl$_4$-CH$_4$-H$_2$ were calculated by Gibbs free energy minimization method. The effects of partial pressure of reactants, temperatures, total pressures, and carbon sources on the final condensed phases were demonstrated. It was found that the CH$_4$ partial pressure is the controlling parameter of SiCl$_4$-ZrCl$_4$-CH$_4$-H$_2$ system. Reasonable reactant concentration ranges, temperatures and pressures for ZrC-SiC system were established according to these deposition phase diagrams. Compared with CH$_4$, lower temperature is required to decompose C$_3$H$_6$ due to its lower chemical stability.

INTRODUCTION

Refractory carbides of zirconium and silicon are major promising coating materials with considerable applications due to their many superior properties, such as great hardness, high melting point, good neutronic property, good thermal shock resistance, and low thermal conductivity[1-4]. However, the thermal decomposition of SiC may occur below its intrinsic melting point, and ZrC readily adsorbs O$_2$ into its lattice and oxidizes at greatly enhanced rates. In addition, ZrO$_2$ undergoes solid-phase transformation[2,5]. If a coating could be prepared by chemical vapor deposition (CVD), since CVD is well employed to prepare ceramic coatings, combining the both advantages of ZrC and SiC, the coating might be able to use in complicated, aggressive and extreme environments. Moreover, CVD SiC and ZrC are investigated intensively[6-13]. However, the codeposition of ZrC and SiC has been rarely reported.

In the last few decades, thermodynamic calculations based on Gibbs free energy minimization of a complex chemical system have been developed in order to predict the vital information of a given system. Therefore, in a new CVD system, a thermodynamic study is generally first used to predict the final compositions of the system at equilibrium in order to understand the influence of deposition parameters such as temperatures, reactant concentrations and pressures on phases, the final compositions and their concentrations. The deposition phase diagrams can be constructed according to the calculation results. They can provide a useful guideline for the selection of processing conditions to obtain specific phases and provide a basic understanding of deposition trends. Hence, they could be used to systematically define deposition parameters for quality coatings and reduce experimental costs.

In the present paper, the thermodynamics of SiCl$_4$-ZrCl$_4$-CH$_4$(C$_3$H$_6$)-H$_2$ system was calculated to understand the Zr-Si-C ternary system deposition process at given conditions. The phase regions constructed by ZrC, SiC, and C will be discussed in detail. The deposition phase diagrams, which demonstrated equilibrium condensed phases as a function of temperatures, reactant concentrations,

[*] Author to whom correspondence should be addressed, Email: lqiaomu@163.com.
Phone: +86-29-88494622, Fax: +86-29-88494620

pressures, and carbon sources, were constructed according to the calculation results. Reasonable reactant concentration ranges, temperatures and pressures for ZrC-SiC system were established according to these deposition phase diagrams.

PROCEDURES

Silicon tetrachloride ($SiCl_4$), zirconium tetrachloride ($ZrCl_4$), methane/propene (CH_4/C_3H_6), and hydrogen (H_2) are the reactant gases for depositing the Zr-Si-C ternary system. Thermodynamic calculations were performed at given conditions by using FactSage programs (Thermfact/CRCT and GTT- Technologies, Canada and Germany) based on Gibbs free energy minimization method. The free energy G of a system consisting of m gaseous species and s solid phases can be described by[14]

$$G = \sum_{i=n_i}^{m} (n_i^g \Delta G_{f_i g}^o + RT \ln P + RT \ln \frac{n_i^g}{N_g}) + \sum_{i=1}^{s} n_i^s \Delta G_{f_i s}^o \qquad (1)$$

Where $\Delta G_{f_i g}^o$ and $\Delta G_{f_i s}^o$ is the formation free energy of the gaseous species and solid species at calculation temperature, respectively. A list of symbols used in this article is in the appendix to aid the reader.

The thermodynamic calculations were conducted at temperatures from 800 to 1200 °C and total pressures from 0.02 to 0.08 atm with continuous varying partial pressures of reactants. H_2 is used to achieve 0.05 atm total pressure. The concentrations of equilibrium species were computed by minimization of the total Gibbs free energy and satisfying the input molar constraints. The CVD phase diagrams for the $SiCl_4$-$ZrCl_4$-$CH_4(C_3H_6)$-H_2 systems were constructed by plotting the equilibrium condensed phases as a function of $ZrCl_4$ and $SiCl_4$ concentrations. Only condensed phases are shown while gaseous phases are ignored for simplicity in CVD phase diagrams.

Thermodynamic data required for the calculation of $SiCl_4$-$ZrCl_4$-CH_4-H_2 system include 61 gaseous species and 34 condensed species. As for $SiCl_4$-$ZrCl_4$-C_3H_6-H_2 system two more gaseous species, C_3H_6 and C_4, should be considered. The gases are nonideal gases. The gaseous species and condensed phases included in the calculation are listed in Table 1. All of the thermodynamic data for these species have already been included in the FactSage internal database.

RESULTS AND DISCUSSION
CH_4 partial pressure effect

The CVD phase diagrams for the $SiCl_4$-$ZrCl_4$-CH_4-H_2 system with a total pressure of 0.05 atm and temperature of 1000 °C at different CH_4 partial pressures are shown in Fig. 1. When the CH_4 partial pressure is 0.00005 atm (Fig.1(a)), the phase diagram includes 10 phase regions, two of which are SiC regions. The phase regions constructed by SiC, ZrC, or C are investigated in detail and other regions will not be investigated and constructed thereafter. By increasing the CH_4 partial pressure to 0.0005 atm (Fig.1(b)), all regions expand and shift to high $ZrCl_4$ and $SiCl_4$ partial pressures. The two SiC regions become one. ZrC+C region appears between ZrC+SiC+C and ZrC, while SiC+C region appears between ZrC+SiC+C and SiC.

The overall chemical reactions for $SiCl_4$-$ZrCl_4$-CH_4-H_2 system are:

$$CH_4(g) \rightarrow ... \rightarrow C(s) + 2H_2(g) \tag{2}$$

$$SiCl_4(g) + CH_4(g) \rightarrow ... \rightarrow SiC(s) + 4HCl(g) \tag{3}$$

$$ZrCl_4(g) + CH_4(g) \rightarrow ... \rightarrow ZrC(s) + 4HCl(g) \tag{4}$$

Table 1. A list of gaseous species and condensed phases included in the calculation

Gaseous species	H_2, $ZrCl_4$, HCl, $ZrCl_3$, H, $SiCl_2$, $ZrCl_2$, Cl, $SiCl_3$, SiH_2Cl_2, $SiHCl_3$, SiH_3Cl, SiH_4, $SiCl_4$, $SiCl$, SiH, Si, $ZrCl$, Cl_2, CH_4, ZrH, Si_2, CH_3, Zr, CH_3Cl, Si_3, Si_2H_6, CH_2, $SiCH_3Cl_3$, Si_2C, CH_2Cl_2, C_2H_2, CH, $CHCl$, C_2H_4, C, SiC, CCl, CCl_2, C_2H_6, C_2H_3, C_2H, $CHCl_3$, C_2HCl, CH_2CHCl, CCl_3, SiC_2, C_2H_5Cl, C_2Cl_2, CH_2CCl_2, CCl_4, C_2, CH_3CHCl_2, $CHClCCl_2$, $CHCl_2CH_2Cl$, C_3, C_2Cl_4, $CHCl_2CHCl_2$, C_2Cl_5H, (C_3H_6, C_4)
Condensed phases	$ZrSi$, Zr_2Si, Zr_5Si_3, Si, $ZrSi_2$, Zr, ZrH_2, ZrC, $ZrCl_3$, $ZrCl_2$, $ZrCl_4$, C, SiC, $SiHCl_3$, $SiCl_4$, Si_2H_6, CH_2Cl_2, $CHCl_3$, C_2H_5Cl, CH_2CCl_2, CCl_4, CH_3CHCl_2, $CHClCCl_2$, C_2Cl_4, $CHCl_2CHCl_2$

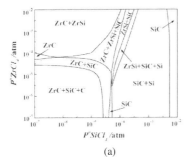 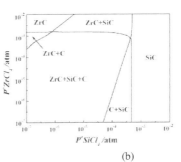

(a)   (b)

Fig.1. Deposition phase diagrams for $SiCl_4$-$ZrCl_4$-CH_4-H_2 system with a total pressure of 0.05 atm and temperature of 1000 °C, the CH_4 partial pressure is (a) 0.00005 atm, (b) 0.0005 atm.

Compared with $ZrCl_4$ and $SiCl_4$, it is easer to decompose for CH_4, and the carbon deposition is relatively easer at 1000 °C (Fig.2). Therefore, there is a ZrC+SiC+C region at low $ZrCl_4$ and $SiCl_4$ partial pressures (Fig.1(a)). According to the van't Hoff isotherm equation and the affinity (A) of a reaction (Eqs. (5)-(9))[15-16], if thermodynamic equilibrium constant is greater than fugacity quotient ($K_f^o > Q_f$), the reaction will proceed forward, in which K_f^o is a function of temperature. By increasing $SiCl_4$ and $ZrCl_4$ partial pressures, Q_f of reactions (3-4) will decrease, both of which will proceed forward. Since more silicon and zirconium consumes more carbon, ZrC+SiC region will appear. Further increasing SiCl$_4$ or ZrCl$_4$ and at low ZrCl$_4$ or SiCl$_4$ partial pressure, Q_f of reaction (3)

or (4) will decrease while that of reaction (4) or (3) will increase. Hence, Q_f of these two reactions

have significant differences. For that reason, reaction (3) or (4) is accelerated by increasing SiCl$_4$ or ZrCl$_4$ partial pressure coupling with retarding of reaction (4) or (3), therefore, there is a monophase SiC or ZrC region. So does the right SiC region.

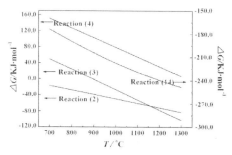

Fig.2. Gibbs free energy of reaction (2), (3), (4), and (14) vs temperature plots

$$A = -\Delta_r G_m = -\sum_i v_i \mu_i = RT \ln \frac{K_f^o}{Q_f} \qquad (5)$$

$$K_f^{\,o} = \prod_i (\frac{f_i}{P^o})^{v_i}_e \qquad (6)$$

$$Q_f = \prod_i (\frac{f_i}{P^o})^{v_i} \qquad (7)$$

$$f_i = P_i \gamma_i \approx f_i^0 n_i \approx \left(\frac{P^2 V_i}{RT} \right) n_i \qquad (8)$$

$$\lim_{P \to 0} \gamma_i = 1 \qquad (9)$$

By increasing the CH$_4$ partial pressure, Q_f of reaction (2) will decrease, and reaction (2) will be

accelerated. Therefore, carbon contained region exists in a wide range (Fig.1(b)). Reactions (3-4) also will be accelerated (Fig.3). Hence, the ZrC region, SiC region, and ZrC+SiC regions shift to higher ZrCl$_4$ and SiCl$_4$ concentrations and expand. Similar to Fig.1(a), further increasing SiCl$_4$ or ZrCl$_4$ and at low ZrCl$_4$ or SiCl$_4$ partial pressures, there is a SiC+C or ZrC+C region.

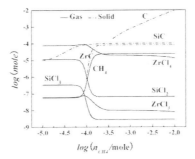

Fig.3. Variation in the amount of solid species and major gas species as a function of CH$_4$ concentration at 1000 °C, in which the partial pressure of ZrCl$_4$ and SiCl$_4$ both are 0.0001 atm, and total pressure is 0.05 atm.

Temperature effect

The CVD phase diagrams for the SiCl$_4$-ZrCl$_4$-CH$_4$-H$_2$ system with a total pressure of 0.05 atm and CH$_4$ partial pressure of 0.00005 atm at different temperatures of 800 °C and 1200 °C are shown in Fig.4. At 1200 °C, the ZrC+SiC+C region and the right SiC region expand, while other interesting regions shrink and shift to low ZrCl$_4$ partial pressure. Moreover, the ZrC region and middle SiC region disappear.

According to the van't Hoff isobar equation (Eq. (10))[15-16], which implies that if the reaction is exothermic ($\Delta H_T < 0$), increasing the temperature will reduce the quantity of products in equilibrium.

Reactions (2-4) are endothermic reactions. The ΔH_T change rates of these three reactions are different (Fig.5). It is indicated that reaction (2) depends strongly on the temperature. The degree of decomposition of CH$_4$, ZrCl$_4$

$$\frac{d \ln K^o}{dT} = -\frac{\Delta_r H_m^o}{RT^2} \tag{10}$$

and SiCl$_4$ increase with increasing temperature, and the deposition of C is easier than that of ZrC and SiC at 1200 °C (Fig.6). Hence, the ZrC+SiC+C region expands at 1200 °C, while ZrC+SiC region shrinks a lot. At 800 °C, the ZrC+SiC+C region disappears. This is due to the low degree of decomposition of CH$_4$ and the increased stability of hydrocarbons in the gas phase at low temperatures.

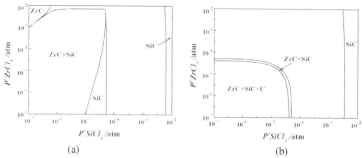

(a)   (b)

Fig.4. Deposition phase diagrams for SiCl$_4$-ZrCl$_4$-CH$_4$-H$_2$ system with a total pressure of 0.05 atm and temperature of (a) 800 °C and (b) 1200 °C, the CH$_4$ partial pressure is 0.00005 atm.

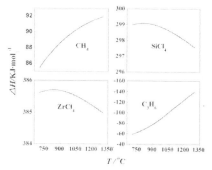

Fig.5. The enthalpy change trend of reaction (2), (3), and (4) as a function of temperatures.

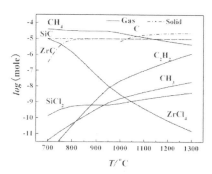

Fig.6. Variation in the amount of solid species and major gas species as a function of temperatures, in which the CH$_4$ partial pressure is 0.00005 atm, and the partial pressure of ZrCl$_4$ and SiCl$_4$ both are 0.00001 atm, and total pressure is 0.05 atm.

Total pressure effect

The CVD phase diagrams for the SiCl$_4$-ZrCl$_4$-CH$_4$-H$_2$ system at temperature of 1000 °C and CH$_4$ partial pressure of 0.00005 atm with different total pressures of 0.02 atm and 0.08 atm are shown in Fig.7. Compared Fig.7(a) with Fig.1(a), it is seen that the interesting regions have insignificant change just ZrC+SiC+C region disappears and ZrC region and SiC regions shrink in size at total pressure of 0.08 atm. By decreasing the total pressure to 0.02 atm (Fig.7(b)), the deposition diagram becomes complicated, the SiC+ZrC region shrinks, while others regions expand and SiC+C region appears.

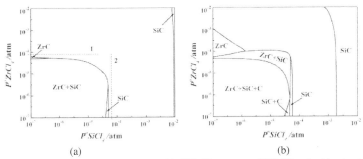

(a)   (b)

Fig.7: Deposition phase diagrams for SiCl$_4$-ZrCl$_4$-CH$_4$-H$_2$ system at 1000 °C and with a total pressure of (a) 0.08 atm, (b) 0.02 atm. The CH$_4$ partial pressure is 0.00005 atm.

According to the Le Châtelier principle (Eqs.(11)-(13))[15-16], if the products consist of more moles of gas than the reactants and the pressure on a system is increased, the system will respond by reacting in the direction that minimizes the effect of the pressure and the mole fractions of the reactants will increase. Reactions (2-4) are accompanied by 100 % volume expansion. K_f^o is a function of temperature. γ_i can be assumed a constant in this article. When the pressure is increased to 0.08 atm, K_x decreases and the reactions (2-4) proceed backward. Therefore, the phase regions shrink. However, the reverse reactions (3-4) are not obvious at low ZrCl$_4$ and SiCl$_4$ concentrations due to their greater K_x. Hence, there is a considerable ZrC+SiC region. When the pressure is decreased to 0.02 atm, K_x increases and these reactions proceed forward. Therefore, the phase regions expand at low pressures.

$$K_f^o = \left(\frac{P}{P^o}\right)^{\sum \nu_i} \cdot K_x \cdot K_\gamma \qquad (11)$$

$$K_x = \prod_i \left(\frac{n_i^g}{N_g} \right)^{v_i} \qquad (12)$$

$$K_\gamma = \prod_i \gamma_i^{v_i} \qquad (13)$$

Carbon source effect

In order to distinguish the effects of different carbon sources on the equilibrium condensed phases, the phase diagrams for the SiCl$_4$-ZrCl$_4$-C$_3$H$_6$-H$_2$ system at 1000 °C with a total pressure of 0.05 atm and C$_3$H$_6$ partial pressure of 0.00005 atm (Fig.8) was also constructed. As can be seen from Fig.8, the deposition diagram at this condition is similar to Fig.7(b), each phase region expand and shift to high ZrCl$_4$ and SiCl$_4$ concentrations.

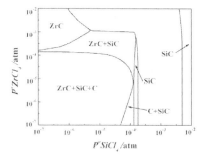

Fig.8. Deposition phase diagrams for SiCl$_4$-ZrCl$_4$-C$_3$H$_6$-H$_2$ system with a total pressure of 0.05 atm and temperature of 1000 °C, the C$_3$H$_6$ partial pressure is 0.00005 atm.

Compared with SiCl$_4$-ZrCl$_4$-CH$_4$ system, the overall chemical reactions for SiCl$_4$-ZrCl$_4$- C$_3$H$_6$-H$_2$ system are:

$$C_3H_6(g) \rightarrow \ldots \rightarrow 3C(s) + 3H_2(g) \qquad (14)$$

$$3SiCl_4(g) + C_3H_6(g) + 3H_2(g) \rightarrow \ldots \rightarrow 6SiC(s) + 12HCl(g) \qquad (15)$$

$$3ZrCl_4(g) + C_3H_6(g) + 3H_2(g) \rightarrow \ldots \rightarrow 6ZrC(s) + 12HCl(g) \qquad (16)$$

Similarly, reactions (14-16) are accompanied by 200 %, 71.43 %, and 71.43 % volume expansion, respectively. Compared with CH$_4$, lower temperature is required to decompose C$_3$H$_6$ due to its lower chemical stability (Fig. 2). Therefore, compared with CH$_4$, all phase regions expand and shift to higher

ZrCl$_4$ and SiCl$_4$ concentrations at the same total pressure, temperature, and carbon source partial pressure.

Controlling parameter

As aforementioned, the boundary line of the middle SiC region (line 2 in Fig. 7(a)), no matter the temperature, total pressure, and carbon source, does not shift remarkably. Thus, the SiCl$_4$ partial pressure is not a controlling parameter. Above 1000 °C, line 1 also does not shift remarkably no matter temperature and total pressure. When the CH$_4$ partial pressure is increased, line 1 and line 2 both shift remarkably. In conclusion, the deposited phases and compositions can be easily varied by adjusting the input CH$_4$ partial pressure. Thus, the CH$_4$ partial pressure is a controlling parameter in this situation. Below 1000 °C, ZrC+SiC region almost exists in the whole ZrCl$_4$ concentration range, which means that the ZrCl$_4$ partial pressure is also a non-controlling parameter. Accordingly, the CH$_4$ partial pressure is the controlling parameter of SiCl$_4$-ZrCl$_4$-CH$_4$-H$_2$ system. At lower temperature, higher total pressure, lower CH$_4$ partial pressure, and lower SiCl$_4$ partial pressure, ZrC+SiC can be gained. The interest filed is preferred at the following condition

$$T \leq 1000^{o}C, P_t \geq 0.05atm, P^{o}_{SiCl_4} \leq 5.4 \cdot 10^{-5}atm, P^{o}_{ZrCl_4} \leq 7 \cdot 10^{-3}atm \text{ and } 1 < \frac{P^{o}_{CH_4}}{P^{o}_{SiCl_4}} < 4.$$

CONCLUSIONS

The CVD phase diagrams for SiCl$_4$-ZrCl$_4$-CH$_4$-H$_2$ and SiCl$_4$-ZrCl$_4$-C$_3$H$_6$-H$_2$ systems were constructed by using the FactSage program based on Gibbs free energy minimization method. The equilibrium condensed phases as a function of reactant concentrations, temperatures, pressures, and carbon sources were studied according to these diagrams. It was found that the CH$_4$ partial pressure is the controlling parameter of SiCl$_4$-ZrCl$_4$-CH$_4$-H$_2$ system. At lower temperature, higher total pressure, lower CH$_4$ partial pressure, and lower SiCl$_4$ partial pressure, specific phases ZrC+SiC can be gained. Compared with CH$_4$, lower temperature is required to decompose C$_3$H$_6$ due to its lower chemical stability. Therefore, all phase regions expand and shift to higher ZrCl$_4$ and SiCl$_4$ concentrations for SiCl$_4$-ZrCl$_4$-C$_3$H$_6$-H$_2$ system.

However, CVD is a complex non-equilibrium system and the thermodynamic calculations and predictions are sensitive to the thermodynamic data. These deposition phase diagrams only give the trends of deposition process. Therefore, complementary experiments are required to obtain more accurate and precise information about SiCl$_4$-ZrCl$_4$-CH$_4$-H$_2$ system.

ACKNOWLEDGEMENT
This work is financially supported by the Chinese Natural Science Foundation (Grant #90176023).

APPENDIX

n_i^g number of moles of gaseous species,

n_i^s number of moles of solid species,

N_g total number of moles of gaseous species,

P total pressure,

R gas constant,

A affinity,

v_i stoichiometric coefficient,

K_f^o thermodynamic equilibrium constant,

Q_f fugacity quotient,

f_i fugacity,

P^o standard-state gas pressure,

P_i partial pressure of gaseous species,

γ_i fugacity coefficient,

f_i^0 the fugacity of pure i at the pressure which is equal to the total pressure in the

     mixture.

REFERENCES

[1]H. O. Pierson, Handbook of Chemical Vapor Deposition (CVD): Principles, Technology, and Applications, Noyes Publications, New York, 1992.

[2]H. O. Pierson, Handbook of Refractory Carbides and Nitrides: Properties, Characteristics, Processing, and Applications, Noyes Publications, New York, 1996.

[3]S. T. Oyama, The Chemistry of Transition Metal Carbides and Nitrides, Blackie Academic &

Professional, Glasgow, 1996.

[4]L. E. Toth, Transition Metal Carbides and Nitrides, Academic Press, New York and London, 1971.

[5]R. F. Voitovich, A. Pugach, High-temperature oxidation of ZrC and HfC, *Powder Metall. Met. C+*, **12**, 916-21 (1973).

[6]Y. Wang, Q. Liu, J. Liu, L. Zhang, L. Cheng, Deposition Mechanism for Chemical Vapor Deposition of Zirconium Carbide Coatings, *J. Am. Ceram. Soc.*, **91**, 1249-52 (2008).

[7]Q. Liu, L. Zhang, L. Cheng, Y. Wang, Morphologies and Growth Mechanisms of Zirconium Carbide Films by Chemical Vapor Deposition, *J. Coat. Technol. Res.*, **6**, 269-73 (2009).

[8]J. J. Glass, J. N. Palmisiano, R. E. Welsh, The Chemical Vapor Deposition of Zirconium Carbide onto Ceramic Substrates, Research report, Bettis atomic power laboratory, West mifflin, Pennsylvania 15122-0079, Operated for the U.S. Department of Energy by Bechtel Bettis, Inc., DE-AC11-98 PN38206, 1999.

[9]Y. S.Won, Y. S. Kim, V. G. Varanasi, O. Kryliouk, T. J. Anderson, C. T. Sirimanne, L. McElwee-White, Growth of ZrC Thin Films by Aerosol-assisted MOCVD, *J. Cryst. Growth*, **304**, 324-32 (2007).

[10]F. Langlais, F. Loumagne, D. Lespiaux, S. Schamm, R. Naslain, Kinetic Processes in the CVD of SiC from CH_3SiCl_3-H_2 in a Vertical Hot-Wall Reactor, *J. Phys. IV*, **5**, 105-12 (1995).

[11]A. Josiek, F. Langlais, Kinetics of CVD of Stoichiometric and Si-excess SiC in the System MTS/H_2 at Medium Decomposition of MTS, *Chem. Vapor. Depos.*, **2**, 141-46 (1996).

[12]F. Langlais, F. Loumagne, R. Naslain, Experimental Kinetic Study of the Chemical Vapour Deposition of SiC-based Ceramics from CH_3SiCl_3/H_2 Gas Precursor, *J. Cryst. Growth*, **155**, 198-204 (1995).

[13]Y. Xu, L. Cheng, L. Zhang, Carbon/Silicon Carbide Composites Prepared by Chemical Vapor Infiltration Combined with Silicon Melt Infiltration, *Carbon*, **37**, 1179-87 (1999).

[14]P. AK, V. K. Sarin, Basic Principles of CVD Thermodynamics and Kinetics in Chemical vapor deposition, J. H. Park, T. S. Sudarshan, Eds.; ASM International Materials Park, OH, Vol. 2, 2001.

[15]R. G. Mortimer, Physical Chemistry, The Benjamin/Cummings Publishing Company, Inc., Redwood City, California, 2000.

[16]B. S. Bokstein, M. I. Mendelev, D. J. Srolovitz, Thermodynamics and Kinetics in Materials Science: A Short Course, Oxford University Press, New York, 2005.

DEBINDERING OF NON OXIDE CERAMICS UNDER PROTECTIVE ATMOSPHERE

Dieter G. Brunner, Gaby Böhm
ANCeram GmbH & Co.KG
Bindlach, Germany

Friedrich Raether, Andreas Klimera*
ISC – Fraunhofer Gesellschaft
Würzburg, Germany

* now Saint Gobain Industrial Ceramics, Rödenthal, Germany

ABSTRACT

Debindering of Non Oxide Ceramics (NOC) like Si_3N_4, $MoSi_2$ or AlN has been investigated using a kinetic model to understand and optimize binder removal in Ar or N_2 atmosphere. The paper focuses on AlN ceramics, especially high volume parts for applications in semiconductor industry, power electronics and space applications. Properties of AlN green parts have been monitored during debindering: wetting behavior of binder, permeability of gaseous species in pore channels, debindering kinetics and mechanical strength. The maximum safe debindering rate was determined allowing debindering without any damage of specimen. A kinetic model was used to calculate optimized heating cycles from the experimental data. Results are shown for an AlN unit dedicated for space application.

INTRODUCTION

Debindering of Non Oxide Ceramics, especially those being sensitive to oxidation like AlN can be done easily and with good results in vacuum. Si_3N_4 can be debindered without major problems in air. Debindering of AlN in air leads to extreme oxygen uptake and reduced thermal conductivity of the final ceramic component. Debindering under vacuum shows good results concerning thermal conductivity, residual carbon content, bending strength and homogeneous color. But cracked organic components contaminate the pumping system causing malfunction and short service intervals. Moreover substituting contaminated waste oil is expensive. Debindering under protective atmospheres, especially in nitrogen, is easier and eco-friendly because exhaust gases can be oxidized thermally. Debindering rates and holding times are usually deduced from TGA curves, measured with small samples. Transfer to bigger components may lead to residual carbon in the part, poor strength, warping [fig. 1], uncontrolled crystallographic phase content and discoloration [fig.2]. Moreover, gaseous and liquid crack products may condense in the waste gas tubes, leading to obstructions or corrode graphite heater and insulation material [fig. 3]. To overcome all these problems, we developed a model for calculating optimum binder burn-out rates in production furnaces based on TG curves measured with small samples at laboratory scale at different heating rates. The model and auxiliary methods are described in the following sections.

Fig. 1: AlN disks after insufficient debindering showing warping

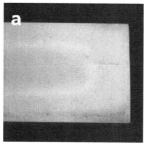

Fig. 2: AlN substrate after sintering in a stack showing zoning due to different yttria/alumina ratios in the secondary phase

Fig. 3: Debris and depositions on used heating graphite elements after debindering and sintering of AlN components

MEASURING METHODES AND RESULTS

Coupled Thermogravimety / Mass Spectroscopy (TG-MS)

TG-MS was used to analyze the carbon remaining in an AlN ceramic sample after debindering. First, samples were debindered under the desired atmosphere e.g. N_2 and with a time/temperature regime to be tested. Thereafter, small samples were extracted from the debindered components and heated in air using TG-MS (STA 449 C, Netzsch, Selb, Germany). Simultaneous measurement of weight loss and continuous registration of mass 44 for carbon dioxide gives the volume and temperature range, where secondary carbon is combusted. Figure 4 shows a typical curve of such a test series. Sensitivity of these measurements for secondary carbon was very high and even different carbon species could be distinguished according to different CO_2 peaks in the mass spectra (compare Fig. 4).

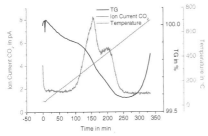

Fig. 4: Weight loss and CO_2 emission after pyrolysis of AlN green sample in N_2 atmosphere during subsequent heating in air

In-situ FTIR Spectroscopy

Evolved gaseous species were detected close to the specimen's surface by using infrared spectroscopy (IRcube, Bruker, Ettlingen, Germany) in a horizontal tube furnace (fig. 5). The sample was placed in a steel tube, closed at both ends by ZnSe windows. The reaction chamber could be evacuated or filled with different types of gases. The IR light beam was oriented parallel to the axis of the furnace and positioned directly above the sample during debindering.

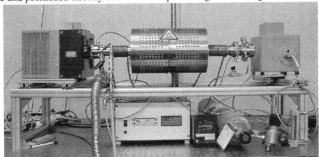

Fig. 5: Setup for FTIR measurement of gaseous species during debindering process

Fig. 6 shows an FTIR spectrum obtained during debindering of green samples with a PVB based binder at 310°. The dominating species evolved from the sample are butyraldehyde, water and CO_2.

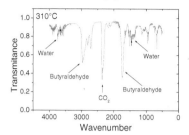

Fig. 6: IR spectrum of the gas atmosphere during debindering an AlN sample in N_2, heating rate 5K/min

Measurement of Youngs modulus and mechanical strength

Youngs modulus of partially debindered samples was measured at room temperature using a ultrasonic method (USIP12, Krautkramers, Hürth, Germany). Fig. 7 shows measured Youngs moduli of AlN samples after previous heating to different peak temperatures. It can be seen that at first Youngs modulus increases with increasing peak temperature. This is attributed to thermally activated additional cross linking within the PVB binder. At higher temperature Youngs modulus decreases due to the decreasing amount of residual binder in the green sample.

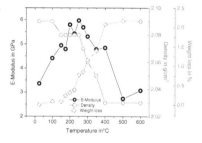

Fig. 7: Youngs modulus E, density and weight loss of AlN green samples after debindering in Ar; heating rate 1 K/min

Mechanical strength of partially debindered samples was measured at room temperature with ball on ring test [1]. For a large number of different ceramics and binders a close correlation between Youngs modulus and mechanical strength was obtained. So, the measurement of Youngs modulus, which can be done much faster, could be used instead of the bending strength measurements. Note that methods have been developed, which allow a measurement of Youngs moduli in situ during the heat treatment of the ceramic probe [2].

Wetting Microscopy

Cylindrical samples of the binder material were placed in a small measuring furnace (TOM, Fraunhofer ISC, Würzburg, Germany) on top of polished AlN disc. During continuous heating, shape changes are monitored (fig. 8a, b). The contact angle was obtained from fitting an ellipse to the contour of the liquid droplet and calculating the angle at the intersection of the ellipse with the contour line of the substrate. It can be seen that the contact angle strongly decreased for the used PVB based binder when temperature was increased. It was assumed that debindering at conditions where the contact angle is above 90° is critical since segregation of the binder would be favorable for thermodynamic reasons [1]. In addition bubbling was observed within the binder droplet at temperatures above 200°C. This was attributed to the vapor pressure of gaseous binder species which cannot diffuse fast enough to the surface of the droplet. This bubbling may cause damages in real components when large clusters of binder are present within the component.

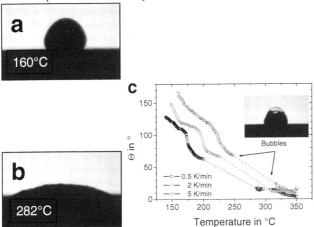

Fig. 8: Wetting angle Θ of binder droplet on AlN substrate monitored at different heating rates

SEM and Cross Section Polishing

With a Cross Section Polisher (Jeol, Tokyo, Japan, fig. 9a) a clean cut was obtained by removing thin surface layers from green specimen using an Argon beam at an incidence angle of 0°, i.e. parallel to the specimen surface (fig.9b). Figure 9c and d present typical SEM images (Ultra 55, Zeiss, Oberkochen, Germany) from this method. The ceramic particles show a light grey color, the organic binder between the particles is dark grey. Pores are black with bright edges. Regions close to the particle contacts and small pores are filled with, larger pores are free of binder. The cross section polishing method was considered superior to other preparation techniques since it allowed a study of the binder distribution in the rather fragile and difficult to handle green samples.

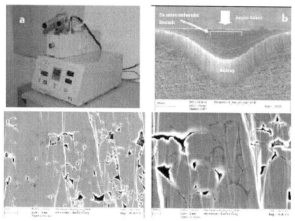

Fig. 9 a: Cross section polishing device, b: area removed by ion beam; c and d: etched area showing binder in smaller pores

Measurement of debindering kinetics and modeling

Kinetics of debindering was described by the so-called Kinetic Field Method. This method has been used successfully since the early nineties to describe sintering shrinkage [3, 4]. The degree of debindering was determined by TG-analysis with different constant heating rates. Fig. 10a, b shows

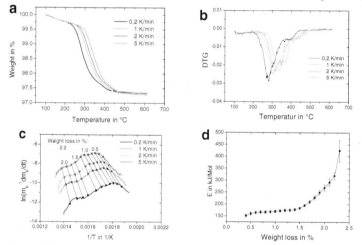

Fig. 10: TG measurement during debindering of AlN samples at different heating rates (a), differential thermogravimetric data DTG (b), kinetic field of the data (c), apparent activation energy E of debindering (d)

TG/DTG results for debindering of AlN green samples containing a PVB based binder. After that the weight loss rate ($\frac{1}{m_b}\frac{dm_b}{dt}$) was plotted logarithmically versus inverse absolute temperature - like an Arrhenius plot (fig. 10c). Due to the inverse plotting, curves must be read from the right to the left. At the beginning the weight loss rate increases with rising temperature, reaches a maximum and approaches zero at the end of the debindering process. Faster heating increases weight loss rate. Connecting all points on the different debindering curves related to the same grade of debindering form a so called iso-weight line. All iso-weight lines measured in all the different debindering tests with different ceramics and binders, could be fitted by straight lines. Assuming first order reaction kinetics for the pyrolysis or evaporation of the organic binder species in inert atmosphere, debindering is described by equation 1:

$$(1)\ \frac{dm_b}{dt} = K(T)\,m_b;\quad \Rightarrow\quad \ln\left(\frac{1}{m_b}\frac{dm_b}{dt}\right) = \ln K(T) = -\frac{E_r}{RT} + C_r$$

With K(T) = reaction constant, m_b = mass fraction of residual binder in the sample
E_r = activation energy for binder evaporation or pyrolysis, R = gas constant, T = absolute temperature, t = time, C_r = constant

From equation 1 it can be seen that the slope m of the iso-weight lines equals to $-\frac{E_r}{R}$. So, the activation energy can be calculated from the slope (see fig. 10d). The activation energy is constant up to a weight loss of 1.5 % and then increases rapidly. It was concluded that the rate controlling step doesn't change up to this weight loss and then other mechanisms become important – probably the pyrolysis of residual hydrated carbons. For some binders, e.g. PVA, the iso-weight lines formed a single line when plotted in the kinetic field diagram. So they followed first order reaction kinetics during debindering according to equation 1. With PVB based binders a shift of the iso-weight lines towards higher temperatures was observed with increasing degree of debindering. This was attributed to the increasing cross linking of the PVB (compare also Figure 7). It is assumed that additional cross linking decreases pyrolysis rate.

In addition it should be noted, that the set of iso-weight lines from one kinetic field allows a prediction of any debindering cycle. Any weight loss and temperature defines a particular point in the kinetic field diagram. The corresponding weight change rate d(M/M$_0$)/dt can be read off. E.g., a weight loss of 1% at a temperature of 400°C leads to an inverse absolute temperature of 0.0015 1/K and a weight loss rate of 1.7 10^{-5} %/min (see fig. 10c). So the grade of debindering at a certain time t+Δt can be calculated according to:

$$(2)\ M/M_0(t+\Delta t) = M/M_0(t) + d(M/M_0)/dt\ \Delta t$$

At this time the temperature will be T+α·Δt, with α = the heating rate in the time interval from t to t+Δt. So another point in the kinetic field diagram can be associated with the time t+Δt. By a step-by-step calculation of the path integral over an arbitrary time – temperature cycle the particular debindering grade can be defined and vice versa. Major advantage is that for a chosen weight loss rate the corresponding temperature-time-regime can be computed.

DETERMINING FASTEST POSSIBLE DEBINDERING CYCLES:

The maximum safe debindering rate is the one at which just no cracks occur in the component. This rate was determined experimentally using large samples and a vacuum tight furnace that was equipped with a special weight sensor (TOM, Fraunhofer ISC, Würzburg, Germany). Samples were heated with fast constant heating rates until fracture was observed (Fig. 11). Cracking of samples was monitored in situ by abrupt changes in the TG-curve. From the slope of the TG-curve immediately before these abrupt changes, the corresponding maximum safe debindering rate was extracted. It was reduced by a safety factor of 1.5. Then a debindering cycle was calculated according to this maximum safe debindering rate where the heating rate was chosen to produce exactly the maximum safe debindering rate during the entire debindering cycle. This calculation was done using the kinetic field method described in the previous paragraph. Small variations of the set debindering rate were tested as well, to consider the requirements described in the preceding sections. Especially the residual carbon after debindering was measured by the TG-MS method and was minimized empirically.

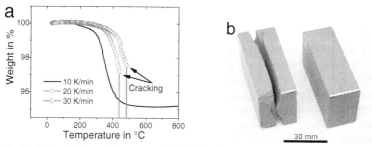

Fig. 11: Determination of fastest possible debindering rate: TG measurement (a) and cracked sample(b)

Optimized debindering conditions were verified with AlN components in a production furnace. They lead to a number of improvements:

- Reduced residual carbon in the components
- Less warping after firing, meaning less "flat firing": energy saving of almost 25 %
- Higher mechanical strength, smaller standard deviation (fig. 12)
- Time optimized debindering saves time and additional 4 % energy
- No discoloration inside larger parts (fig. 13).

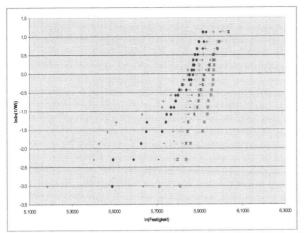

Fig. 12a: 3-point bending strength of various samples debindered under standard conditions

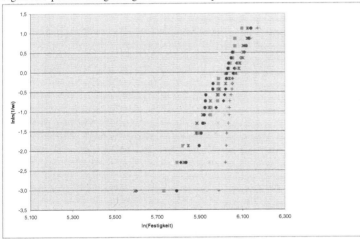

Fig. 12b: 3-point bending strength of various samples debindered according to our kinetic field model

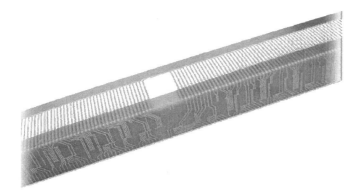

Fig. 13: AlN header bar for CCD camera, showing extremely homogeneous color; length about 10"

CONCLUSIONS

Using the kinetic field method for binder burnout calculation, combined with the determination of the maximum safe debindering rate, lead to a computer program which can predict optimum debindering curves for NO ceramics. Verification was done with AlN-, MoSi2 - and a mixed oxide ceramic used for cutting tools. It showed excellent experimental results in series production. Debindering of NOC can be done under vacuum conditions, air or protective atmospheres. Vacuum is preferred in small furnaces showing good results concerning residual C but liquid organic waste causes problems when entering the vacuum pump system or being released to environment. Debindering in air is common, shows good results concerning residual C but is critical due to damage which can be introduced during subsequent handling of the components. Uncontrolled oxygen uptake may cause detrimental chemical constitutions of sintered NOC. Debindering in flowing N_2 avoids additional handling. Exhaust gases can be burned under oxidizing conditions in a subsequent heat treatment to avoid environmental pollution.

REFERENCES

[1]   Klimera, A.: Festigkeitssteigerung bei Aluminiumnitridkeramiken. Dissertation, Bayerische Julius-Maximilians-Universität Würzburg (2007)
[2]   Roebben, G., Bollen, B., Brebels, A., Van Humbeeck, J., Van der Biest, O.: Impulse excitation apparatus to measure resonant frequencies, elastic moduli, and internal friction at room and high temperature. Rev. Sci. Instrum., 68(1997) 4511–4515
[3]   Palmour, H., Hare, T. M.: Sintering 85, Plenum Press, New York 1987
[4]   Raether, F.: Current state of in-situ measuring methods for the control of firing processes. J. Am. Ceram. Soc. 92 (2009) 146-152

SOFTENING OF RARE EARTH ORTHOPHOSPHATES BY TRANSFORMATION PLASTICITY: POSSIBLE APPLICATIONS TO FIBER-MATRIX INTERPHASES IN CERAMIC COMPOSITES

R. S. Hay, G. Fair
Air Force Research Laboratory
Materials and Manufacturing Directorate
WPAFB, OH

E. E. Boakye, P. Mogilevsky, T. A. Parthasarathy
Air Force Research Laboratory
UES, Inc., Dayton, OH

J. Davis
Wright State University
Fairborn, OH

ABSTRACT

Rare-earth orthophosphate interphases made from nanoparticle precursors have been successfully demonstrated for dense matrix oxide-oxide CMCs. For these interphases the major concern is high fiber pull-out stresses, typically ~80 - 200 MPa. Plastic deformation mechanisms in a 10 – 100 nm thick zone of rare-earth orthophosphate adjacent to the fiber govern pullout friction. For lower fiber pull-out stresses, rare-earth orthophosphates and vanadates that soften by transformation plasticity during the martensitic xenotime → monazite phase transformation were investigated. Predictive methods developed for prediction of deformation twinning in orthophosphates were extended to transformation plasticity. Nano-indentation testing was used to develop and test materials suitable for transformation plasticity weakened fiber coatings. Transformation plasticity significantly softens $TbPO_4$ and $(Gd,Dy)PO_4$ solid-solutions in the xenotime phase. Transformed regions were characterized by TEM; some evidence suggests that the phase transformation may reverse with time. Preliminary attempts to coat single-crystal alumina fibers with these materials were made. The potential to tailor fiber-matrix interphase friction in CMCs is discussed.

INTRODUCTION

Rare earth orthophosphates are stable at high temperatures in oxidizing and high vacuum environments.[1-4] Oxide-oxide CMC's with rare-earth orthophosphate interphases such as monazite and xenotime have been successfully demonstrated by a number of different research groups.[5-10] However, these materials are much less mature than other CMCs. The available information suggests that performance limiting factors for oxide-oxide CMCs are: 1) crack deflection and fiber-pullout shear stress (friction) of the rare-earth orthophosphate fiber-matrix interfaces,[11] 2) creep and high temperature strength of the nano-grain size (60 – 100 nm) oxide fibers, particularly in humid or combustion environments,[12,13] and 3) environmental effects, particularly environmentally assisted subcritical crack growth (EASCG).[14-17]

The major concern for rare-earth orthophosphate interphases is the high fiber pull-out stresses, typically ~80 - 200 MPa, which may be near the borderline of acceptable values.[1,18,19] These stresses contrast with 5 – 20 MPa typically measured or inferred for carbon and BN interphases.[20,21] High pullout stress biases fiber failure towards shorter pullout lengths, which may slightly increase strength, but at the expense of toughness (strain to failure) and flaw tolerance, the most desirable CMC attributes.[22,23] The optimal pull-out stresses are not known, and depend on the particular structural application. Pullout friction is governed by the plastic deformation mechanisms of a thin layer (10 – 100 nm) of the interphase material adjacent to the fiber (Fig. 1).[24] These mechanisms include dislocation slip, microfracture, formation of deformation twin nano-lamellae, and cataclastic flow of deformed nanoparticles (Fig. 1).

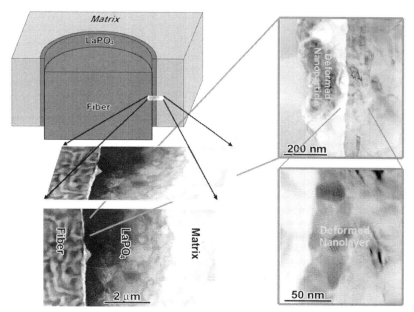

Fig. 1 TEM image sequence showing deformation in the monazite fiber-matrix interphase during room-temperature fiber pushout.[24] Deformation is concentrated in a ~100 nm thick layer adjacent to the fiber. An intensely deformed cataclastic nanoparticle and a deformed, recrystallized layer are marked.

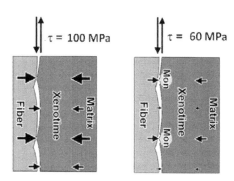

Fig. 2 An illustration of idealized transformation plasticity weakening by a xenotime coating on a fiber. Local high pressures and shear stress from fiber roughness trigger the xenotime → monazite phase transformation,n which weakens the interphase material and lowers push-out friction.

Some approaches to lowering pullout stress are use of softer ABO_4 phases such as tungstates, vanadates, or niobates, or nanoscale texturing of such materials so that cleavage or slip planes are in the plane of the fiber-matrix interface. However, many of these phases are unlikely to be stable with CMC constituents at high temperatures.[25] Another approach to lowering pullout stress involves use of interphases that undergo a $-\Delta V$ martensitic phase transformation. The concept is illustrated for the xenotime → monazite phase transformation in figure 2. The phase transformation can be driven by local high pressures and shear stresses caused by accommodation of fiber roughness during pullout. There are two possible effects; friction reduction by transformation plasticity, or by local reduction of normal stress from contraction of small volumes of the interphase material during the phase transformation (Fig. 2).

TRANSFORMATION PLASTICITY

Transformation plasticity occurs when the atomic rearrangements during a phase transformation simultaneously accommodate stress and therefore weaken the material. It has been known to metallurgists for over 80 years,[26-30] but the phenomena has seen little application in ceramics (a brittle-field mechanism, called "transformation weakening", was proposed for interphase crack deflection,[31,32] but not for reduction of pullout friction). Transformation plasticity has also been proposed to weaken earth materials.[28,33]

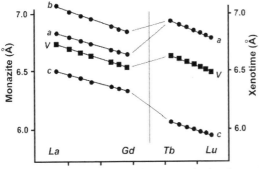

Fig. 3. *Lattice parameter of the rare earth phosphates. The stable phase changes from monazite to xenotime between GdPO4 and TbPO4. The xenotime phase has ~5.8% larger molar volume than the monazite phase.*

Large rare-earths, such as La and Ce, form orthophosphates with the monazite structure. Small rare earths, such as Y and Lu, form orthophosphates with the xenotime structure. The change between the structures occurs between the $GdPO_4$ and $TbPO_4$ compositions (Fig. 3, 4).[34] Similar observations have been made for rare-earth orthovanadates; here the change from monazite to xenotime structures occurs between $LaVO_4$ and $CeVO_4$.[35] The transformation between the two structures should be martensitic and kinetically facile, since only small atomic shuffles, half the magnitude of those involved in (100) and (001) monazite deformation twinning, are involved (Fig. 5).[36] The shear accompanying the xenotime → monazite transformation does not have a shear sense, or sign, unlike deformation twinning,[37] and because of the tetragonal xenotime symmetry, the transformation can occur on (100), (010), and (001) in xenotime. Pressures that induce the transformation can be estimated from thermodynamic data,[38-42] and are ~ 1 GPa for $TbPO_4$ (Fig. 6), but approach zero for $(Dy_x,Gd_{1-x})PO_4$ solid-solutions near monazite-xenotime equilibrium. Thermodynamic calculations suggests monazite-xenotime equilibrium for a $(Dy_{0.8}Gd_{0.2})PO_4$ solid-solution, with little temperature dependence.[43,44] It should be possible to shift the monazite-xenotime equilibrium to larger rare-earth cations, which are softer because

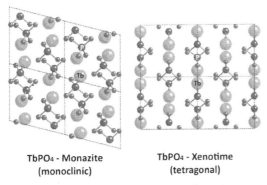

TbPO₄ - Monazite (monoclinic)　　**TbPO₄ - Xenotime (tetragonal)**

Fig. 4. *Crystal structures for monazite and xenotime TbPO4 polymorphs.*

of longer rare-earth – oxygen bonds, by making $RE(P_xV_{1-x})O_4$ solid-solutions. The trade-off for orthovanadates is that these compounds are less refractory and less stable, particularly with respect to reduction,

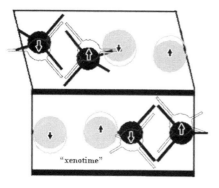

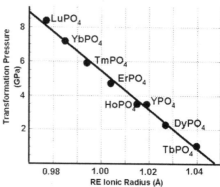

Fig. 5 Shuffles for rare-earth cations (gray) and PO₄ tetrahedra during deformation twinning on the (100) plane in monazite. The maximum shuffle required for twinning is 0.07 nm, which is less than ½ the La-O bond distance so diffusion is not required. The atoms shuffle through the "xenotime" structure, which requires only 0.035 nm shuffles for a martensitic transformation.

Fig. 6 Calculated xenotime → monazite transformation pressures at room temperature for different rare-earth orthophosphates.

in comparison with orthophosphates.[42]   The volume loss of the xenotime → monazite transformation (~5.8 vol%) is equivalent to ~2% linear shrinkage, and should in principle relax normal stress induced by interface roughness of 5 nm magnitude, and therefore the pullout friction, if the transformation occurs through an interphase 250 nm thick. The actual reduction in friction will depend on nanostructural details, such as the extent of and mechanisms by which the phase transformation is accommodated in the interphase. The ease of inducing the transformation, and the subsequent effect on mechanical properties through transformation plasticity, can be assessed by the stress-strain signature and hysteresis during instrumented nano-indentation, using a method that is widely applied to silicon.[45-48]

We present and discuss preliminary experiments on rare-earth orthophosphates that soften by transformation plasticity during the xenotime → monazite phase transformation. The effort involves identification of soft rare-earth orthophosphate compositions and solid-solutions, characterization of mechanical properties by instrumented indentation, and characterization/verification of fiber push-out/pullout friction of fibers coated with the identified compositions.

EXPERIMENTS
    Rare-earth orthophosphate powders with GdPO₄, TbPO₄, and DyPO₄ compositions were prepared and characterized for phase and particle morphology by methods used previously.[49,50] A series of powders with $(Gd_x Dy_{1-x})PO_4$ solid-solutions were also prepared. The powders were cold-pressed and sintered at 1600°C for 20 hours. Phase presence was checked using X-ray. The sintered pellets were sectioned and polished with diamond laps to < 1 μm surface finish. Indentation of the polished pellets was done using a MTS Nano Indenter XP system with Berkovich indenter, a 10 μm tip, and a 1:7 depth-to-side ratio. The indentation depth limit was 1.5 μm. Indentations were characterized by optical microscopy (reflected light) and SEM. Focused ion beam (FIB) sections were milled out beneath some TbPO₄ indentations and thinned to electron transparency. These sections were characterized in a Phillips CM200 FEG TEM operating at 200 kV.

Preliminary fiber coating experiments were conducted using TbPO$_4$ and (Gd,Dy)PO$_4$ solid-solution precursors described elsewhere.[50] Single-crystal alumina fibers (Saphikon™) were coated using heterogenous nucleation and growth from rare-earth citrates; the methods are described in other publications.[50-53] TEM specimens were prepared of the coated fibers by published methods. The coatings were characterized for microstructure, phase and composition in a Phillips CM200 FEG TEM operating at 200 kV using EDS and selected area electron diffraction. Coating thicknesses of 300 – 400 nm were used with a final coating heat-treatment of 5 minutes at 900°C. Coated fibers were slip cast in Sumitomo alumina powder and hot-pressed at 1400°C, 15 MPa for 1 hour to make minicomposites.

RESULTS AND DISCUSSION

Abnormal grain growth hindered densification of the TbPO$_4$ and (Gd,Dy)PO$_4$ xenotime pellets. Polishing of TbPO$_4$ and (Gd,Dy)PO$_4$ pellets with xenotime compositions near the monazite-xenotime phase boundary was also very difficult (Fig. 7). These material removal rate was significantly faster for these compositions than GdPO$_4$ and DyPO$_4$ compositions. This was attributed to poor densification, transformation plasticity, and to grain pullout driven by stress generated from local xenotime → monazite phase transformations during polishing. Pellets with GdPO$_4$ and DyPO$_4$ compositions were not difficult to polish.

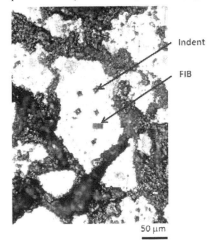

Indentation experiments were done on the sporadically dispersed grains or clumps of grains that were well polished (Fig. 7). Indentation load-displacement curves for GdPO$_4$ (monazite), DyPO$_4$ (xenotime), and TbPO$_4$ (xenotime) are shown in figure 8. GdPO$_4$ is the rare-earth orthophosphate monazite phase closest to the xenotime stability field (Fig. 3). TbPO$_4$ is the rare-earth orthophosphate xenotime phase closest to the monazite stability field, and DyPO$_4$ is the xenotime phase that is next closest to the monazite stability field (Fig. 3).

Fig. 7 Optical micrograph (reflected light) of a polished TbPO$_4$ surface. Indentations, and the location of a FIB section through an indentation are marked.

Indentation load-displacement curves for GdPO$_4$ (monazite), DyPO$_4$ (xenotime), and TbPO$_4$ (xenotime) are shown in figure 8. The TbPO$_4$ was much softer than either DyPO$_4$ or GdPO$_4$. This is consistent with the presence of transformation plasticity in TbPO$_4$, but not in DyPO$_4$. There was much more variation in load for a 1500 nm displacement for different indentations in the TbPO$_4$ than either DyPO$_4$ or GdPO$_4$, suggesting significant anisotropy for transformation plasticity. Grain orientations with maximum shear stress along <100>{010} and [010](001) are expected to deform most easily by transformation plasticity (Fig. 5).

The indentation load-displacement curves for (Gd$_{0.5}$,Dy$_{0.5}$)PO$_4$ (xenotime) are shown with the DyPO$_4$ (xenotime) load-displacement curve from figure 8 in figure 9. The (Gd$_{0.5}$,Dy$_{0.5}$)PO$_4$ is even softer than TbPO$_4$, and there were even larger variations in load for a 1500 nm displacement for different indentations. Some load-displacement curves exhibit up to 75% rebound hysteresis. This is diagnostic of a ferroelastic or shape memory material, and may be due to either phase transformation reversal or elastic deformation twinning.[54]

Several focused-ion beam (FIB) sections were cut directly beneath the bottoms of indentations in TbPO$_4$. Most show only intense plastic deformation upon examination by TEM. A couple did show conclusive evidence for the xenotime → monazite phase transformation. The best example is shown in

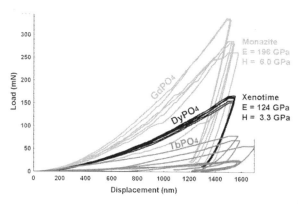

Fig. 8. Indentation – displacement curves for GdPO₄ (monazite), DyPO₄ (xenotime),and TbPO₄ (xenotime). TbPO₄ is by far the softest material. Variation in load displacement curves may reflect anisotropy of transformation plasticity mechanisms related to the crystallography of the martensitic xenotime → monazite transformation.

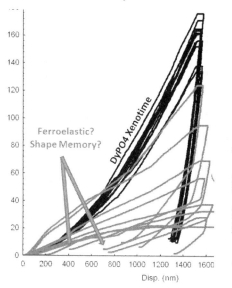

Fig. 9. Indentation – displacement curves for DyPO₄ (xenotime), and (Gd₀.₅,Dy₀.₅)PO₄ (xenotime). Some (Gd₀.₅,Dy₀.₅)PO₄ indentations show very large displacement rebounds. These are arrowed as ferroelastic or shape memory transformations.

figure 10. In this figure a TEM micrograph montage is shown along with the relationship of the imaged area to the indenter tip. An area several microns in extent directly beneath the indenter tip has transformed to monazite, as shown by selected area electron diffraction patterns from an area outside (1) and inside (2) the transformed area. This area was imaged exactly one week after the indentation experiments were performed. Subsequent TEM examination months after the indentation experiments found that some of the monazite region had transformed back to xenotime. Time dependent transformation reversal has been observed for other pressure induced martensitic phase transformations under indentation such as β-eucryptite, and in some cases the effect has been correlated with environmental effects such as subcritical crack growth from moisture.[55] The lack of a transformed region in some FIB sections may then be due to either time-dependent reversal of a transformed area between the time of indentation and TEM examination, or lack of any original transformed area. The high variance in load-displacement curves (Fig, 8 & 9) diagnostic of transformation plasticity anisotropy suggests transformation plasticity may not occur under all indentations. We did not track particular load-displacement curves with the indentation FIB sections, so we cannot confirm the anisotropy effect. Work is in progress to identify the transformation reversal mechanisms and time dependence.

Single crystal alumina fibers were coated with TbPO₄ and with (Gd₀.₅,Dy₀.₅)PO₄ for fiber pushout experiments. Temperatures of at least 1400°C were required to deposit these compositions as the xenotime phase in thin film form; this was at least 200°C higher than required for the powders. As observed with the sintered pellets (Fig. 7),

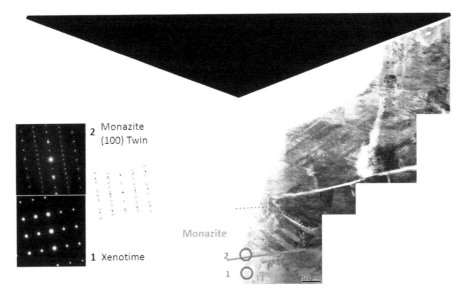

Fig. 10. TEM micrograph of indented TbPO₄ (xenotime), showing transformation to monazite underneath the indent with selected area electron diffraction patterns of the circled areas (1) & (2). The TEM cross-section was prepared by FIB.

there were significant densification problems that appear to be associated with exaggerated grain growth and loss of phosphorous. The porous coatings were easily infiltrated by the Sumitomo alumina powder and were not suitable for high quality fiber pushout specimens. These processing difficulties are currently under study; we will report more thorough description, characterization, and analysis in a future publication.

If the processing obstacles to making dense coatings of transformable rare-earth orthophosphates are solved, the potential to tailor the fiber pullout friction in CMCs suggested by the large variation in indentation load-displacement for rare-earth orthophosphates and their solid-solutions is particularly attractive for CMC fiber-matrix interphase engineering. Reversal of the transformation with time may also be a desirable feature; the fiber-matrix interphase material resets for applications where load is intermittently applied.

SUMMARY AND CONCLUSIONS

Transformation plasticity of TbPO₄ and (Gd$_{0.5}$,Dy$_{0.5}$)PO₄ rare-earth orthophosphates with the xenotime phase was demonstrated by nano-indentation and TEM characterization. The martensitic xenotime → monazite phase transformations are mechanistically similar to deformation twinning in monazite and require only very small shuffles. There is some evidence that these transformations may reverse with time; this complicates TEM characterization of transformation extent and mechanisms. The TbPO₄ and (Gd$_{0.5}$,Dy$_{0.5}$)PO₄ compositions are difficult to process as dense pellets or coatings; this currently hinders their use as fiber-matrix interphases in CMCs. Work is in progress to solve the processing difficulties, and to thoroughly characterize the transformation anisotropy and reversal mechanisms.

REFERENCES

[1] Morgan, P. E. D. and Marshall, D. B., Ceramic Composites of Monazite and Alumina, *J. Am. Ceram. Soc.* 78 (6), 1553-63, 1995.

[2] Morgan, P. E. D., Marshall, D. B., and Housley, R. M., High Temperature Stability of Monazite-Alumina Composites, *Mater. Sci. Eng.* A195, 215-222, 1995.

[3] Davis, J. B., Marshall, D. B., Oka, K. S., Housley, R. M., and Morgan, P. E. D., Ceramic Composites for Thermal Protection Systems, *Composites* A30, 483-488, 1999.

[4] Marshall, D. B., Morgan, P. E. D., Housley, R. M., and Cheung, J. T., High-Temperature Stability of the Al_2O_3-$LaPO_4$ System, *J. Am. Ceram. Soc.* 81 (4), 951-956, 1998.

[5] Keller, K. A., Mah, T., Parthasarathy, T. A., Boakye, E. E., Mogilevsky, P., and Cinibulk, M. K., Effectiveness of Monazite Coatings in Oxide/Oxide Composites After Long Term Exposure at High Temperature, *J. Am. Ceram. Soc.* 86 (2), 325-332, 2003.

[6] Lee, P.-Y., Imai, M., and Yano, T., Fracture Behavior of Monazite-Coated Alumina Fiber-Reinforced Alumina-Matrix Composites at Elevated Temperature, *J. Ceram. Soc. Japan* 112 (12), 628-633, 2004.

[7] Marshall, D. B. and Davis, J. B., Ceramics for Future Power Generation Technology: Fiber Reinforced Oxide Composites, *Curr. Opin. Solid State Mater. Sci.* 5, 283-289, 2001.

[8] Kaya, C., Butler, E. G., Selcuk, A., Boccaccini, A. R., and Lewis, M. H., Mullite (Nextel™ 720) Fibre-Reinforced Mullite Matrix Composites Exhibiting Favourable Thermomechanical Properties, *J. Eur. Ceram. Soc.* 22, 2333-2342, 2002.

[9] Davis, J. B., Marshall, D. B., and Morgan, P. E. D., Oxide Composites of $LaPO_4$ and Al_2O_3, *J. Eur. Ceram. Soc.* 19, 2421-2426, 1999.

[10] Davis, J. B., Marshall, D. B., and Morgan, P. E. D., Monazite Containing Oxide-Oxide Composites, *J. Eur. Ceram. Soc.* 20 (5), 583-587, 2000.

[11] Zok, F. W., Developments in Oxide Fiber Composites, *J. Am. Ceram. Soc.* 89 (11), 3309-3324, 2006.

[12] Ruggles-Wrenn, M. B. and Laffey, P. D., Creep Behavior of Nextel™ 720/Alumina Ceramic Composite at Elevated Temperature in Air and in Steam, *Compos. Sci. Tech.* 68, 2260-2266, 2008.

[13] Ruggles-Wrenn, M. B., Siegert, G. T., and Back, S. S., Creep Behavior of NextelTM 720/Alumina Ceramic Composite with +/-45 Fiber Orientation at 1200 C, *Compos. Sci. Tech.* 68, 1588-1595, 2008.

[14] Boakye, E., Hay, R. S., and Petry, M. D., Continuous Coating of Oxide Fiber Tows Using Liquid Precursors: Monazite Coatings on Nextel 720, *J. Am. Ceram. Soc.* 82 (9), 2321-2331, 1999.

[15] Boakye, E. E., Hay, R. S., Mogilevsky, P., and Douglas, L. M., Monazite Coatings on Fibers: II, Coating without Strength Degradation, *J. Am. Ceram. Soc.* 84 (12), 2793-2801, 2001.

[16] Hay, R. S., Boakye, E. E., and Petry, M. D., Effect of Coating Deposition Temperature on Monazite Coated Fiber, *J. Eur. Ceram. Soc.* 20, 589-97, 2000.

[17] Hay, R. S. and Boakye, E., Monazite Coatings on Fibers: I, Effect of Temperature and Alumina Doping on Coated Fiber Tensile Strength, *J. Am. Ceram. Soc.* 84 (12), 2783-2792, 2001.

[18] Kuo, D.-H., Kriven, W. M., and Mackin, T. J., Control of Interfacial Properties through Fiber Coatings: Monazite Coatings in Oxide-Oxide Composites, *J. Am. Ceram. Soc.* 80 (12), 2987-2996, 1997.

[19] Chawla, K. K., Liu, H., Janczak-Rusch, J., and Sambasivan, S., Microstructure and Properties of Monazite ($LaPO_4$) Coated Saphikon Fiber/Alumina Matrix Composites, *J. Eur. Ceram. Soc.* 20, 551-559, 2000.

[20] Cao, H. C., Bischoff, E., Sbaizero, O., Ruhle, M., Evans, A. G., Marshall, D. B., and Brennan, J. J., Effect of Interfaces on the Properties of Fiber-Reinforced Ceramics, *J. Am. Ceram. Soc.* 73 (6), 1691-99, 1990.

[21] Curtin, W. A., Eldredge, J. I., and Srinivasan, G. V., Push-Out Test on a New Silicon Carbide/Reaction Bonded Silicon Carbide Ceramic Matrix Composite, *J. Am. Ceram. Soc.* 76 (9), 2300-2304, 1993.

[22] Kerans, R. J., Hay, R. S., Parthasarathy, T. A., and Cinibulk, M. K., Interface Design for Oxidation Resistant Ceramic Composites, *J. Am. Ceram. Soc.* 85 (11), 2599-2632, 2002.

23. Curtin, W. A., Ahn, B. K., and Takeda, N., Modeling Brittle and Tough Stress-Strain Behavior in Unidirectional Ceramic Matrix Composites, *Acta mater.* 46 (10), 3409-3420, 1998.

24. Davis, J. B., Hay, R. S., Marshall, D. B., Morgan, P. E. D., and Sayir, A., The Influence of Interfacial Roughness on Fiber Sliding in Oxide Composites with La-Monazite Interphases, *J. Am. Ceram. Soc.* 86 (2), 305-316, 2003.

25. Morgan, P. E. D. and Marshall, D. B., Functional Interfaces for Oxide/Oxide Composites, *Mater. Sci. Eng.* A162, 15-25, 1993.

26. Sauveur, A., What is Steel? Another Answer, *The Iron Age* 113, 581-583, 1924.

27. Wassermann, G., Untersuchungen an einer Eisen-Nickel Legierung uber die Verformbarkeit Wahrend der g-a Unwandlung, *Archiv Fur der Eisenhutt* 7, 321-325, 1937.

28. Poirier, J.-P., *Creep of Crystals* Cambridge University Press, Cambridge, 1985.

29. Stringfellow, R. G., Parks, D. M., and Olson, G. B., A Constitutive Model for Transformation Plasticity Accompanying Strain-Induced Martensitic Transformations in Metastable Austenitic Steels, *Acta metall. mater.* 40 (7), 1703-1716, 1992.

30. Fischer, F. D., A Micromechanical Model for Transformation Plasticity in Steels, *Acta metall. mater.* 38 (6), 1535-1546, 1990.

31. Kriven, W. M. and Lee, S.-J., U.S.A 6,361,888, 2002, Toughening of Ceramic Composites by Transformation Weakening of Interphases.

32. Kriven, W. M. and Lee, S.-J., Toughening of Mullite/Cordierite Laminated Composites by Transformation Weakening of b-Cristobalite Interphases, *J. Am. Ceram. Soc.* 88 (6), 1521-1528, 2005.

33. Poirier, J. P., On Transformation Plasticity, *J. Geophys. Res.* 87 (B8), 6791-6798, 1982.

34. Kolitsch, U. and Holtsam, D., Crystal Chemistry of REEXO$_4$ Compounds (X = P, As, V). II. Review of REEXO$_4$ Compounds and their Stability Fields, *Eur. J. Mineral.* 16, 117-126, 2004.

35. Jia, C.-J., Sun, L.-D., You, L.-P., Jiang, X.-C., Luo, F., Pang, Y.-C., and Yan, C.-H., Selective Synthesis of Monazite- and Zircon-type LaVO$_4$ Nanocrystals, *J. Phys. Chem. B* 109, 3284-3290, 2005.

36. Hay, R. S. and Marshall, D. B., Deformation Twinning in Monazite, *Acta mater.* 51 (18), 5235-5254, 2003.

37. Wenk, H.-R., Plasticity Modeling in Minerals and Rocks, in *Texture and Anisotropy*, Kocks, U. F., Tome, C. N., and Wenk, H.-R. Cambridge University Press, Cambridge, UK, 1998, pp. 561-596.

38. Ushakov, S. V., Helean, K. B., Navrotsky, A., and Boatner, L. A., Thermochemistry of Rare-Earth Orthophosphates, *J. Mater. Res.* 16 (9), 2623-2633, 2001.

39. Thiriet, C., Konings, R. J. M., Javorsky, P., and Wastin, F., The Heat Capacity of Cerium Orthophosphate CePO$_4$, the Synthetic Analogue of Monazite, *Phys. Chem. Minerals* 31, 347-352, 2004.

40. Popa, K., Sedmidubsky, D., Benes, O., Thiriet, C., and Konings, R. J. M., The High Temperature Heat Capacity of LnPO$_4$ (Ln = La, Ce, Gd) by Drop Calorimetry, *J. Chem. Thermo.*, 2005.

41. Thiriet, C., Konings, R. J. M., Javorsky, P., Magnani, N., and Wastin, F., The Low Temperature Heat Capacity of LaPO$_4$ and GdPO$_4$, the Thermodynamic Functions of the Monazite-Type LnPO$_4$ Series, *J. Chem. Thermo.* 37, 131-139, 2005.

42. Dorogova, M., Navrotsky, A., and Boatner, L. A., Enthalpies of Formation of Rare Earth Orthovanadates, REVO$_4$, *J. Solid State Chem.* 180, 847-851, 2007.

43. Mogilevsky, P., Boakye, E. E., and Hay, R. S., Solid Solubility and Thermal Expansion in LaPO$_4$-YPO$_4$ System, *J. Am. Ceram. Soc.* 90 (6), 1899-1907, 2007.

44. Keller, K. A., Mogilevsky, P., Parthasarathy, T. A., Lee, H. D., and Mah, T.-I., Monazite coatings in dense ($\geq$90%) alumina-chromia minicomposites, *J. Am. Ceram. Soc.* submitted, 2007.

45. Bradby, J. E., Williams, J. S., Wong-Leung, J., Swain, M. V., and Munroe, P., Mechanical Deformation in Silicon by Micro-Indentation, *J. Mater. Res.* 16 (5), 1500-1507, 2001.

46. Domnich, V., Gogotsi, Y., and Dub, S., Effect of Phase Transformations on the Shape of the Unloading Curve in the Nanoindentation of Silicon, *Appl. Phys. Lett.* 76 (16), 2214-2217, 2000.

47. Zarudi, I., Zhang, L. C., and Swain, M. V., Behavior of Monocrystalline Silicon Under Cyclic Microindentations with a Spherical Indenter, *Appl. Phys. Lett.* 82 (7), 1027-1029, 2003.

[48.] Zhang, L. and Zarudi, I., Towards a Deeper Understanding of Plastic Deformation in Mono-crystalline Silicon, *Int. J. Mech. Sci.* 43, 1985-1996, 2001.

[49.] Boakye, E. E., Hay, R. S., Mogilevsky, P., and Cinibulk, M. K., Two Phase Monazite/Xenotime $30LaPO_4$-$70YPO_4$ Coating of Ceramic Fiber Tows, *J. Am. Ceram. Soc* 91 (1), 17-25, 2008.

[50.] Boakye, E. E., Fair, G. E., Mogilevsky, P., and Hay, R. S., Synthesis and Phase Composition of Lanthanide Phosphate Nanoparticles $LnPO_4$ (Ln = La, Gd, Tb, Dy, Y) and Solid Solutions for Fiber Coatings, *J. Am. Ceram. Soc.* 91 (12), 3841-3849, 2008.

[51.] Fair, G. E., Hay, R. S., and Boakye, E. E., Precipitation Coating of Monazite on Woven Ceramic Fibers – I. Feasibility, *J. Am. Ceram. Soc.* 90 (2), 448-455, 2007.

[52.] Fair, G. E., Hay, R. S., and Boakye, E. E., Precipitation Coating of Rare-Earth Orthophosphates on Woven Ceramic Fibers- Effect of Rare-Earth Cation on Coating Morphology and Coated Fiber Strength, *J. Am. Ceram. Soc* 91 (7), 2117-2123, 2008.

[53.] Fair, G. E., Hay, R. S., and Boakye, E. E., Precipitation Coating of Monazite on Woven Ceramic Fibers – II. Effect of Processing Conditions on Coating Morphology and Strength Retention of Nextel[TM] 610 and 720 Fibers, *J. Am. Ceram. Soc.* 91 (5), 1508-1516, 2008.

[54.] Salje, E. K. H., *Phase Transitions in Ferroelastic and Co-Elastic Crystals* Cambridge University Press, 1990.

[55.] Reimanis, I. E., Seick, C., Fitzpatrick, K., Fuller, E. R., and Landin, S., Spontaneous Ejecta from b-Eucryptite Composites, *J. Am. Ceram. Soc.* 90 (8), 2497-2501, 2007.

INFLUENCE OF FIBER ARCHITECTURE ON IMPACT RESISTANCE OF UNCOATED SIC/SIC COMPOSITES

Ramakrishna T. Bhatt
US Army Vehicle Technology Directorate
NASA Glenn Research Center
21000 Brookpark Road, Cleveland, Ohio 44135

Laura M. Cosgriff
Cleveland State University, Cleveland, OH 44115

Dennis S. Fox
NASA Glenn Research Center, Cleveland, OH 44135

ABSTRACT

2-D and 2.5D woven SiC/SiC composites fabricated by melt infiltration (MI) method were impact tested at ambient temperature and at 1316^0C in air using 1.59-mm diameter steel-ball projectiles at velocities ranging from 115 m/s to 300 m/s. The extent of substrate damage with increasing projectile velocity was imaged and analyzed using optical microscopy, pulsed thermography, and computed tomography. Results indicate that both types of composites impact tested at ambient temperature and at 1316^0C showed increased surface or internal damage with increased projectile velocity. At a fixed projectile velocity, the extant of impact damage caused at ambient temperature is nearly the same as that at 1316^0C. Predominant impact damage mechanisms in 2-D SiC/SiC composites are fiber ply delamination, fiber fracture and matrix shearing, and in 2.5D SiC/SiC composites are fiber fracture and matrix shearing with no evidence of delamination cracks. Under similar testing conditions, the depth of projectile penetration into 2.5D SiC/SiC composites is significantly lower than that in 2D SiC/SiC composites.

INTRODUCTION

The efficiency of engines used for aero propulsion and for land-based power generation will depend strongly on the upper use temperature and life capability of the structural materials used for the hot-section components and the cooling requirements of these components. Components with improved thermal capability and longer life between maintenance cycles will allow improved system performance by reducing cooling requirements and life-cycle costs. The structural materials used for current turbine components are limited to 1100^0C because of their poor creep resistance, poor oxidation performance, and lifing issues. Further improvements in power efficiency in the engines can be achieved by replacing metallic components with a new class of material typically referred to as fiber-reinforced ceramic matrix composites (CMCs) of which silicon carbide fiber reinforced silicon-carbide matrix composites (SiC/SiC) are of particular interest [1]. These materials are not only ~40% lighter and capable of 200 to 300^0C higher use temperatures than state-of-the-art metallic alloys and oxide matrix composites (~1100^0C), but also capable of providing significantly better static and dynamic toughness than un-reinforced silicon-based monolithic ceramics. However, these materials also show limited strain capability compared to metals and require an environmental barrier coating to survive in the combustion environment. Various environmental barrier coating systems have been developed for CMCs which show potential for limited time applications at temperatures as high as 1450^0C [2, 3]. Because of their potential advantages over metallic materials, SiC/SiC CMCs are actively being pursued for high-temperature structural applications such as engine combustor liners, turbine components, and exhaust nozzles [4, 5].

From a durability and lifing point of view, the CMC materials for turbine blade and vane applications should not only have high design properties and oxidation resistance, but also should have adequate erosion, corrosion, and foreign object damage (FOD) resistance. In general, foreign objects moving with the gas stream of the engine could vary in size from sub-micron to several centimeters in diameter. Depending on where the object strikes and its velocity, impact damage can result in chipping and spalling of the coating and damage to the substrate. This can lead to internal oxidation by recession and loss of mechanical properties, or in the worst situations, failure of the blade from bending and shearing [6, 7]. Sometimes the secondary damage may be much greater than the primary damage. For example, the major damage in the blade may not be due to the initial impact but due to damage created by intensity of the shock wave propagating back and forth from the impact site to the root of the blade, causing it to buckle or break. The extent of impact damage can vary depending on substrate and projectile parameters: for the substrate this includes hardness, thickness, support, coating, fiber architecture, stiffness, and temperature, and for the projectile this includes the size, hardness, shape, velocity and angle of incidence relative to the substrate.

Previous single particle impact studies indicate that although the damage is local, the 2-D woven uncoated and environmental barrier coated MI SiC/SiC composites are prone to delamination and that the presence of a thicker EBC is more effective at withstanding damage than a thinner EBC[8, 9]. The current study was conducted to understand basic impact damage mechanisms and the role played by the fiber architecture and test temperature on the impact damage of uncoated MI SiC/SiC composites as well as to categorize and quantify the damage by NDE methods.

EXPERIMENTAL PROCEDURE

For the impact study, the 2-D woven Sylramic-iBN SiC fiber-reinforced SiC matrix composite panels were purchased from GE Composite Ceramic Products (GECCP), Newark, Delaware, and the 2-D and 2.5D woven Hi-Nicalon-S SiC/SiC composite panels were provided by Goodrich, Brecksville, Ohio. The SiC fiber tows (Sylramic^{TM}) produced by Dow Corning Corporation, Midland, Michigan, and the Hi-Nicalon-S SiC fiber tows were produced by Nippon Carbon, Japan. The fibers can be woven into 2-D cloth, or into a wide variety of 2.5D architectures depending on the segment of fiber required in the through-the-thickness direction and the degree of interlocking that is required between fiber plies. In this study, two different 2.5D woven fiber preforms designated as Type-I and Type-II were used. The fibers were woven into 2-D, 0/90, 5-harness satin fabric or into 2.5D fiber architecture by Albany Engineered Composites, New Hampshire. The Sylramic fiber cloth was further treated to convert Sylramic to Sylramic-iBN fibers by a proprietary process [10]. For composite fabrication, 2-D and 2.5D woven fiber preforms are first compressed into predetermined thickness in a graphite fixture, and then coated with a thin BN layer (<1 μm) followed by layer of SiC (<3 μm) both by chemical vapor deposition. At this stage, the fiber preforms are rigidized and contain open porosity between 20 to 40 vol% depending on the thickness of the woven fiber stack, and the amounts of BN and SiC deposited. The rigidized preforms are first infiltrated with SiC slurry and then with molten silicon. The fabrication details are described in reference [11]. According to the vendor, the as-fabricated SiC/SiC composites contained ~35 vol% SiC fibers, ~5 vol% BN coating, ~25vol% CVI SiC coating, ~16 vol% SiC particles, and ~16 vol% silicon. The remaining ~3 vol% is closed porosity.

The as-fabricated composite panels had nominal dimensions of ~ 230-mm (L) x 150-mm (W) x 2.4-mm (T). The panels were first cut into rectangular specimens and the machined into tensile dog-boned specimens of dimensions 152-mm (L), 13-mm (W), and 2.4-mm (T) with a reduced gage section as shown in figure 1.

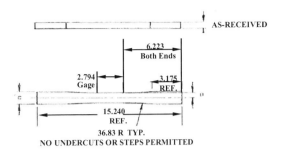

Figure1. Geometry of a typical impact test specimen (all dimensions are in centimeters).

Impact testing

A gas propelled impact gun unit similar to that shown in figure 2 was used for impact testing. Detailed description of unit can be found in reference [12]. The target specimen was mounted in a test fixture that was placed in front of and at a distance ~ 25 to 30 cm away from the impact gun. The top and bottom ends of the specimen were held in a fixture using "C" clamps. For high temperature testing the gage section of the specimen was heated by an atmospheric pressure burner rig. The specimen temperature was monitored by a 2 color optical pyrometer.

At both ambient and high temperatures, the specimens were impacted with hardened (HRC≥60) chrome steel-balls (diameter ~1.59-mm, density 7.8 gm/cc) at velocity ranging from 110 m/s to 300 m/s. At each test condition only one or two specimens were tested. The gage section of the specimen before and after impact testing was imaged by optical microscopy, pulsed thermography (PT), and computed tomography (CT) to determine extent of surface and volumetric damage. The PT and CT imaging procedures similar to those discussed in references [13] and [14] were used.

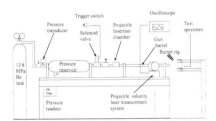

Figure 2. Impact testing apparatus with a specimen holder and a burner rig.

EXPERIMENTAL RESULTS AND DISCUSSION

Representative SEM photographs of the transverse cross-sections of 2-D and two types of 2.5D woven MI SiC/SiC composites are shown in figure 3. Figure 3(a) represents 2-D woven MI SiC/SiC composites which consist of 8 layers of 2-D woven SiC fiber mats. In this figure, the dark grey regions are BN/SiC coated SiC fiber tows, and the light grey and white regions are SiC particles and silicon metal, respectively. Figure 3(b) and 3(c) represent two types of 2.5-D woven MI SiC/SiC composites. In these figures, the undulating dark and grey oblong regions are fiber tows, and the surrounding whitish regions are SiC matrix which consists of SiC particles and silicon metal. The undulating fibers (also referred to as warp weaver fibers) are along the tensile loading direction and the oblong bundle of fibers (also referred to as pick fibers) are

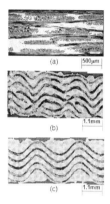

Figure 3. SEM photographs of the cross-section of as-fabricated MI SiC/SiC composites showing fiber architecture and SiC matrix: (a) 2-D, (b) 2.5D Type-I, and (c) 2.5D Type-II composites.

transverse to the loading direction. All three photographs in figure 3 are taken at different magnifications therefore in some photographs the features within the tows are clearly delineated and in others are not.

Characterization results indicate that at a fixed projectile velocity, the resulting damaged zone width and depth of into the substrate for 2-D woven MI SiC/SiC composites impact tested at ambient and at 1316^0C were similar. Also the impact behaviors of 2-D MI SiC/SiC composites fabricated by GECCP and Goodrich at the tested temperatures were similar although the reinforcing fibers were different. In addition, these composites showed fiber ply delamination at low impact velocities (~115m/sec) followed by fiber fracture and matrix shearing, which is consistent with literature data on similar 2-D MI SiC/SiC composites[3]. The delamination between the fiber ply and the matrix is primarily due to low through-the-thickness tensile strength. The impact behavior with increase in projectile velocity and extent of volumetric damage at a fixed projectile velocity for two types of 2.5D MI SiC/SiC composites impact tested at ambient and at 1316^0C were also similar. Therefore, for brevity only optical, PT and CT images of 2.5D Type-II MI SiC/SiC composites impact tested at 1316^0C are shown in figures 4, 5 and 6, and characterization data of all three types of composites are discussed in other sections of the paper.

Figure 4 shows the optical photographs of the gage section of the 2.5-D Type-II woven MI SiC/SiC composite specimens impact tested at 1316^0C from the impacted and backsides with increasing projectile velocities. On the impacted side, the surface damage is minimal and increased with increase in projectile velocity. Even on specimens tested at 300 m/sec projectile velocity visible damage zone is approximately two to three times the width the fiber tow. On the back side of the specimens, no damage was detected under the tested conditions.

Figure 5 shows the thermal image of corresponding specimens shown in figures 4. Each individual thermal image represents the thermal response at a specific time and was chosen based on the amount of contrast to illustrate the defects. Surface conditions such as roughness and reflectivity, internal volumetric flaws and thermal conductivities of the constituents of the specimen will affect the thermal response. Examination of Fig 5 indicates that at the four projectile velocities both surface or internal damages were noticed both on the impacted and back side of the specimen. The damage zone width is larger than that observed by optical microscopy. Some of the contrast variations in the images represent the fiber weave. These variations were caused by the surface undulations or possible differences in thermal conductivities of the fiber, the interface and the matrix. The dull to intense white spot at the center of the image indicates the presence of subsurface or internal damage. Figure 5 also shows that for the same projectile velocity, the damage zone width measured by optical method is always smaller than that detected with thermal method. This indicates a greater amount of subsurface damage surrounding the crater. Also noticed is that the width of the damage zone measured in the thermal image is greater on the back side than in the impacted side possibly due to bending of specimen during impact testing.

Figure 6 shows CT images of the impacted specimen shown in figures 4 and 5 with increasing projectile velocity. The dark arrows in the figure indicate the impacted site. The CT images in general can detect internal flaws such as delamination, shear band and pores and their locations. In this case, CT images reveal that damage is limited and occurs only on the impacted side of the specimen, and that damage increased with increase in projectile velocity. Few isolated pores within the specimen are also detected. However, the internal damage as recorded by the PT images captured from the back side

Impacted side   Back side

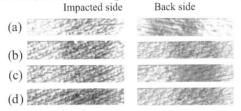

Figure 4 Optical photographs of 2.5-D Type-II MI SiC/SiC composite specimens impact tested at 1316^0C showing impact damage with projectile velocity: (a) 115 m/s, (b) 160 m/s, and (c) 220 m/s (d) 300 m/s.

Impacted side Back side

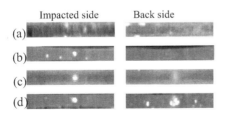

Figure 5 PT images of 2.5-D SiC/SiC composite specimens impact tested at 1316^0C showing impact damage with projectile velocity: (a) 115 m/s, (b) 160 m/s, and (c) 220 m/s (d) 300 m/s.

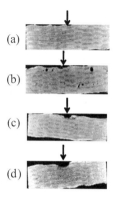

Figure 6.   CT images of the cross-sections of the 2.5D Type-II MI SiC/SiC composites impact tested at 1316°C showing impact damage with projectile velocity: (a) 115 m/s, (b) 160 m/s, (c) 220 m/s, and (d) 300 m/s. The black arrows indicate impacted sites.

of the impacted specimens at 220 and 300 m/s (Fig. 5) is not detected in the CT images. The fiber architecture may have masked the backside damage that was detected with pulsed thermography. Examination by optical microscopy of the transverse cross-sections is planned in the future in order to confirm the nature of the damage.

Figures 7 and 8 show variation of maximum damage zone width with increasing projectile velocity on the front side (i.e. impacted side) and the backside of the 2D and two types of 2.5-D MI SiC/SiC composites impact tested at ambient and 1316°C, respectively. The damage zone or crater size was measured from pulsed thermography. In these figures, no distinction was made in plotting the data of the impact tested 2-D MI SiC/SiC composites from GECCP and Goodrich since both exhibited similar damage zone width at a fixed projectile velocity. Comparison of figure 7 and 8 indicates that either on the impacted side or on the back side of all three types of composites, the damage zone width increases linearly with increasing projectile velocity, but the test

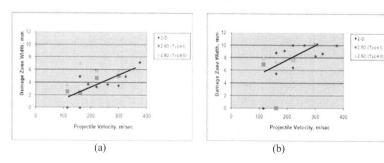

(a)   (b)

Figure 7.   Variation of damaged zone width with projectile velocity for 2D and 2.5D MI SiC/SiC composites impact tested at ambient temperature: (a) Impacted side (b) Back side. The damage zone width was measured from thermographic images.

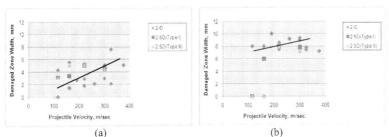

(a)   (b)

Figure 8. Variation of damaged zone width with projectile velocity for 2D and 2.5D MI SiC/SiC composites impact tested at 1316°C: (a) Impacted side (b) Back side. The damage zone width was measured from thermographic images.

temperature or the fiber architecture had relatively no significant effect on impact damage zone width. However, at a fixed projectile velocity, damage zone width recorded for all three composites from the backside is always greater than on the impacted side.

From the CT images of impact tested 2D and 2.5D composites at ambient temperature and at 1316°C, the depth of penetration of the projectile was measured. Figure 9 shows this data which indicate that depth of penetration of the projectile is significantly lower in 2.5D composites than that in 2D composites at ambient temperature as well as 1316°C and at projectile velocities greater than 160m/sec.

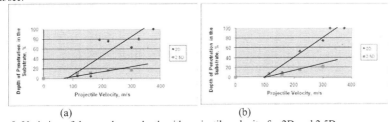

(a)   (b)

Figure 9. Variation of damaged zone depth with projectile velocity for 2D and 2.5D Type-II MI SiC/SiC composites impact tested at (a) ambient temperature, (b) 1316°C.The damage zone depth was measured from CT image.

SUMMARY OF RESULTS

The influence of fiber architecture and test temperature on impact behavior of uncoated MI SiC/SiC composites has been studied. The partially supported 2-D and two types of 2.5D woven composites composite specimens were impact tested at ambient and at 1316°C using 1.59-mm diameter hardened steel balls at projectile velocities up to 300 m/s. The surface and internal damage created by impact testing was analyzed by optical microscopy, PT and CT. Results show that both 2-D and 2.5D woven MI SiC/SiC composites show increase in impact damage with increase in projectile velocity. In general, for the fiber architecture investigated, the PT images of the impacted specimens indicate that the damage on the impacted side is much lower than that on the back side. However both optical microscopy and CT do not show any indication of damage on the back side. Also for a given architecture and at a fixed projectile velocity, the width and depth damage created at ambient and at 1316^0C is nearly the same. Depth of penetration of the projectile is significantly lower in 2.5D compared to 2D MI SiC/SiC composites at a fixed projectile velocity. Delamination cracks are not observed in 2.5D MI SiC/SiC composites at the tested projectile velocities, but such cracks are prevalent in 2D MI SiC/SiC composites at projectile velocity as low as 115 m/sec.

CONCLUSION

The 2-D woven MI SiC/SiC composites are prone to delamination under impact conditions. By reinforcing fibers in the through-the-thickness direction, such as 2.5D fiber architecture, delamination can be mitigated or completely avoided, but other modes of impact damage do exist.

ACKNOWLEDGEMENTS

The authors would like to thank Dr Vijay Pujar, Goodrich Technical Center for providing composite specimens for this study, R.W. Rauser for conducting CT scans and M. Cuy for impact testing. This work was supported by the Fundamental Aeronautics (FA) Program, NASA Glenn Research Center, Cleveland, Ohio.

REFERENCES

1. J.A DiCarlo and Mark Van Roode, Proceedings of IGTI conference, Paper#90151, (2006).
2. K.N. Lee, D.S. Fox, J.I. Eldridge, D. Zhu, R.C. Robinson, N.P. Bansal, and R.A. Miller, *J. Am. Ceram. Soc.*, **86** (8): p1299-1306, (2003).
3. K.N. Lee, D.S. Fox, and N.P. Bansal, *J. Eur. Ceram. Soc.*, **25** (10): p1705-1715, 2005.
4. R. Naslain, Comp. Sci. Tech., **64,** p155-170 (2004).
5. L. Zawada, G. Richardson, and P. Spriet, High Temperature Ceramic Matrix Composites- 5 (HTCMC-5), eds., M. Singh, R.J. Kerans, E. Lara-Curzio, and R. Naslain. pp. 491-498 (2004)
6. D.L. McDanels and R.A. Signorelli, NASA TN D-8204, (1976).
7. H. Tsuruta, M. Masuda, T. Soma, and M. Matsui, *J. Am. Ceram. Soc.*, **73** (6), p1714-1718, (1990).
8. R.T. Bhatt, S.R. Choi, L.M. Cosgriff, D.S. Fox, and K.N. Lee, Material Science and Engineering A, **476,** p20-28(2008)
9. R.T. Bhatt, S.R. Choi, L.M. Cosgriff, D.S. Fox, and K.N. Lee ,Materials Science and Engineering A, **476**, p8-19 (2008)
10. H.M. Yun and J.A. DiCarlo, *Ceram. Eng. Sci. Proc.*, **20**[3], p259-272 (1999).
11. S. K. Lau, S. J. Calandra, and R. W. Ohnsorg, US Patent #5,840,221, (1998).
12. S. R. Choi, J. M. Pereira, L. A. Janosik, and R. T. Bhatt, *Ceram. Eng. Sci. Proc.*, **23**[3], p193-202 (2002).
13. L. M. Cosgriff, R.T. Bhatt, S. R. Choi, and D.S. Fox, in *Proc. SPIE*, **5767**, p363-372, 2005.
14. P.O. Moore, "Nondestructive Testing Handbook: Radiographic Testing" Vol. 4. 3rd ed. Columbus: American Society for Nondestructive Testing, Inc, (2002).

OXIDATION KINETICS AND STRENGTH VERSUS SCALE THICKNESS FOR HI-NICALON™-S SiC FIBER

R. S. Hay, G. Fair
Air Force Research Laboratory
Materials and Manufacturing Directorate
WPAFB, OH

E. Urban
Appalachian State University
Boone, NC

J. Morrow, J. Somerson
U. Cincinnati
Cincinnati, OH

M. Wilson
Wright State University
Fairborn, OH

ABSTRACT
    The strength of Hi-Nicalon™-S SiC fibers was measured after oxidation in dry air between 800° and 1200°C. Fiber strength increases by approximately 10% for scale thickness up to ~100 nm, and decreases for thicker scales. Weibull modulus had no systematic variation with either fiber strength or SiO_2 scale thickness. The fiber strength increase is about the same as the calculated thermal residual compressive stress in the SiO_2 scale. Surface flaw healing may also be important. The strength decrease may be associated with stress concentrations in the scale caused by crystallization to α-cristobalite. Ceramic matrix composites made with SiC fibers with thin SiO_2 scales may benefit from the increased strength, improved resistance to fiber strength degradation associated with active oxidation, and better environmental stability in combustion environments.

INTRODUCTION
    Oxidation of SiC-SiC ceramic matrix composites (CMCs) has been thoroughly studied, but there are still aspects that are poorly understood. Oxidation of the BN or carbon fiber-matrix interphase at high and intermediate temperatures (pesting), and loss of SiC at high temperature in the presence of water by $Si(OH)_4$ formation are topics that have received intensive study.[1-9] Phenomena that are less well understood include 1) catastrophic loss of strength associated with rapid porosity formation and microstructure change in SiC fibers after active oxidation at low pO_2,[10,11] and 2) SiC fiber strength as a function of silica scale thickness, and the resistance to active oxidation that this scale imparts.[12] Fiber strength defines the ultimate attainable CMC strength – any strength degradation is therefore critical.[13] Complications include presence of impurities in the fiber, particularly alkali and alkali earths, that locally drastically increase oxidation rates, reduce scale viscosity, and lower temperatures for scale crystallization to α-cristobalite.[14,15] The general consensus is that oxidation of SiC fibers reduces fiber strength.[16-18] However, recent work in our laboratory has shown that thin silica scales (< 100 nm) actually increase SiC fiber strength, as might be expected from the residual compressive stress in the scales and possible surface flaw healing. Ambiguous effects of SiC oxidation on strength have also been observed for bulk material.[19]
    Passive oxidation kinetics for SiC are well known to be parabolic for thick SiO_2 scales and linear (interface controlled) for thin scales. They obey the "Deal-Grove" kinetics first described for silicon.[20] "Deal-Grove" kinetics have been reformulated for cylindrical geometry for Nicalon™ fibers.[21] Unfortunately this treatment assumes a one for one replacement of SiC by amorphous SiO_2; this is clearly not correct for stoichiometric SiC fibers such as Hi-Nicalon™-S, Tyranno™-SA, and

Sylramic™, where the much higher molar volume of amorphous silica (v_m = 27.34 cm³) causes the oxidized fiber to expand more than the remaining SiC (v_m = 12.46 cm³ for β-SiC) contracts. Controversy still exists about the possible presence of a "reactive" layer at the SiC-SiO$_2$ interface that converts diffusing O$_2$ molecules to individual O atoms in the SiO$_2$ network,[22] and a "carbon-condensed" layer associated with inhibited CO transport.[23,24] Oxidation kinetics change if the scale crystallizes, which in turn depends on impurity concentrations and environmental effects such as water or other combustion environment species.[25,26]

Passive oxidation can increase strength by healing surface flaws, and by inhibition of surface cracking from the compressive residual stresses present in the scale. Calculation or measurement of stress in the scale can be problematic. Thermal residual stress can be estimated from the CTE mismatch between SiO$_2$ and SiC and the temperature at which this stress is locked in and not relieved by viscous SiO$_2$ flow. For transversely isotropic materials systems with cylindrical symmetry the thermal residual stress can be calculated using exact analytical solutions to the elasticity problem in NDSANDS.[27,28] For isotropic systems simpler solutions are available. For very thin films the thermal residual stress (σ) must approach:

$$\sigma = \Delta\alpha\ \Delta T\ E/(1-v) \qquad [1]$$

Where E is modulus, v is Poisson's ratio, $\Delta\alpha$ is the CTE difference between the SiC phase and the SiO$_2$ phase, and ΔT is the temperature difference over which thermal stress is preserved. For a ΔT of ~1000°C, the in-plane residual stress will be about 300 MPa in the SiO$_2$ scale.

The other stress present in the scale is the growth stress, which forms from viscoelastic constraint of the volume expansion of oxidation of SiC to SiO$_2$. This thermal residual stress is superimposed on this pre-existing stress. Expressions for this stress have been developed elsewhere for cylindrical substrates and are complex.[29] In general these stresses are significantly smaller than the thermal residual stress. A final consideration is the effect of the viscoelastic growth stress on SiC oxidation kinetics. Analysis done elsewhere suggest that in some cases the effect can be significant.[29]

This paper presents preliminary data and analysis of the strength of oxidized Hi-Nicalon™-S SiC fiber as a function of SiO$_2$ scale thickness. Hi-Nicalon™-S fiber was chosen because it has near-stoichiometric SiC composition, and the smoothest surface of currently available SiC fibers.[30] The properties of Hi-Nicalon™-S fiber are described in several publications.[17,31-35] Possible strengthening and degradation mechanisms will be discussed. It is hoped that the data will be useful in designing optimal fiber coating and matrix processing methods for SiC-SiC CMCs. More thorough data and analysis will be presented in a forthcoming publication.

EXPERIMENTS

The fiber sizing was removed by dissolution in boiling distilled deionized water in a glass beaker for one hour. This process was repeated a second time with fresh distilled deionized water. Fibers were then dried in a drying oven for 20 minutes at 120°C. Great care was taken to avoid contamination of the fibers during handling. This procedure was found to be necessary to form a smooth, uniform SiO$_2$ scale during oxidation that was not contaminated by residual inorganics from the sizing. The desized fiber tows were oxidized in flowing dry air (< 10 ppm H$_2$O) using an alumina muffle tube furnace and an alumina boat dedicated to these experiments. Both the muffle tube and boat were baked-out at 1600°C for 1 hour in laboratory air prior to use. Fiber oxidation was done with temperatures from 800 to 1200°C for times up to 100 hours. A total of 36 different heat-treatments were done. Fiber tows were heated at 10°C/minute to these temperatures. Experiments with no hold time and just a 10°C/minute ramp up to temperature were also done.

The uniformity of the SiO$_2$ oxidation product was characterized using reflected light interference fringes with optical microscopy. Surface morphology and cracking of the oxidation product were characterized by SEM. Cross-section TEM specimens were prepared from some of the oxidized fibers by published methods.[36,37] TEM sections were ion-milled at 5 kV and examined using a 200 kV Phillips LaB$_6$-filament TEM. SiO$_2$ oxidation product thickness, SiC grain growth, and cracking and crystallization of the SiO$_2$ scale were characterized.

The strengths of the oxidized fibers were measured by tensile testing of at least 30 filaments using published methods.[38] The average and Weibull characteristic value for the failure stress were calculated, along with the Weibull modulus. The average and standard deviation of the fiber diameter were measured by optical microscopy and SEM. The average fiber diameter was 12.1 μm.

RESULTS AND DISCUSSION

TEM micrographs of SiO$_2$ scales formed after oxidation for various times at 1000°C in dry air are shown in figure 1. All scales formed at 1000°C are amorphous. Scales formed for longer times at 1100 and 1200°C were at least partially crystallized to α-cristobalite. This is shown in figure 2 in a TEM micrograph of fiber oxidized for 30 hours at 1200°C.

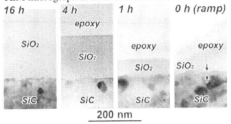

Fig. 1 Cross-sectional TEM micrographs of Hi-Nicalon™-S fibers oxidized in dry air at 1000°C.

The time and temperature dependence for oxidation kinetics for Hi-Nicalon™-S fiber are shown in figure 3. The thicknesses are corrected for scale formation that occurs during heat-up. Crystallized or partially crystallized scale is denoted by open circles; others (solid circles) are completely amorphous. The time exponents for growth are slightly larger than the n = ½ value expected for parabolic growth for 800 - 1000°C. This is likely due to interface control of oxidation kinetics for thin scales, where n = 1. This is particularly evident for the 900°C data, which shows distinctly higher slope at low thickness. The time exponent for growth is slightly less than ½ for 1200°C data; this is most likely due to retardation of kinetics by scale crystallization. An activation energy of ~225 kJ/mol°K was calculated assuming parabolic growth kinetics. This activation energy is slightly higher than other results for stoichiometric SiC fibers. Collection of more data in progress and may slightly modify our result, particularly after analysis using "Deal-Grove"

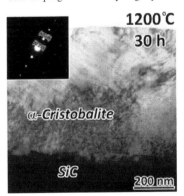

Fig. 2 TEM micrograph of SiO$_2$ scale that has crystallized to α-cristobalite on Hi-Nicalon™-S fiber oxidized for 30 hours at 1200°C.

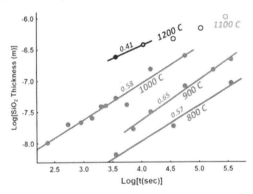

Fig. 3 Oxidation kinetics of Hi-Nicalon™-S fiber in dry air. Open circles are crystallized scale; solid circles are amorphous scale. The time exponent for scale growth is listed above each line.

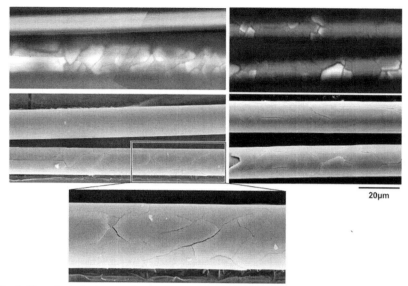

20µm

Fig. 4. Top: Optical micrographs in reflected light of Hi-Nicalon^{TM}-S fiber surfaces oxidized for 81 hours at 1100°C. Through-thickness cracks are evident, and interference fringes are diagnostic of debonding between the SiO$_2$ and SiC. Bottom: SEM micrographs of the same areas, showing through thickness cracks. Bottom, Inset: Higher magnification SEM image of the scale.

kinetics for cylindrical substrates discussed in the Introduction.

Scales that were thick and crystallized were cracked. This is illustrated in optical (reflected light) and SEM micrographs in figure 4 for fibers heat-treated for 81 hours at 1100°C. Through thickness cracks are evident in both reflected light and by SEM, and interference fringes in reflected light are diagnostic of debonding between the scale and the underlying SiC. Cracking is expected from the $-\Delta V$ of ~ 10% accompanying crystallization to α-crisotbalite.

The relationship between scale thickness and fiber strength is shown in figure 5. The average strength and Weibull characteristic strength are both plotted. As-received fibers had average strengths of 2.85 GPa (3 separate measurements). Data from all temperatures (800 -1200°C) are plotted. A distinct relationship between oxidation temperature and fiber strength was not observed; a 100 nm thick scale that formed in a short time at high temperature had about the same effect on fiber strength as a 100 nm scale that formed in a long time at low temperature. Fiber strength increased about 10%, with a maxima near SiO$_2$ thickness of about 100 nm, and decreased for thicker scales. Crystallized scales are denoted by open circles. The data are plotted for strengths calculated from the original fiber diameter (12.1 µm) and for the oxidized fiber diameter calculated from the molar volumes (v_m) of SiC and amorphous SiO$_2$. These calculations assume that the SiO$_2$ scale carries no load. The maximum strength increase of about 250 MPa is close to the compressive thermal residual stress in a thin SiO$_2$ scale that is about 300 MPa.

If a new population of strength governing flaws starts to form for SiO$_2$ thicknesses greater than 100 nm, it might be expected that that the Weibull modulus of the fibers would change above those thicknesses. This was checked – no correlation was found. The observed relationship between Weibull modulus and SiO$_2$ thickness is shown in figure 6. The 10% volume loss during crystallization

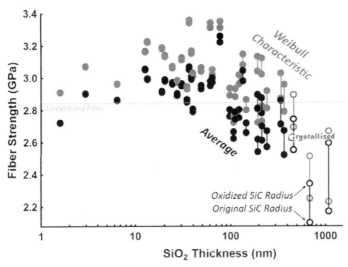

Fig. 5. Fiber strength of Hi-Nicalon^{TM}-S fiber as a function of SiO$_2$ oxidation product thickness. Both the average and Weibull characteristic strength are plotted, and strengths calculated from the original and the oxidized fiber diameters are both plotted.

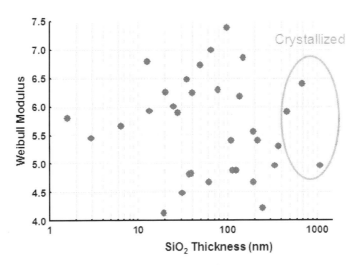

Fig. 6. The measured Weibull modulus of Hi-Nicalon^{TM}-S fibers as a function of SiO$_2$ scale thickness formed after oxidation at 800 to 1200°C.

of α-cristobalite should have a dramatic effect on residual stress in the scale and drive the scale cracking seen in figure 4. However, Weibull modulus also did not correlate with the limitied data available for crystallization of the scale (Fig. 6). Further data collection, characterization, and analysis are in progress to establish the strength degradation mechanisms that operate for thick SiO_2 scales on SiC fibers.

SUMMARY AND CONCLUSIONS

The strength of Hi-Nicalon™-S fiber oxidized in dry air increases about 10% for SiO_2 scales up to ~100 nm thick. This strength increase is about the same as the residual compressive thermal stress expected in the scale. The fiber strength decreases for scales thicker than 100 nm, but an obvious change in Weibull modulus does not accompany this strength change. Thick SiO_2 scales were crystallized to α-cristobalite and were cracked and partially debonded from the SiC fibers. The volume loss accompanying α-cristobalite crsyatllization may drive this cracking; it may also be at least partially responsible for fiber strength degradation. Further data collection and analysis are in progress to define the degradation mechanisms.

REFERENCES

[1] Kerans, R. J., Hay, R. S., Parthasarathy, T. A., and Cinibulk, M. K., Interface Design for Oxidation Resistant Ceramic Composites, *J. Am. Ceram. Soc.* 85 (11), 2599-2632, 2002.

[2] Ogbuji, L. U. J. T., Pest-Resistance in SiC/BN/SiC Composites, *J. Eur. Ceram. Soc.* 23, 613-617, 2003.

[3] Morscher, G. N., Hurst, J., and Brewer, D., Intermediate-Temperature Stress Rupture of a Woven Hi-Nicalon, BN-Interphase, SiC-Matrix Composite in Air, *J. Am. Ceram. Soc.* 83 (6), 1441-1449, 2000.

[4] Jacobson, N. S., Morscher, G. N., Bryant, D. R., and Tressler, R. E., High-Temperature Oxidation of Boron Nitride: II, Boron Nitride Layers in Composites, *J. Am. Ceram. Soc.* 82 (6), 1473-1482, 1999.

[5] Ogbuji, L. U. J. T., A Pervasive Mode of Oxidative Degradation in a SiC-SiC Composite, *J. Am. Ceram. Soc.* 81 (11), 2777-2784, 1998.

[6] Opila, E. J., Oxidation and Volatilization of Silica Formers in Water Vapor, *J. Am. Ceram. Soc.* 86 (8), 1238-1248, 2003.

[7] Opila, E. J., Variation of the Oxidation Rate of Silicon Carbide with Water Vapor Pressure, *J. Am. Ceram. Soc.* 82 (3), 625-636, 1999.

[8] Opila, E. J. and Hann, R. E., Paralinear Oxidation of CVD SiC in Water Vapor, *J. Am. Ceram. Soc.* 80 (1), 197-205, 1997.

[9] Opila, E. J., Fox, D. S., and Jacobson, N. S., Mass Spectrometric identification of Si-O-H(g) Species from the Reaction of Silica with Water Vapor at Atmospheric Pressure, *J. Am. Ceram. Soc.* 80 (4), 1009-1012, 1997.

[10] Shimoo, T., Okamura, K., and Morisada, Y., Active-to-Passive Oxidation Transition for Polycarbosilane-Derived Silicon Carbide Fibers Heated in Ar-O_2 Mixtures, *J. Mater. Sci.* 37, 1793-1800, 2002.

[11] Shimoo, T., Morisada, Y., and Okamura, K., Oxidation Behavior of Si-M-C-O Fibers Under Wide Range of Oxygen Partial Pressures, *J. Mater. Sci.* 37, 4361-4368, 2002.

[12] Shimoo, T., Morisada, Y., and Okamura, K., Suppression of Active Oxidation of Polycarbosilane-Derived Silicon Carbide Fibers by Preoxidation at High Oxygen Pressure, *J. Am. Ceram. Soc.* 86 (5), 838-845, 2003.

[13] Curtin, W. A., Ahn, B. K., and Takeda, N., Modeling Brittle and Tough Stress-Strain Behavior in Unidirectional Ceramic Matrix Composites, *Acta mater.* 46 (10), 3409-3420, 1998.

[14] Doremus, R. H., Viscosity of Silica, *J. Appl. Phys.* 92 (12), 7619-7629, 2002.

[15] Pezzotti, G. and Painter, G. S., Mechanisms of Dopant-Induced Changes in Intergranular SiO_2 Viscosity in Polycrystalline Silicon Nitride, *J. Am. Ceram. Soc.* 85 (1), 91-96, 2002.

[16] Takeda, M., Urano, A., Sakamoto, J., and Imai, Y., Microstructure and Oxidation Behavior of Silicon Carbide Fibers Derived from Polycarbosilane, *J. Am. Ceram. Soc.* 83 (5), 1171-1176, 2000.

[17] Shimoo, T., Takeuchi, H., and Okamura, K., Oxidation Kinetics and Mechanical Property of Stoichiometric SiC Fibers (Hi-Nicalon-S), *J. Ceram. Soc. Japan* 108 (1264), 1096-1102, 2000.

[18] Kim, H.-E. and Moorhead, A. J., Strength of Nicalon Silicon Carbide Fibers Exposed to High-Temperature Gaseous Environments, *J. Am. Ceram. Soc.* 74 (3), 666-669, 1991.

[19] Easler, T. E., Bradt, R. C., and Tressler, R. E., Strength Distributions of SiC Ceramics After Oxidation and Oxidation Under Load, *J. Am. Ceram. Soc.* 64 (12), 731-734, 1981.

[20] Deal, B. E. and Grove, A. S., General Relationships for the Thermal Oxidation of Silicon, *J. Appl. Phys.* 36 (12), 3770-3778, 1965.

[21] Zhu, Y. T., Taylor, S. T., Stout, M. G., Butt, D. P., and Lowe, T. C., Kinetics of Thermal, Passive Oxidation of Nicalon Fibers, *J. Am. Ceram. Soc.* 81 (3), 655-660, 1998.

[22] Mott, N. F., Rigo, S., Rochet, F., and Stoneham, A. M., Oxidation of Silicon, *Philos. Mag. B* 60 (2), 189-212, 1989.

[23] Cooper, R. F. and Chyung, K., Structure and Chemistry of Fiber-Matrix Interfaces in Silicon Carbide Fibre-Reinforced Glass-Ceramic Composites: An Electron Microscopy Study, *J. Mater. Sci.* 22, 3148-3160, 1987.

[24] Akashi, T., Kasajima, M., Kiyono, H., and Shimada, S., SIMS Study of SiC Single Crystal Oxidized in Atmosphere Containing Isotopic Water Vapor, *J. Ceram. Soc. Japan* 116 (9), 960-964, 2008.

[25] Choi, D. J. and Scott, W. D., Devitrification and Delayed Crazing of SiO_2 on Single Crystal Silicon and Chemically Vapor Deposited Silicon Nitride, *J. Am. Ceram. Soc.* 70 (10), C269-C272, 1987.

[26] Zhou, W., Fu, H., Zhang, L., Sun, X., She, S., and Ma, J., Effects of impurities and Manufacturing Methods on the Devitrification of Silica Fibers, *J. Am. Ceram. Soc.* 74 (5), 1125-1128, 1991.

[27] Pagano, N. J. and Tandon, G. P., Elastic Response of Multidirectional Coated-Fiber Composites, *Compos. Sci. Tech.* 31, 273, 1988.

[28] Pagano, N. J. and Tandon, G. P., Thermo-Elastic Model for Multidirectional Coated-Fiber Composites: Traction Formulation, *Compos. Sci. Tech.* 38, 1, 1990.

[29] Hsueh, C. H. and Evans, A. G., Oxidation Induced Stresses and Some Effects on the Behavior of Oxide Films, *J. Appl. Phys.* 54 (11), 6672-6686, 1983.

[30] Hinoki, T., Snead, L. L., Lara-Curzio, E., Park, J., and Kohyama, A., Effect of Fiber/Matrix Interfacial Properties on Mechanical Properties of Unidirectional Crystalline Silicon Carbide Composites, *Ceram. Eng. Sci. Proc.* 23 (3), 511-518, 2002.

[31] Sauder, C. and Lamon, J., Tensile Creep Behavior of SiC-Based Fibers With a Low Oxygen Content, *J. Am. Ceram. Soc.* 90 (4), 1146-1156, 2007.

[32] Bunsell, A. R. and Piant, A., A Review of the Development of Three Generations of Small Diameter Silicon Carbide Fibres, *J. Mater. Sci.* 41, 823-839, 2006.

[33] Ishikawa, T., Advances in Inorganic Fibers, *Adv. Polym. Sci.* 178, 109-144, 2005.

[34] Sha, J. J., Nozawa, T., Park, J. S., Katoh, Y., and Kohyaman, A., Effect of Heat-Treatment on the Tensile Strength and Creep Resistance of Advanced SiC Fibers, *J. Nucl. Mater.* 329-333, 592-596, 2004.

[35] Tanaka, T., Shibayama, S., Takeda, M., and Yokoyama, A., Recent Progress of Hi-Nicalon Type S Development, *Ceram. Eng. Sci. Proc.* 24 (4), 217-223, 2003.

[36] Hay, R. S., Welch, J. R., and Cinibulk, M. K., TEM Specimen Preparation and Characterization of Ceramic Coatings on Fiber Tows, *Thin Solid Films* 308-309, 389-392, 1997.

[37] Cinibulk, M. K., Welch, J. R., and Hay, R. S., Preparation of Thin Sections of Coated Fibers for Characterization by Transmission Electron Microscopy, *J. Am. Ceram. Soc.* 79 (9), 2481-2484, 1996.

[38.] Petry, M. D., Mah, T., and Kerans, R. J., Validity of Using Average Diameter for Determination of Tensile Strength and Weibull Modulus of Ceramic Filaments, *J. Am. Ceram. Soc.* 80 (10), 2741-2744, 1997.

CERAMIC MATRIX COMPOSITES DENSIFICATION BY ACTIVE FILLER IMPREGNATION FOLLOWED BY A P.I.P. PROCESS

S. Le Ber, M.-A. Dourges, L. Maillé, R. Pailler, A. Guette
Université de Bordeaux 1
Laboratoire des Composites Thermo-Structuraux, UMR 5801 CNRS-SAFRAN-CEA-UB1
3 allée de la Boétie, 33600 Pessac, France

ABSTRACT

The nitridation of a micro-sized $TiSi_2$ powder was investigated by thermogravimetry over a temperature range of 1000 to 1300°C under nitrogen atmosphere. A two-step nitridation mechanism was observed: the formation of TiN and Si in a first step and beyond 1200°C the nitridation of Si into Si_3N_4 in a second step. The raw $TiSi_2$ powder was milled by different methods in order to obtain submicronic powders. The effects of milling on particle size and on nitridation kinetics have been examined. This study shows the importance of specific surface area on the nitridation rate: planetary milling improved significantly the nitridation rate, and experimental results showed that it is possible to produce both nitrides TiN and Si_3N_4 at a temperature as low as 1100°C. The interest of $TiSi_2$ as active filler has been evaluated: manufacturing of monolithic ceramics was performed by pyrolysis of a preceramic polymer (polysiloxane) / active filler ($TiSi_2$) blend. During pyrolysis under nitrogen atmosphere, nitridation of $TiSi_2$ produced volume expansion and compensated for polymer shrinkage during pyrolytic conversion. A feeble carburization of the active filler was also observed.

1. INTRODUCTION

In the preparation of Ceramic Matrix Composites (CMCs), densification of fiber preforms can be performed via different routes with different fibers and different techniques, such as chemical vapour infiltration (CVI), polymer impregnation and pyrolysis (PIP), reactive melt infiltration (RMI) or slurry infiltration and hot processing (SI-HP).[0] Using complementary methods of densification as slurry impregnation with filler powder and liquid polymer impregnation enables to obtain an effective process with a low price/performance ratio. Adding fillers to the polymer allows to modulate certain properties of the final ceramic, such as mechanical behaviour, electrical or thermal properties. However, large volume shrinkages of up to 80% can be measured during the pyrolytic conversion of polymers to ceramic.[0,0] P. Greil suggested overcoming this problem with active fillers, which react during pyrolysis to form oxides, carbides or nitrides.[0,0,0,0] These reactions occur with a volume expansion that can compensate for polymer shrinkage. Some active fillers present high potential volume changes; for instance silicon is oxidized with a 97% volume increase, and boron is nitrided with a 142% volume increase. However, most active fillers react only at high temperatures (T>1400°C) and/or in oxidizing atmosphere. This can be a major drawback if the fibers are damaged at high temperatures, or if the ceramic derived from the precursor faces oxidation. Titanium disilicide ($TiSi_2$, density=4.01g/cm³) has been identified as an interesting active filler. Indeed, under nitrogen atmosphere, $TiSi_2$ can form TiN (d=5.43g/cm³) and Si_3N_4 (d=3.19g/cm³) with a 57% volume increase; besides this nitridation starts in a temperature range around 1000°C. This study explores the influence of temperature on the nitridation rate of a micronic $TiSi_2$ powder. This powder was milled in different conditions; reactivity of milled powders was then studied. Behaviour of active filler was also explored in pellets containing $TiSi_2$ powder mixed with a liquid polysiloxane. A ceramic material was thus obtained in the system Ti-Si-C-N-O.

2. MATERIALS AND PROCEDURE

A high purity micro-sized $TiSi_2$ powder (C-54 stable phase, 99.95% in purity, -325mesh, Neyco) was used in this work. This raw powder was milled with a vibratory mixer mill (Retsch MM200) and with planetary ball mills (Retsch PM200 and Fritsch pulverisette 7).

The resin used as a preceramic polymer is a phenyl-containing polysiloxane, provided by Snecma Propulsion Solide. It is a solvent-free liquid silicone resin, which cures by the effect of heat and an included platinum catalyst. In this work samples were cured by a thermal treatment of 2h at 200°C in air.

Liquid resin was mechanically mixed with $TiSi_2$ powder in a 1/1 volume ratio, in order to obtain a homogeneous polymer/powder mixture. This mixture was poured in aluminum cylinder-shaped molds, and subsequently heated during 2h at 200°C for curing. Solid cured samples (diameter=12mm, thickness=3mm) were obtained. They were heated under argon or nitrogen atmosphere at a ramping rate of 5°C/min up to 1100°C, and maintained at this temperature for 1h. Masses and volumes were measured before and after pyrolysis.

Nitridations and pyrolyses were performed in a thermogravimetric (TG) analyser (Setaram TAG24). Samples (~100mg in Al_2O_3 crucibles) were heated under a flow of pure argon or nitrogen gas (50mL/min) from 20°C to the chosen temperature at a ramping rate of 1 or 5°C/min.

Volumes were measured with a Micromeritics AccuPyc II 1340 helium pycnometer ($1cm^3$ model) using helium gas. Materials structure was identified by X-ray diffraction (XRD). XRD patterns were obtained with a Siemens D5000 diffractometer using Cu Kα radiation. They were recorded using a step size of 0.04° for the 2θ range 10-90°, and a counting time of 4s per step. The microstructure of powders was observed by scanning electron microscopy (SEM). Experiments were carried out using a Quanta 400 FEG microscope. Specific surface areas were determined by the BET method with an ASAP 2010 (Micromeritics); samples were degassed by heating at 220°C during 4h immediately prior to measurements.

3. TITANIUM DISILICIDE NITRIDATION

3.1. Nitridation of micro-sized $TiSi_2$ powder

SEM micrographs of the starting powder are shown in Figure 1. They indicate a wide particle size distribution, ranging from 1 to 50µm. A low specific surface area of 0.33m²/g was measured by the BET method.

Figure 1. SEM micrographs of the $TiSi_2$ starting powder.

A non-isothermal nitridation of this powder was first performed. The sample was heated from 20 to 1300°C at a low rate of 1°C/min in flowing pure nitrogen gas. Curve for weight gain plotted as a function of temperature is shown in Figure 2.

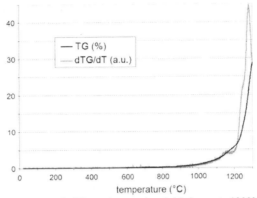

Figure 2. TG curve of TiSi₂ starting powder nitridation up to 1300°C in N₂.

No significant weight gain was measured until the temperature reached 900°C. Two peaks of dTG/dT were observed at T= 1160°C and at T=1280°C, revealing two reaction steps. These results were used to choose the suitable temperature range for this study. The TiSi₂ powder was heated under a flow of nitrogen at a rate of 5°C/min up to 1000°C and maintained at this temperature for 5h. The same experiment was performed at higher temperatures: 1100, 1200 and 1300°C, in order to understand the different reactions that occurred. Curves presenting weight gain and temperature plotted as a function of time are shown in Figure 3. X-ray diffraction analyses were performed after every experiment (Figure 4).

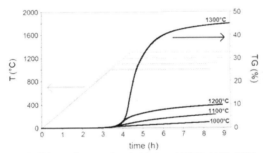

Figure 3. TG curves of TiSi₂ starting powder nitridation at 1000, 1100, 1200 and 1300°C in N₂.

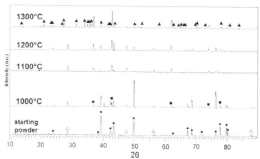

Figure 4. XRD patterns of the starting powder and after TGA experiments in N_2.
($\bullet$)$TiSi_2$; (Δ)Si; ($\blacksquare$)TiN; ( )Ti; ($\blacktriangle$)β-Si_3N_4.

These results indicate a two-step nitridation of $TiSi_2$, as previously reported in literature.[0,0] After 5h at 1000°C, the weight gain was only of 2.3%; XRD patterns showed the presence of $TiSi_2$ (JCPDS 71-0187), TiN (38-1420) and Si (27-1402) phases. The equation (1) formulates this partial nitridation, which comes with a maximum weight gain of 13.5% and a maximum volume gain of 36.9%.

$$TiSi_2 + \tfrac{1}{2} N_2 \rightarrow TiN + 2\ Si \qquad (1)$$

$$\Delta W/W_0 = 13.5\% \qquad\qquad \Delta V/V_0 = 36.9\%$$

After 5h at 1100°C the weight gain reached 5.7%, and the peaks of TiN and Si phases were more intense. From 1200°C a second step was observed, with a total nitridation (2) leading to the formation of both nitrides TiN and Si_3N_4. The weight gain after 5h at 1200°C: 9.8%, was still above 13.5% but XRD peaks corresponding to the β-Si_3N_4 phase (JCPDS 33-1160 and 71-6479) started to emerge.

$$TiSi_2 + 11/6\ N_2 \rightarrow TiN + 2\ Si_3N_4 \qquad (2)$$

$$\Delta W/W_0 = 49.4\% \qquad\qquad \Delta V/V_0 = 56.9\%$$

This reaction was accelerated at 1300°C, with a higher conversion rate: indeed a weight gain of 44.8% was measured after 5h at 1300°C, close to the theoretical maximum value of 49.4%. $TiSi_2$ peaks disappeared whereas β-Si_3N_4 peaks became more intense.

Some thermodynamic studies in the Ti-Si-N ternary system indicate that $TiSi_2$ can be nitrided into TiN and Si_3N_4 at 1100°C.[0,0] This assertion was not confirmed by our first experiments, even after a long time of nitridation at 1100°C. It was therefore decided to mill the starting powder presented above, with the intention of getting close to a nanometric scale and explore the nitridation of milled powders.

3.2. Milling study
The starting powder is the finest $TiSi_2$ powder commercially available. It was therefore milled in order to obtain sub-micronic particles. A conventional milling was performed with a vibratory ball

mill; high-energy planetary millings were also carried out with the help of two companies specialized in milling (Retsch and Fritsch). Many different parameters control a high-energy milling, such as the solvent or the milling balls size and material.[0] Milling conditions used in this work are reported in Table I.

Table I. Milling conditions

	mixer milling	planetary milling 1	planetary milling 2	planetary milling 3
mill	Retsch MM200	Retsch PM200	Retsch PM200	Fritsch pulverisette 7
vibrational frequency / operating speed	30Hz	480 rpm	480 rpm	550 rpm
jar	5mL, steel	50mL, ZrO_2	50mL, WC	45mL, ZrO_2
balls	Ø=7mm, steel, 2 balls (4g)	Ø=2mm, ZrO_2, 30mL (110g)	Ø=3mm, WC, 30mL (305g)	Ø=3mm, ZrO_2, 20mL (70g)
solvent	none	15mL heptane	15mL isopropanol	10mL isopropanol
mass of $TiSi_2$	3g	18g	18g	20g
milling time	3h	3h	3h	2h

SEM micrographs of milled powder are shown in Figure 5. They show a decrease of particle size after all millings, but diameters of several micrometers can still be observed after mixer milling and planetary milling 1. Planetary millings 2 and 3 performed with isopropanol were more efficient: SEM micrographs show a homogenization of particle size distribution with diameters below 1μm.

(a)   (b)

(c)   (d)

Figure 5. SEM micrographs of TiSi₂ milled powders after: (a) mixer milling;
(b) planetary milling 1; (c) planetary milling 2; (d) planetary milling 3.

Specific surface areas were measured by the BET method. Results are listed in Table II as well as granulometry data provided by Retsch and Fritsch.

Table II. Specific surface area and granulometry of raw and milled powders

	raw powder (R)	mixer milling (MM)	planetary milling 1 (PM1)	planetary milling 2 (PM2)	planetary milling 3 (PM3)
S (m²/g)	0.3	2.0	7.3	41	36
d_{50}	10µm	4µm	1.7µm	550nm	380nm
d_{90}	-	-	-	1.1µm	1.4µm

Sub-micronic powders were obtained with planetary millings 2 and 3; a specific surface area of 41m²/g was notably achieved with planetary milling 1, which was performed using tungsten carbide jar and balls. X-ray diffractions were performed on every milled powder (Figure 6).

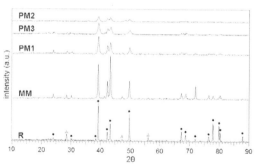

Figure 6. XRD patterns of raw and milled powders. (●)TiSi₂; (Δ)Si.

X-ray diffraction revealed no contamination by oxygen, iron or carbon. The same peaks corresponding to the C-54 $TiSi_2$ stable phase were observed for milled and raw powders, but a general trend could be observed: a decrease of diffraction peak intensities with the average particle size, accompanied by an increase of the FWHM due to the refinement of the crystallite size in powder particles, and possibly to internal stresses.

3.3. Nitridation of milled powders

TG analyses were performed on milled powders at 1100°C. TG curves are shown in Figure 7.

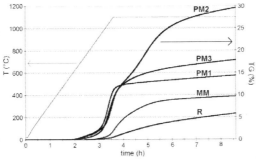

Figure 7. TG curves of $TiSi_2$ milled and raw powders heated to 1100°C in N_2.

Mixer milling improved significantly the nitridation rate, with a weight gain of 8.6% after 5h at 1100°C. Nitridation of planetary milled powders began at a clearly lower temperature of 650°C, with much steeper slopes during the heating stage. Furthermore the weight gain exceeded the value of 13.5%, especially after planetary milling 2 for which the weight gain was of 29.5% after 5h at 1100°C. This indicated the formation of Si_3N_4. X-ray diffraction analyses were performed after every experiment (Figure 8).

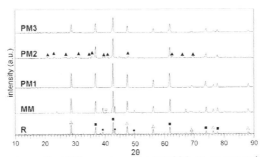

Figure 8. XRD patterns after TGA experiments at 1100°C in N_2 on raw and milled powders.
(●)$TiSi_2$; (Δ)Si; (■)TiN; ( )Ti; (▲)β-Si_3N_4.

XRD patterns of all nitrided products indicated the presence of TiN and Si. After 5h at 1100°C in N_2, only raw and mixer-milled nitrided powders still indicated peaks of $TiSi_2$. The XRD patterns of powders nitrided after planetary millings 1 and 3 exhibited only peaks corresponding to the TiN and Si phases. The XRD pattern of the $TiSi_2$ powder nitrided after planetary milling 2 exhibited also small peaks corresponding to the β-Si_3N_4 phase.

Planetary milling 2, performed with isopropanol and with tungsten carbide jar and balls, enabled to obtain a powder with the highest specific surface area; this powder was nitrided at 1100°C. X-ray diffraction showed the formation of TiN and Si_3N_4; the weight gain measured after 5h at 1100°C indicated that 45% of silicon was nitrided. Further studies are in process in order to understand the efficiency of planetary milling 2, and the enhanced reactivity of the resulting milled powder. It can first be noticed that tungsten carbide has a much higher density (d=15.6g/cm³) than zirconium dioxide (d=5.7g/cm³). As a result milling balls with the same diameter and the same speed will have a much higher energy, enabling a more performing milling.

4. ACTIVE-FILLER CONTROLLED PYROLYSIS

4.1 Polymer-ceramic conversion

A sample of polysiloxane was cured during 2h at 200°C in air. This thermal treatment led to a solid, transparent and brittle material. After curing it was heated up to 1100°C at a rate of 5°C/min under nitrogen atmosphere, in a thermogravimetric analyser (Figure 9). A weight loss was observed starting from 250°C; the dTG/dT curve exhibited three peaks at T=450°C, 530°C and 720°C. A total weight loss of 26.9% was measured at 1100°C. The same experiment was performed in argon and no difference was observed. Different gases (H_2O, CO_2, H_2, C_6H_6 and CH_4) are released during the pyrolytic decomposition, which leads to an amorphous SiO_xC_y black glass.[0,0,0,0]

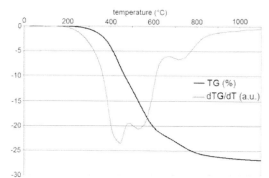

Figure 9. TG curves of cured resin heated up to 1100°C in N_2.

The volume of the sample was measured by helium pycnometry before and after pyrolysis. An important volume shrinkage of 55% was calculated.

4.2 Pyrolysis of active filler/polymer mixtures

Mixtures containing resin and raw or planetary-milled (PM1) $TiSi_2$ in a 1/1 volume ratio were cured and pyrolysed to manufacture ceramic residues as a model for a CMC matrix. Mass and volume variations are reported in Table III.

Table III. Mass and volume variations measured for 1h-long pyrolyses at 1100°C

sample	composition	atmosphere	$\Delta W/W_0$ (%)	$\Delta V/V_0$ (%)
0	100%V resin	N_2	-27.1	-55
1	50%V resin + 50%V raw $TiSi_2$	N_2	-3.9	-23
2	50%V resin + 50%V planetary-milled $TiSi_2$	N_2	-0.2	-17
3	50%V resin + 50%V planetary-milled $TiSi_2$	Ar	-6.0	-27

As expected the highest weight loss and volume shrinkage among samples containing $TiSi_2$ were measured for the sample 3, which was pyrolysed in argon and within which the active filler could therefore not be nitrided. Among samples 2 and 3, pyrolysed under nitrogen atmosphere, the lowest values were measured for the sample containing planetary-milled $TiSi_2$. Its reactivity remains better in the presence of the polymer. It can be noticed that a higher initial volume fraction of $TiSi_2$ appears necessary to achieve zero shrinkage. X-ray diffractions were performed on samples 2 and 3 after pyrolysis (Figure 10).

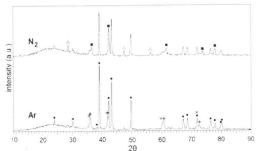

Figure 10. XRD patterns after pyrolysis at 1100°C of samples 2 and 3.
(●)$TiSi_2$; (Δ)Si; (■)TiN; (+)TiC; (×)SiC.

The XRD pattern of sample 3 indicated the presence of $TiSi_2$, and revealed the formation of carbides TiC (JCPDS 32-1383) and SiC (JCPDS 89-2222). The XRD pattern of sample 2 exhibited the same peaks plus those corresponding to TiN and Si phases. The $TiSi_2$ was not as well nitrided in this sample as the powder alone was: after nitridation by the same thermal treatment of planetary-milled (PM1) powder alone, the XRD pattern indicated no peaks of $TiSi_2$.

In order to determine whether the carbides were formed by reaction of $TiSi_2$ with carbon from solid or from gaseous decomposition products of the polymer, pyrolyses under argon atmosphere were performed at different temperatures on cured polymer/active filler 1/1 mixtures. Temperatures of 550°C and 750°C were chosen for these pyrolyses. They respectively correspond to the temperature range where benzene and methane are released. X-ray diffractions were performed after these pyrolyses (Figure 11).

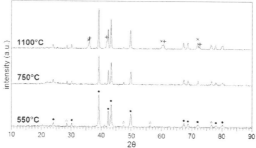

Figure 11. XRD patterns after pyrolyses under argon atmosphere at 550, 750 and 1100°C.
(●)TiSi₂; (Δ)Si; (+)TiC; (×)SiC.

XRD patterns of samples pyrolysed at 550°C and 750°C exhibit only the same peaks than those of the TiSi₂ powder. Peaks of carbides phases appear only for the sample pyrolysed at 1100°C. It can therefore be inferred that TiSi₂ did not react with the hydrocarbon species, but with the solid carbon of the Si-O-C phase.

5. CONCLUSION

The nitridation of a micro-sized TiSi₂ powder under nitrogen atmosphere has been explored by X-ray diffraction and thermogravimetry in the 1000-1300°C temperature range. Up to 1100°C this nitridation produced TiN and Si; from 1200°C Si₃N₄ was also produced.

Different millings were performed on this powder; by reducing the particle size and increasing the specific surface area, higher nitridation rates were achieved. A planetary milling enabled to obtain a sub-micronic powder with a specific surface area of 41m²/g. The nitridation of this powder produced both nitrides TiN and Si₃N₄ at a temperature of only 1100°C.

The behaviour of TiSi₂ as active filler was explored by pyrolyses of TiSi₂/polysiloxane mixtures, as a model for a CMC matrix. Nitridation of TiSi₂ enabled to compensate significantly for the volume shrinkage occurring during polymer pyrolysis. A carburization of TiSi₂ was also observed at 1100°C due to its reaction with solid carbon contained in the Si-O-C phase.

ACKNOWLEDGEMENTS

This work was supported by Snecma Propulsion Solide (Safran Group) and by the French national project NaCoMat. The authors are grateful to Dr. Gerhard Beckers from Retsch and M. Walter de Oliveira from Fritsch, and to Jacques Thébault from Snecma Propulsion Solide for helpful discussions.

REFERENCES
[1]R.R. Naslain, SiC-Matrix Composites: Non-brittle Ceramics for Thermo Structural Application, *Int. J. Appl. Ceram. Technol.*, **2**(2), 75–84 (2005).
[2]P. Greil, Active-filler-controlled pyrolysis of preceramic polymers, *J. Am. Ceram. Soc.*, **78**(4), 835-48 (1995).
[3]P. Greil, Near net shape manufacturing of polymer derived ceramics, *J. Eur. Ceram. Soc.*, **18**(13), 1905-14 (1998).
[4]D. Suttor, T. Erny, and P. Greil, Fiber-reinforced ceramic-matrix composites with a polysiloxane / boron-derived matrix, *J. Am. Ceram. Soc.*, **80**(7), 1831-40 (1997).
[5]P. Greil, Near net shape manufacturing of ceramics, *Mat. Chem. Phys.*, **61**(1), 64-68 (1999).

[6]J.M. Cordoba, M.D. Alcalà, M.J. Sayagués, M.A. Avilés, C. Real, and F.J. Gotor, Nitriding study of titanium silicide intermetallics obtained by mechanical alloying, *Intermetallics*, **16**(8), 948-54 (2008).

[7]R. Beyers, R. Sinclair, and M.E. Thomas, Phase equilibria in thin-film metallizations, *J. Vac. Sci. Technol.*, B **2**(4), 781-84 (1984).

[8]M. Paulasto, J.K. Kivilahti, and F.J.J. van Loo, Interfacial reactions in Ti/Si_3N_4 and TiN/Si diffusion couples, *J. Appl. Phys.*, **77**(9), 4412-16 (1995).

[9]X. Ma, C. Li, and W. Zhang, The thermodynamic assessment of the Ti-Si-N system and the interfacial reaction analysis, *J. Alloys Compd*, **394**(1-2), 138-47 (2005).

[10]C. Suryanarayara, Mechanical alloying and milling, *Prog. Mat. Sci.*, **46**(1-2), 1-184 (2001).

[11]J. Cordelair, and P. Greil, Electrical conductivity measurements as a microprobe for structure transitions in polysiloxane derived Si-O-C ceramics, *J. Eur. Ceram. Soc.*, **20**(12), 1947-57 (2000).

[12]H.D. Akkas, M.L. Oveçoglu, and M. Tanoglu, Silicon oxycarbide-based composites produced from pyrolysis of polysiloxanes with active Ti filler, *J. Eur. Ceram. Soc.*, **26**(15), 3441-49 (2006).

[13]T. Erny, M. Seibold, O. Jarchow, and P. Greil, Microstructure development of oxycarbide composites during active-filler controlled polymer pyrolysis, *J. Am. Ceram. Soc.*, **76**(1), 207-13 (1993).

HIGH POTENTIAL OF COMPOSITES WITH CARBON FIBERS AND A SELF-SEALING CERAMIC MATRIX IN MOIST ENVIRONMENTS UNDER HIGH PRESSURES AT 600°C

L. Quémard[a], F. Rebillat[a], A. Guette[a], H. Tawil[b] and C. Louchet-Pouillerie[b]
[a]University bordeaux1, Laboratoire des Composites Thermostructuraux (LCTS),
Pessac, France
[b]Snecma Propulsion Solide,
Le Haillan, France

ABSTRACT
    Self-healing ceramic matrix composites $C_f/[Si,C,B]_m$ are potential candidates to replace the current nickel-based alloys in hot sections of aeronautic turbine engines. Their development is a huge issue due to their much lower cost than $SiC_f/[Si,C,B]_m$, actual candidates for high temperature applications. The principles of self-healing approach are to consume part of the incoming oxygen and limit access of residual oxygen to the carbon interphase by filling the matrix micro-cracks with a B_2O_3-SiO_2 oxide phase. The durability of these composites was studied at 600°C for exposure duration up to 1000 hours in moist environments under high pressures. Their oxidation/corrosion behavior is investigated from (i) post-exposure tensile tests at room temperature, (ii) regular weight changes with time, (iii) post-exposure observations by microscopy and (iv) chemical and structural characterizations. Thus, at 600°C, all of the post-exposure characterizations demonstrate the ability of the sequenced [Si,C,B] matrix to protect the C fibers from environmental attacks.

INTRODUCTION
    Non-oxide ceramic matrix composites such as $C_{(f)}/SiC_{(m)}$ consist of SiC matrix reinforced with C fibers. They are among the low cost composites. These composites exhibit a low density associated with high thermomechanical properties and are potential candidates to replace the current nickel-based alloys for a variety of long-term applications in the aerospace field. $C_{(f)}/SiC_{(m)}$ components can be subjected to service conditions in the aerospace field that include mechanical loading under very high temperature. However, in complex environment containing oxygen and steam, the oxidation of C fibers can occur at a temperature lower than 500°C and leads to a decrease of the mechanical resistance of composites. $C_{(f)}/[Si,C,B]_{(m)}$ composites with a sequenced self-healing matrix have been developed[1,2] and investigated[3-7] to protect the C fibers against oxidation effects up to 1400°C. The principles of the self-sealing approach are to consume part of the incoming oxygen and to limit access of residual oxygen to the C fibers by sealing the matrix microcracks with a SiO_2-B_2O_3 oxide phase.
The aim of this study is to evaluate, at intermediate temperature(600°C), the effects of both oxygen and water vapor on the self-healing process of $C_{(f)}/[Si,C,B]_{(m)}$ composites subjected to high pressure environments.

MATERIALS AND TEST SPECIMENS
The composite
    The material investigated is the SEPCARBINOX®A500[1-3] manufactured by Snecma Propulsion Solide (France) via chemical vapor infiltration (CVI). It is a woven- C fiber reinforced [Si,C,B] sequenced matrix composite (Fig.1). The different matrix layers are crystallized SiC, amorphous B_4C and a SiC-B_4C phase named Si-B-C.
The mainly tested specimen, are small bars, 60 mm long, with a gauge section width of 16 mm, and a thickness of 5 mm. A few tested dog-bone specimens is 200 mm long, with a grip section width of 24 mm, a reduced gauge section width of 16 mm, and a thickness of 4.4 mm. All specimens are seal-coated with CVI layers of SiC, B_4C and Si-B-C.

The constituents

At intermediate temperature, the reactivity of crystallized SiC is low and the self-sealing process involves the B_4C and Si-B-C layers. The efficiency of the self-healing process under environments containing both oxygen and water vapor results from the competition between the oxidation of these matrix layers and the volatilization of the resulting oxide phase.

Under dry air, B_4C undergoes oxidation below 600°C, leading to the formation of $B_2O_{3(l)}$ (equations 1-2).

$$B_4C + \tfrac{7}{2} O_2 \rightarrow 2 B_2O_{3(l)} + CO_{(g)} \qquad (1)$$
$$B_4C + 4 O_2 \rightarrow 2 B_2O_{3(l)} + CO_{2(g)} \qquad (2)$$

Under water vapor-containing environments, $B_2O_{3(l)}$ may react significantly to form gaseous hydroxides species, $H_xB_yO_{z(g)}$[4,5]. The competition between the oxidation (equations 1-2) and volatilization reactions (equations 3-5) can result in the recession of B_4C.

$$3/2 \, B_2O_{3\,(l)} + 3/2 \, H_2O_{\,(g)} \rightarrow H_3B_3O_{6\,(g)} \qquad (3)$$
$$3/2 \, B_2O_{3\,(l)} + 1/2 \, H_2O_{\,(g} \rightarrow H_3BO_{3\,(g)} \qquad (4)$$
$$1/2 \, B_2O_{3\,(l)} + 1/2 \, H_2O_{\,(g)} \rightarrow HBO_{2\,(g)} \qquad (5)$$

The Si-B-C matrix layer can be described as a mixture of SiC nanocrystals in an amorphous B_4C phase[6]. The SiC nanocrystals can oxidize significantly at 600°C and at atmospheric pressure to form silica[4].

In high temperature environments, the SiC matrix layers can oxidize significantly under dry air above 1000°C to form a protective SiO_2 scale (equations 6-7).

$$SiC + \tfrac{3}{2} O_2 \rightarrow SiO_2 + CO_{(g)} \qquad (6)$$
$$SiC + 2 O_2 \rightarrow SiO_2 + CO_{2(g)} \qquad (7)$$

In environments containing O_2 and H_2O, the formation of silica is dramatically enhanced. Under steam-containing environments, the SiO_2 scale may also volatilize, by formation of $Si(OH)_{4\,(g)}$ species[8] (equation 8).

$$SiO_{2\,(s)} + 2 \, H_2O_{\,(g)} \rightarrow Si(OH)_{4\,(g)} \qquad (8)$$

Finally, at 600°C, mainly the B_2O_3[4,5] volatilization under water-vapor-containing environments may lead to a reduction of the quantity of the protective oxide phase in the $C_{(f)}/[Si,C,B]_{(m)}$.

TEST PROCEDURES

Inside these materials, a regular crack network is generated in the matrix, with the relaxation of the residual, thermal stresses during the cooling after their process. This crack network facilitates the ingress of the corrosive species. The pre-existing micro-cracks are present across all the composite section (Figure 1-a)). At room temperature (RT), their mean spacing distance is between 350 and 500 μm in the seal coat. In the bulk, one or two micro-cracks cross each tow. Their widths are 4 μm in the seal-coat and 1 μm inside a tow.

The material mechanical behavior shows, from the very first loading, a non linear stress-strain behavior (Figure 1-b)). These materials have a relatively low value of initial modulus (65 GPa). Along the non linear stress-strain behavior, no plateau is observed up to the ultimate rupture of the specimens. This behavior, induced by matrix cracking, indicates a progressive damaging. Moreover, the hysteresis loops are widely opened and the residual strains after unloading are high. This indicates a low fiber-matrix load transfer, thus a weak interfacial shear stress and matrix cracks widely opened. All the samples have gage failures with a non-brittle rupture characterized by fiber pull-out. The size of the tested specimens, small bar or dog-bone, does not influence the mechanical behavior in tension.

The corrosion test conditions are reported on figure 2. High pressure corrosion tests are conducted in a High Pressure - High Temperature Furnace developed for these researches[9]. High pressure moist air is provided by a pressurized gas supply system then mixed with water in an evaporator; the air and water flows are independently controlled by mass flow meters and the Air/H_2O gas mixture is injected in the test tube. In an over hand, at atmospheric pressure, the dry air flows through a heated water column in order to be saturated in steam before its introduction in the tube of the furnace.

In both corrosion test equipments, the SEPCARBINOX®A500 specimens are oriented parallel to the gas flow and placed on alumina sample holders (purity: 99.7 %, OMG, France) specially designed. The heating and cooling rates, used at atmospheric pressure under ambient air, are respectively 150°C.h^{-1} and 100°C.h^{-1}. The exposures are regularly interrupted to weigh the specimens using a scale (Precisa Instruments AG, Switzerland) with an accuracy of 1.10^{-2} mg.

Then, post-exposure cyclic tensile tests up to failure are performed at RT on the dog-bone specimens in a servo controlled testing machine (INSTRON 1185) equipped with self aligning grips at a cross-head speed of 0.40 ± 0.05 %.min^{-1}.

After exposures, the test specimens are cut perpendicularly and parallel to the loading axis then polished for examination using an optical microscope. Furthermore, the fractured surfaces are analyzed by scanning electron microscopy (SEM). The oxides are characterized by raman microspectrometry (Labram 10 spectrometer from Jobin Yvon, France), infra-red spectroscopy (Bruker IFS66) and electron probe micro-analysis (EPMA).

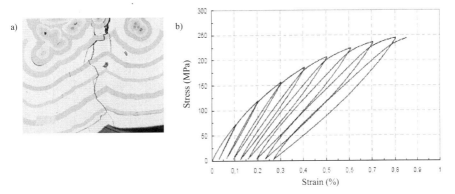

Figure1. SEPCARBINOX®A500 material : a) polished cross-section of the as-received material and b) a stress-strain curve in tension obtained with loading-unloading hysteresis loops at room temperature.

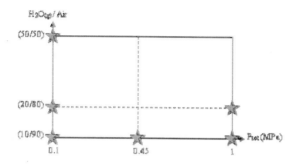

Figure 2. Summary of test conditions for CERASEP®A410 specimens exposed
in various environments.

DURABILITY OF THE SEPCARBINOX®A500 at 600°C
Mechanical properties

The post-exposure mechanical results, determined at room temperature, show that a non linear material mechanical behavior is conserved after all the corrosion tests at 600°C for durations up to 1000 hours (Figure 3). The exposed materials have an initial modulus between 25 % and 70 % higher than the modulus of the as-received material. At the beginning of the strain-stress curves, a linear elastic domain appears. At room temperature, the sealing borosilicate glassy phase (in a solid state), filling the matrix cracks, may bond the damaged matrix blocs together and be responsible to the increase of the initial Young Modulus and to a new matrix cracking process along this pre-existing crack network.

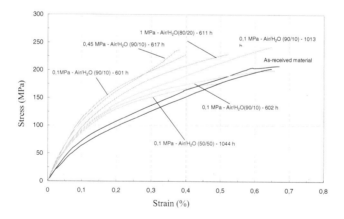

Figure 3. Stress-strain curves of the CERASEP®A410 specimens obtained at RT after exposure for 600 hours at 600°C in various environments.

In general, the values of ultimate tensile strength (UTS) are not affected, except for aging in the air/steam (50/50) gas mixture, for which a decrease of around 15 % is measured on small bars after 1000 h and 20 % on dog-bone specimens after 600 h (Figure 4). Further, the values of strains to failure are often 20 % smaller than that measured for the as received materials. This decrease appears as soon as 50 h of aging and it doesn't evaluate up to 1000 h.

During a tensile test, the exposed composite materials present the same variations of width of the hysteresis loops and residual strains after unloading than the as-received materials (Figure 1).

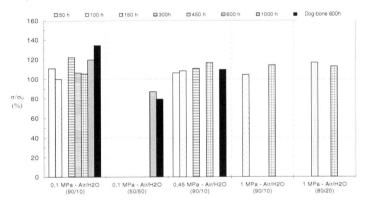

Figure 4. Effect of the corrosion environments on the ultimate tensile strength of A500 materials.

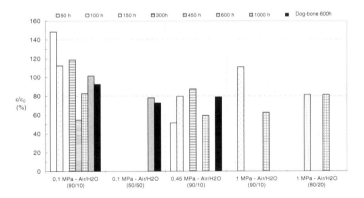

Figure 5. Effect of the corrosion environments on the strain to failure of A500 materials.

Weight changes

The SEPCARBINOX®A500 weight changes are important in specific conditions, with global variations ranging between -10 to +5 % (Figure 6). Nevertheless, the weight changes and the morphological analyses highlight effects of the total pressure. Thus, composite weight gain rates are obtained at high total pressures while weight loss rates are mainly obtained in lower pressure environments. These weight changes are essentially due to the boron matrix layers recession and oxidation processes and to the carbon fibers consumption. Indeed, only the SiC matrix layers are assumed to be relatively inert to oxygen and steam at 600°C.

Respectively for lowest moisture ratio (Figure 8), a very limited oxidation of fibers is observed around the tip of cracks.

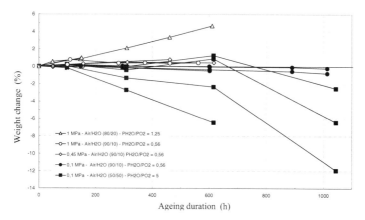

Figure 6. Weight changes of SEPCARBINOX®A500 specimens exposed in various environments at 600°C up to 1000 hours.

According to the post-exposure observations (Figure 7) and the weight changes of the SEPCARBINOX®A500 composites (Figure 6), the weight loss rates at atmospheric pressure result from : (i) a local recession of a part of the boron matrix layers which is due to a competition between the kinetics of oxidation and volatilization and (ii) to a simultaneous important oxidation of carbon fibers poorly protected due to this local absence of the liquid sealing boria-containing phase (Figure 7). Furthermore, at atmospheric pressure with the higher moisture ratio, the recession of the boron matrix layers and the fibers oxidation are enhanced. Increasing the moisture content by a factor 5 leads to a weight lost of 2.5 % after 600 h against no significant change noticed for a 90/10 air/moisture gas mixture.

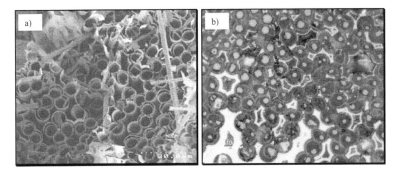

Figure 7. Observations of a SEPCarbinox®A500 specimen, exposed at 600°C and 0.1 MPa in an Air/Steam (50/50) gas mixture for 1000 hours, then tested in tension at room temperature, showing in different places across the composite : a) the recession of the boron-containing matrix layers, by scanning electron microscopy on a fracture surface in tension, b) the oxidized carbon fibers embedded in a sealing boria-containing glassy phase, by optic microscopy, on a polished cross section.

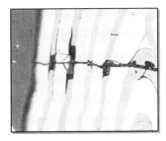

Figure 8. Polished cross section of a SEPCARBINOX®A500 specimen, exposed at 600°C and 0.1 MPa in an air/steam (90/10) gas mixture for 600 hours, then tested in tension at room temperature, showing the recession of the boron-containing matrix layers in matrix cracks.

With the increase of the total pressure, the filling of matrix cracks by a residual glassy phase is evidenced. Across woven close to the seal-coat, the carbon fibers are: (i) only oxidized in very narrow zones surrounding these cracks and (ii) systematically embedded in this sealing glassy phase.

For the high total pressures, the quantity of the liquid sealing boria-containing phase is more important, according to the more important measured weight gain rates. Thus, increasing the moisture content from 10 to 20 % in the gas mixture leads to an increase of the weight gain by a factor 11 after 600 h. No corrosion seems to affect the bulk of the composite (Figure 9).

Increasing of the moisture content leads to a reverse effect at high pressure and at atmospheric pressure, with the enhancement of the weight loss.

 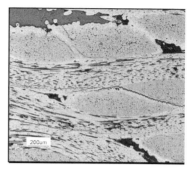

Figure 9. SEPCARBINOX®A500 specimen, exposed at 600°C and 1 MPa in an air/steam (90/10) gas mixture for 600 hours, then tested in tension at room temperature, showing a typical non brittle fractured surface and an image of the bulk very similar to the as-received material.

Characterization of the oxides scale

Auger Electron Spectroscopy analyses were carried out on the glassy phase formed in the seal-coat and surrounding the carbon fibers (Figure 10). With increasing the moisture content in the gas mixture or for higher total pressure, the oxide scale has high silica content. The boria content is limited by its volatility in steam environments. Respectively, the large amount of silica comes from the accelerated oxidation of the Si-B-C phase by both O_2 and H_2O oxidizing species in excess in the gas mixture. Thus, the faster generation of the liquid boria-containing glass facilitates the silicon carbide dissolving, and the enrichment in silica[8].

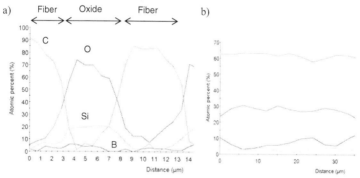

Figure 10. AES line profile realized in zones surrounding C fibers in the bulk of SEPCARBINOX®A500 specimens, exposed at 600°C for 600 hours: a) under 0.1 MPa in an Air/Steam (50/50) gas mixture and b) under 1 MPa in an air/steam (80/20) gas mixture

Discussion

Finally, a scenario of the corrosion progression through a matrix crack in the [Si,C,B] matrix is proposed. At atmospheric pressure with high moisture content, a weight loss is measured and a filling of the gap around the fiber by oxidation of the inner boron-containing layers cannot occur. Processes of localised sealing are caused by a low oxide volatilization kinetic due to the decrease of P_{H2O} through the crack. At the opposite, at higher total pressure, the rapid Si-B-C oxidation kinetic allows to quickly fill the pre-damaging crack with a borosilicate glass. Then, a reduced part of the oxygen continues to diffuse toward the B_4C layer (through the superficial borosilicate, more stable against moisture) and causes the amount of B_2O_3 to increase. The generated B_2O_3 inside the composite may diffuse through the borosilicate up to the surface and is volatilized. With the boria volatilization, the oxide scale has higher silica content and is in an enough important quantity to fill the cracks. This phenomenon results in a weight gain rate of the SEPCARBINOX[®]A500.

Those results are in agreement with recent studies[3,10,11] conducted at Pratt and Whitney (Florida, USA) and at Arnold Engineering Development Center (AEDC, Arnold Air Force Base, Tennessee, USA). These works consisted in testing SEPCARBINOX[®]A500 seals in Pratt and Whitney engines that power the F-16 and F-15 fighters. The seals were tested in ground engines for nearly 1300 hours including 100 hours in after burner conditions that corresponds to a number of Total Accumulated Cycles (TAC) of 5000 approximately. The maximal stresses in the seals, calculated taking into account the worst-case flight point corresponding to the after burner conditions, were 63 MPa (760°C) in the 1-1 axis (exit end) and 55 MPa (430°C) in the 2-2 axis (forward hinge). The post-exposure mechanical results showed a decrease around 20 % of the ultimate tensile properties at RT. The oxidation of carbon fibers is limited to the woven close to the seal-coat along the exposed surfaces.

CONCLUSIONS

The durability of a $C_{(f)}/[Si,C,B]_{(m)}$ composite with a sequenced self-sealing matrix and C fibers was investigated at 600°C for exposure duration up to 1000 hours. The specimens are aged in a variety of slow-flowing air/steam gas mixtures and total pressures, ranging from atmospheric pressure with a 10-50% water content to 1 MPa with 10-20% water content, with a flowing velocity between 5 and 10 $cm.s^{-1}$.

All of the post-exposure characterizations demonstrate the ability of the sequenced [Si,C,B] matrix to protect the C fibers from environmental attacks. Two different degradation modes of the matrix, depending on the total pressure are discussed in terms of the reactivity of the boron-containing layers, and their relative positions in the sequenced matrix.

ACKNOWLEDGEMENTS

This work has been supported by the Centre National de la Recherche Scientifique (CNRS) and Snecma Propulsion Solide (SPS) through a grant given to L. Quémard. The authors are grateful to M. Cataldi, J. Lamon and G. Falguieres for fruitful discussions, Snecma for the production of the samples, R. Bouvier for assistance with corrosion tests, P. Ophele and F. Labarriere for assistance with mechanical tests and M. Lahaye for assistance with AES analyses.

REFERENCES
[1]F.Lamouroux, R.Pailler, R.Naslain and M. Cataldi,French Patent, n° 95 14843 (1995).
[2]L. Vandenbulcke and S. Goujard, Multilayer systems based on B, B_4C, SiC and SiBC for environmental composite protection, Progress in Advanced Materials and Mechanics, Progress in Advanced Materials and Mechanics, 1198-1203 (1996).
[3] E. Bouillon, P. Spriet, G. Habarou, C. Louchet, T. Arnold, G. C. Ojard, D. T. Feindel, C. P. Logan, K. Rogers and D. P. Stetson, Engine test and post engine test characterization of self sealing ceramic matrix composites for nozzle applications in gas turbine engines,, ASME TURBO EXPO 2004, Power for Land, Sea and Air, Vienna, Austria, ID GT-2004-53976 (2004).
[4]X. Martin, Oxydation/corrosion de matériaux composites ($SiC_f/SiCB_m$) à matrice auto-cicatrisante, PhD Thesis 2749, University of Bordeaux 1, France, (2003).

[5]L. Quémard, X. Martin, F. Rebillat and A. Guette, Thermodynamic study of B_2O_3 reactivity in $H_2O_{(g)}N_{2(g)}/O_{2(g)}$ atmospheres at high pressure and high temperature, Proceedings of High Temperature Ceramic Matrix Composites 5 (HTCMC5), edited by Singh M., Kerans R., Lara-Curzio E. and Naslain R., published by the Am. Ceram. Soc. (Westerville, Ohio, USA), 327-332 (2004).

[6]G. Farizy, Mécanisme de fluage sous air de composites $SiC_f/SiBC_m$ à matrice auto-cicatrisante, Ph.D. Thesis, University of Caen, France, (2002).

[7]L. U. J. T. Ogbuji, A pervasive mode of oxidative degradation in a SiC-SiC composite, J. Am. Ceram. Soc., **81/11**, 2777-2784 (1998).

[8]E. J. Opila and R. E. Jr. Hann, Paralinear Oxidation of CVD SiC in Water Vapor, J. Am. Ceram. Soc., **80/1**, 197-205 (1997).

[9]L. Quemard, F. Rebillat, A. Guette and H. Tawil, Development of an original design of high temperature - high pressure furnace,Proceedings of High Temperature Ceramic Matrix Composites 5 (HTCMC5), edited by Singh M., Kerans R., Lara-Curzio E. and Naslain R., published by the Am. Ceram. Soc. (Westerville, Ohio, USA), 543-548 (2004).

[10]L. Zawada, G. Richardson and P. Spriet, Ceramic matrix composites for aerospace turbine engine exhaust nozzles, Proceedings of High Temperature Ceramic Matrix Composites 5 (HTCMC5), edited by M. Singh, R. Kerans, E. Lara-Curzio and R. Naslain, published by the Am. Ceram. Soc. (Westerville, Ohio, USA), 491-498 (2004).

[11]E. Bouillon, P. Spriet, G. Habarou, C. Louchet, T. Arnold, G. C. Ojard, D. T. Feindel, C. P. Logan, K. Rogers and D. P. Stetson,, Engine test experience and characterization of self sealing ceramic matrix composite for nozzle appications in gas turbine engines, IGTI 2003: ASME TURBO EXPO 2003, Atlanta, Georgia, USA, ID GT-2003-38967 (2003).

QUANTIFICATION OF HIGHER SIC FIBER OXIDATION RATES IN PRESENCE OF B_2O_3
UNDER AIR

F. Rebillat, E. Garitte & A. Guette
University of Bordeaux, Laboratoire des Composites Thermostructuraux (LCTS),
Pessac, France

ABSTRACT
    Ceramic Matrix Composites, such as SiC/SiC, exhibit excellent thermomechanical properties.
To enhance the lifetime of these composites, a self-healing matrix with boron bearing species has been
developed in order to resist in a large range of temperatures under oxidant atmosphere. The chemical
and thermal stabilities of the generated B_2O_3 are enhanced through the formation of borosilicate by
dissolving SiO_2. However, the dissolving process should be avoided as soon as it acts over the silica
protective layer on the SiC fiber surface: the oxidation of SiC fiber may be highly enhanced. A
quantification of the silica dissolution rate is done, through the development of a simple global
modelling taking into account a competition between the silica formation rate by oxidation and its
disappearance by dissolution in B_2O_3. This scenario is checked by many experiments in oxidizing
environment with various oxygen partial pressures at different temperatures ranging from 600°C to
900°C.

INTRODUCTION
    Ceramic Matrix Composites (CMC) such as SiC/SiC composites are proposed for thermo
structural applications in the aeronautical and space domains because of their thermo mechanical
properties and their resistance to corrosive atmosphere [1,2]. Under dry air, the composite is able to
protect itself by the formation of a protective silica layer at temperature higher than 1000°C [3]. That's
why it is necessary to introduce in the matrix a ceramic material which is able to generate an oxide at
lower temperature. The incorporation of boron bearing species such as B_4C in the matrix allows the
self-healing of the materials at lower temperature, starting around 500°C. The filling up of matrix
microcracks is made by formation of B_2O_3, a liquid oxide. A borosilicate is formed when SiO_2 is
dissolved in B_2O_3 [4]. Nevertheless, the boron oxide is very corrosive against SiC fibers : (i) the boron
oxide is able to dissolve a large amount of SiC, depending on the solubility between SiC and B_2O_3 [5]
and (ii) a chemical reaction can occur between SiC and B_2O_3 under argon at high temperature (1200°C)
[6]. Thus, « oxidation » of SiC is favoured in presence of B_2O_3. This phenomenon has been punctually
observed in a composite at temperature as low as 600°C [7]. It provokes a reduction of fiber diameters
(figure 1), eventually their premature rupture and a drastic reduction of the lifetime of the composite.
The aim of this work is to determine the rate of SiC fiber degradation in presence of B_2O_3 liquid in
order to predict the lifetime of these composites. For this, the oxidation rates are measured for silicon
carbide with or without boron oxide and kinetic laws are explicated. Then, by comparing the results,
the acceleration phenomenon will be highlighted. The effect of several parameters (such as the
temperature, the amount of boron oxide) will be taken into account. Finally, a scenario of the
dissolution phenomenon is proposed.

Figure 1. fibers sections close to the tip of a matrix crack, containing boron bearing species, oxidized at 1200°C under a total pressure of 1 MPa containing 200 kPa of moisture during 600h [7].

EXPERIMENTAL PROCEDURE

The materials

The silicon carbide fibers used are Hi-Nicalon® (14 μm of diameter).

Vitreous borosilicate oxide is prepared from commercial boron oxide powder (Chempur, 99%, amorphous) and silica powder (Chempur, 99,9%, quartz α). After low heating at 300°C for 3 hours in order to dehydrate powders, the mixture is heated at 900°C for one hour, and then undergoes a quenching. The variation of composition of borosilicate glassy phases allows pointing out the impact of B_2O_3 on its capability to dissolve silica (figure 2) and respectively, to contribute to enhance the silicon carbide oxidation rate by dissolving its silica protective scale, in function of temperature.

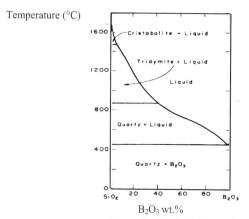

Figure 2. SiO_2-B_2O_3 binary diagram [8].

The oxidation device

The thermo gravimetric analyses (TGA) are performed using an apparatus (SETARAM, France) equipped with a vertical alumina tube and a balance with a sensibility of 1μg. Hi-Nicalon fibers are placed in a platinum crucible for each experiment. The crucible is hanging from the beam using platinum suspensions. The oxidant phase is introduced and exhausted respectively through the bottom and the top of the furnace below the balance. A helium flows is on the other hand introduced in the balance to protect it from the oxidant gases.

The tests are performed at different temperatures ranging from 600 to 1000°C. The working pressure is always equal to atmospheric pressure. The atmosphere consists of dry gas mixtures. In order to determine the influence of the oxygen quantity on the oxidation rate of SiC Hi-Nicalon fibers, the oxidation tests are conducted for different partial pressures of oxygen. The flows of O_2 and N_2 are controlled by mass flow meter and the total gas flow is fixed to 2 l/h. The gas velocity is then 0.2 cm/s in the cold zone of the furnace. Thus, the oxygen partial pressure is regulated from 0 to 100 kPa and the nitrogen one is fixed by $100 - P(O_2)$.

The oxidation tests
    All the oxidation tests were performed at fixed temperatures ranging between 600 and 1000°C. The mass variation is measured with time and expressed per unit of surface area. The test duration depends on the temperature, but it does not exceed 20 hours. The sample is made of about 6 cm of fibers (cut into pieces to cover uniformly the bottom of the platinum crucible; m = 39 mg). The oxidation tests are performed on fibers covered or not with boron oxide. In the mixture, the mass ratio of B_2O_3 is fixed at 43,4 % (the volume of SiC is around 3 times higher than this of B_2O_3) (figure 3). The mixture is preheated at 300°C during 30 minutes in order to dehydrate the boron oxide. The oxidizing gas mixture is introduced in the furnace only when the experiment temperature is reached (in the same time, a homogenous mixture is gotten).

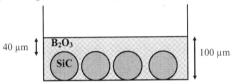

Figure 3. Schematic of the tested sample, once B_2O_3 surrounds SiC fibers.

Theoretical model for data exploitation
    During the oxidation process, the chemical reactions of our specimens of Hi-Nicalon fibers, according to its excess in carbon, under dry environment are [9,10]:

$$SiC_{(s)} + 1/2\ C_{(s)} + 7/4\ O_{2(g)} = SiO_{2(s)} + 3/2\ CO_{(g)} \tag{1}$$
$$SiC_{(s)} + 1/2\ C_{(s)} + 5/2\ O_{2(g)} = SiO_{2(s)} + 3/2\ CO_{2(g)} \tag{1'}$$

For the both reactions, the molar quantity of formed SiO_2, n_{SiO_2} is:

$$n_{SiO_2} = \frac{\Delta m}{\Delta M} \tag{2}$$

where Δm and ΔM are, respectively, the mass variation measured by a thermo gravimetric method and the mass balance (3) during the oxidation of Hi-Nicalon fibers, according to reaction (1).

$$\Delta M = 2\ M_O - 3/2\ M_C = 14\ \text{g.mol}^{-1} \tag{3}$$

where M_O and M_C are respectively the molar mass. To calculate the thickness of formed SiO_2, the determination of the volume variation of Hi-Nicalon fiber is necessary. The formed SiO_2 volume, V_{SiO_2}, is:

$$V_{SiO_2} = \pi N L \left[\left(r_0 - e_{SiC} + e_{SiO_2} \right)^2 - \left(r_0 - e_{SiC} \right)^2 \right] = \frac{n_{SiO_2} M_{SiO_2}}{\rho_{SiO_2}} \tag{4}$$

where N is the fiber number, L the initial fiber length, r_0 the average fiber radius, ρ_{SiO2} the SiO_2 volume mass and M_{SiO2} the SiO_2 molar mass. From Eq. (4), the thickness of SiO_2 formation is:

$$e_{SiO_2} = (r_0 - e_{SiC}) + \sqrt{(r_0 - e_{SiC})^2 + \frac{n_{SiO_2} M_{SiO_2}}{\pi N L \rho_{SiO_2}}} \qquad (5)$$

Moreover, the volume of SiC consumption, V_{SiC} is:

$$V_{SiC} = \pi N L \left[r_0^2 - (r_0 - e_{SiC})^2 \right] = \frac{n_{SiC} M_{SiC}}{\rho_{SiC}} \qquad (6)$$

Thus, the thickness of consumed SiC is:

$$e_{SiC} = r_0 - \sqrt{r_0^2 - \frac{n_{SiC} M_{SiC}}{\pi N L \rho_{SiC}}} \qquad (7)$$

where r_0 is the fiber radius, ρ_{SiC} the SiC volume mass, M_{SiC} the SiC molar mass, n_{SiC} the molar number of consumed SiC, deduced from the whole mass variation through the relation (6).

RESULTS
Oxidation of silicon carbide fibers in dry air
    The weight changes resulting from oxidation of SiC fibers (weight changes divided by the initial weight of the sample) under dry air between 800 and 1000°C are shown in figure 4. The SiC fiber oxidation is limited by oxygen diffusion through the oxide layer, and the weight changes follow a parabolic law[18]. The oxidation phenomenon is thermally activated and is described by an Arrhenius law (table 1).

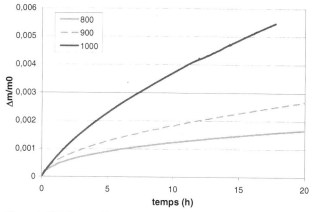

Figure 4. SiC fibers oxidation in an ATG under dry air (P_{O_2} = 20 kPa).

Table 1. oxidation of SiC Hi-Nicalon fibers under dry air (total flow 2 $l.h^{-1}$) :
a) experimental oxidation rate in function of temperature and b) oxidation kinetic law,
(respectively, expressed in mass variation, in % of mass, in formed silica thickness and in consumed SiC thickness)

a)

OXYDATION DES FIBRES DE SiC HI-NICALON				
Température	k_p [$mg^2.cm^{-4}.h^{-1}$]	k_p'' [h^{-1}]	k_p' [$\mu m^2.h^{-1}$] of SiO_2	k_p' [$\mu m^2.h^{-1}$] of SiC
800°C	$8,11.10^{-8}$	$1,19.10^{-7}$	$2,45.10^{-5}$	$9,18.10^{-6}$
900°C	$3,18.10^{-7}$	$4,08.10^{-7}$	$9,99.10^{-5}$	$3,65.10^{-5}$
1 000°C	$1,36.10^{-6}$	$1,86.10^{-6}$	$4,86.10^{-4}$	$1,65.10^{-4}$
Ea (kJ/mol)	159	155	169	163

b)

Hi-Nicalon SiC fibers	
$k_p = 4,53 . e^{(-19198/T)}$	[$mg^2.cm^{-4}.h^{-1}$]
$k_p = 3,99 . e^{(-18680/T)}$	[h^{-1}]
$k_p = 3,86.10^{+3} . e^{(-20326/T)}$	[$\mu m^2.h^{-1}$](SiO_2)
$k_p = 7,75.10^{+2} . e^{(-19647/T)}$	[$\mu m^2.h^{-1}$](SiC)

Oxidation of silicon carbide fibers in contact with B_2O_3

Weight changes resulting from oxidation of SiC fibers, under dry air and in contact with B_2O_3 between 600 and 1000°C, are shown in figure 5. From the global shape of these curses, the SiC fiber oxidation process appears to be limited by oxygen diffusion through an oxide scale. This process is thermally activated.

Respectively, using relations (6) and (7), the SiC consumed thicknesses, during oxidation of Hi-Nicalon SiC fibers in the same oxidation conditions, with and without being embedded in B_2O_3, are calculated. They are compared in figure 6.

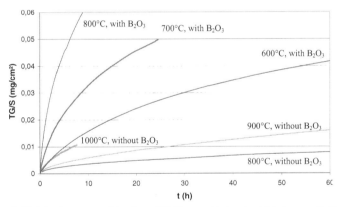

Figure 5. Weight changes resulting from SiC fibers oxidation (weight changes divided by the sample surface area) under dry air and in contact with B_2O_3 between 600 and 1000°C in a TGA ((v(gaz) = (0.25 ± 0.05) cm/s) $P_{O2} = 13$ kPa, (56.6 wt.% mass. de SiC))

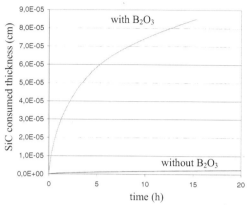

Figure 6. SiC consumed thickness during SiC Hi-Nicalon fibers
with and without being embedded in B_2O_3 at 900°C under $P_{O_2} = 13$ kPa

Considering an initial similar amount of SiC fibers, a much higher weight gain is measured during a shorter oxidation tests in presence of boron oxide: this gain is around 15 times higher after 5 hours. Further, in contact with B_2O_3, the oxidation rate at 600°C becomes higher than that at 1000°C, without B_2O_3. Boron oxide seems to act on the SiC fiber oxidation process, by reducing the efficiency of the protective silica oxide layer.

Oxidation behaviour in function of oxygen partial pressure
        The same mass ratio SiC/B_2O_3 is maintained and the tests are performed under dry atmosphere (with $P(O_2) = 1.2$ to 80 kPa) at a fixed temperature, 900°C. The weight changes are shown in figure 3. First of all, variations of oxygen partial pressure neatly modify the relative oxidation rate and certainly the phenomena involved in presence of B_2O_3.
In this configuration, as long as the oxygen partial pressure is lower or equal to 6.4 kPa, the weight gains increase with the oxygen partial pressure.
By increasing the oxygen partial pressure, the oxygen diffusion through the boron oxide appears to be much more limited; a lower weight gain is recorded. This may be induced by a progressive formation of a more protective oxide scale, certainly rich in SiO_2, around fibers. The parabolic shape of the weight variation curve has an enhanced curvature, related to slower oxygen diffusion toward SiC fibers, as soon as this protective scale is able to be formed.

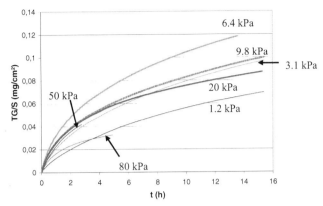

Figure 7. Effect of P(O$_2$) on the oxidation of SiC fibers embedded in B_2O_3 at 900°C under dry atmosphere (56.6 wt.% of SiC)

Oxidation behaviour in function of the composition of the borosilicate glassy phase
Since in a composite, the healing oxide should be a borosilicate glassy phase, the oxidation rate of SiC fiber in borosilicate has to be also considered (figure 8).

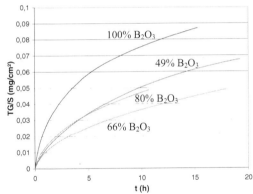

Figure 8. Effect of the SiO$_2$ content in the borosilicate, embedding SiC fibers, on the oxidation rate at 900°C under P$_{O_2}$ = 13 kPa

The oxidation rate depends on the thickness of the oxide layer and the nature of this oxide. Oxidation tests are done at 900°C, under dry air (with a weight of SiC fiber always fixed to 56,6% of the total weight of the mixture (SiC fibers + borosilicate)). To form a borosilicate, silica is initially introduced in the boron oxide glassy phase, embedding fibers. The weight gains are lower by increasing the SiO$_2$ content in borosilicate. The O$_2$ diffusion coefficient in borosilicate is smaller than in pure B_2O_3 and the O$_2$ access toward fibers is much more limited. The enhanced curvature of parabolic weight gain curves reveals that the formation of a more protective scale is favoured around fibers.

For a composition of borosilicate close to the limit of solubility of SiO_2 in B_2O_3, the development of a much more protective scale (as expected) is not favoured since a high SiC oxidation rate is maintained. The silica scale formed at the fiber surface should be dissolved, before precipitating far from the SiC consumed surface (solubility of SiO_2 in B_2O_3 is fixed at each temperature).

Quantity of B_2O_3 embedding fibers
　　The quantity of diffusing oxygen is also affected by the increase of the B_2O_3 thickness layer to cross over fibers. Thus, oxidation tests are done SiC fibers (with the same quantity as previously) embedded in a more important amount of boron oxide: the mass ratio of the SiC is changed from 0.566 to 0.167. In the case, B_2O_3 is introduced in a large excess (respectively, the volume ratio of B_2O_3 on SiC evolves from 0.33 and 2.5). The weight changes obtained in a TGA system are shown in figure 9.
The increase of the B_2O_3 amount, surrounding the fibers, leads to a decrease of the oxidation rate of SiC fibers. As expected, in presence of a large amount of boron oxide, the oxygen diffusion through this oxide is slowed down. However, this oxidation rate of SiC fibers remains higher than that measured for oxidized fibers without B_2O_3.

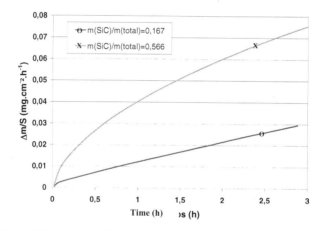

Figure 9. Effect of the B_2O_3 amount embedding fibers, on their oxidation rate at 1000°C under dry air.

Chemical characterization of oxidized fibers
　　The cross sections of oxidized fibers embedded in borosilicate (with various boron oxide content), are observed by scanning electron microscopy (figure 10).
The observation of a silica layer around each fiber shows that the surrounding glassy phase is not able to dissolve the entire silica quantity formed by oxidation of SiC fibers. Across the tow, the thicknesses of silica scales are not uniform. The phenomena of mass transfers in such a liquid environment (around and inside fibers packing) should be extremely complex.
As expected, in same oxidation conditions, the protective silica scale is thicker when the surrounding borosilicate contents more silica.
To complete these results, EDX analyses are realized on the oxide layer embedding SiC fibers, oxidized at 900°C, in borosilicates with various compositions (Figure 11). Far from the fiber surface (expected at 7 μm), silicon is detected as shown by the elevation of the intensity of Si peak. Its ratio in

the glassy phase decreases progressively. This evolution has to be related to a gradient of concentration in the borosilicate (and not to a clearly defined silica scale). Far from the shouted fiber, interactions of the electron spot with surrounding fibers may cause perturbations in signal levels: reverse signal levels to these expected in function of the silica content in borosilicate (the curve for 49 % B$_2$O$_3$ (51 % SiO$_2$) is below that of 80 % B$_2$O$_3$ (20 % SiO$_2$)).

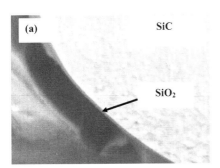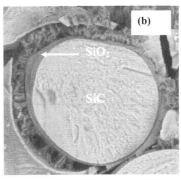

Figure 10. Observation by back scattering electron microscopy of oxidized SiC fibers at 900°C, under atmospheric pressure with P$_{O2}$ = 13 kPa, during 15 hours (v(gaz) = 0,3 cm/s), in : (a) B$_2$O$_3$ and (b) a borosilicate (66 wt.% of B$_2$O$_3$), (56,6 wt.% de SiC).

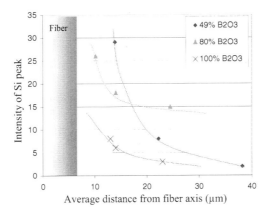

Figure 11. EDX analyses realized on the oxide layer embedding SiC fibers, oxidized at 900°C, under atmospheric pressure with P$_{O2}$ = 13 kPa, in borosilicates with various compositions.

DISCUSSION

Reactivity

The presence of boron oxide near SiC fibers leads to an increase of the fibers oxidation rate. Indeed, boron oxide has the capability to dissolve high quantities of silica [8]. This chemical affinity is illustrated by the spontaneous wetting of silica surfaces, formed over SiC [16].

SiC and B_2O_3 cannot co-exist, according to the phase equilibrium in the Si-B-C-O system [17]. Along the interface between SiC, or SiO_2, and B_2O_3, few phases should spontaneously be formed as silica and/or boron carbide and/or carbon (Figure 12).

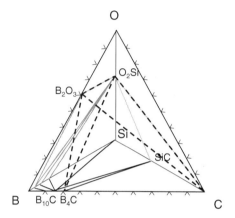

Figure 12. Phase equilibrium in the Si-B-C-O system [17] :
(dotted line to insist on the *"$B_2O_3/SiO_2/C/B_4C$"* tetrahedron).

Oxidation kinetic

The generic law of scale growth is described by equation (8) where Δm is the weight gain, k a growth rate constant and n an empirical exponent characteristic of the process limiting oxide growth [11].

$$\Delta m = k.t^{n}$$

(8)

The relationship remains valid when weight gain is replaced with oxide scale thickness. In the simplest cases n takes the value of 1.0 (representing linear kinetics) or 0.5 (parabolic kinetics), according as growth is limited by a reaction or by diffusion. The linear kinetics implies that the underlying chemical reaction is the step which limits the overall oxidation process. This is the case when oxidant diffusion through the oxide scale is so fast as to impose no limitation; for instance, the oxide liquefied (as the formed silica is dissolved in the melted glassy phase) is poorly consolidated short-circuit paths for diffusion of gases. Partial melting of the oxide (as an occasional burst of gas bubbles, or periodic spallation) causes transient linear excursions in an otherwise parabolic oxidation kinetic. The thickening of the scale by generation of new protecting oxide at the substrate interface is offset by a concomitant depletion by dissolution at the B_2O_3-containing glassy phase interface. On the other hand, parabolic kinetics implies that the process is controlled by diffusion (of oxidant to inward to react with the substrate, or of an oxidation product outward to the environment) and the slowing down the oxide scale growth rate with time is associated with the monotonic increase of diffusion distance. While the foregoing is conceptually clear, its application to real systems is not straightforward. Luthra[12] has lighted the difficulty of making reliable distinctions between prevailing mechanisms for values of n between 0.5 and 1.0. For n values outside the interval $0.5 < n < 1.0$, simple analytical models do not exist. For example, cubic and quadratic kinetics (with $n = 1/3$ and $\frac{1}{4}$, respectively) are sometimes suggested, but those relationships are strictly empirical and do not correspond to any known physical processes in oxide growth. They are easily shown to reflect competition between mechanisms, especially when such mechanisms operate simultaneously in different strata of the oxide scale [5].

By reporting the logarithm of the mass variation in function of time, it is highlighted that n may take the value of 0.9 for short times and later 0.5 (Figure 13). Initially, the oxidation behavior can be represented by linear kinetic and after by a parabolic kinetic, according as growth is successively limited by a reaction (value very close to 1) or by diffusion

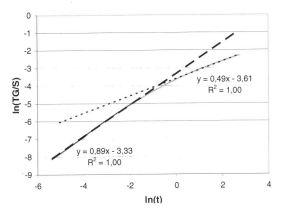

Figure 13. Logarithm of the mass variation in function of time, during oxidation of SiC fibers at 900°C under atmospheric pressure with $P_{O2} = 9.8$ kPa, during 15 h (v(gaz)=0,3 cm/s), deep in B_2O_3 (56,6%mass. de SiC)

The diffusion of oxygen up to fibers through the surrounding B_2O_3 is not a limiting phenomenon in the oxidation process; the oxidation rate, in presence of B_2O_3, is much quicker than those measured for fibers in a dry atmosphere. During experiment, the quantity of oxygen diffusing up to fibers through the surrounding B_2O_3 should remains almost constant since the thickness of B_2O_3 is not significantly modified.

Thus, in a first step, the oxidation rate of SiC fibers is mainly controlled by the dissolution rate of SiO_2 by B_2O_3 (interfacial phenomenon). In a second step, the thickening of a protective scale by generation of silica at the substrate interface allows getting a limiting diffusion of oxygen through this oxide layer formed at the fiber surface.

Scenario

The basis of a « model » can be established in order to predict the degradation of SiC fibers in presence of boron oxide. Two competitive phenomena have to be considered : (1) the dissolution of silica (formed by direct oxidation of SiC fibers) by boron oxide leading to the formation of a borosilicate and (2) the formation of a silica layer as soon as its formation rate is higher than its dissolution one (equivalent with a saturation in silica of the borosilicate). V_D represents the initial silica dissolution rate and V_F the initial silica formation rate (figure 14 a)).

Boron oxide is able to dissolve its own weight in SiC and a large amount of SiO_2 leading to the formation of a borosilicate [5, 7, 8]. Without the creation of a protective silica layer, the reaction remains interfacial.

With the increase of dissolved silica in function of time, the dissolution rate starts to be limited by diffusion of SiO_2 in B_2O_3 : transition between the linear part and the parabolic part in the Deal and Grove's model[18]. As soon as the borosilicate will be saturated in silica, a solid silica layer is formed between SiC fibers. Further the oxygen diffusion starts to be limited through the thin silica scale, according to its more efficient protection [13-15]. A simple calculation of oxygen concentration across a multilayered oxide scale (by a Fick diffusion process) shows that as soon as a silica layer is formed, the oxygen concentration in the surrounding B_2O_3 is almost constant, equal to the value imposed in the gas environment. A gradient of oxygen concentration is mainly present inside the silica layer at the fiber surface.

The thickness of the silica scale, able to be formed, is limited by the dissolution process and the consumption rate of SiC is more important (figure 14 b)). Further, a limited part of formed silica is dissolved and a growth of a protective layer is favored.

This SiO_2 layer can be formed earlier by replacing the surrounding boron oxide by a borosilicate or by increasing the oxygen partial pressure (the silica formation rate is enhanced).

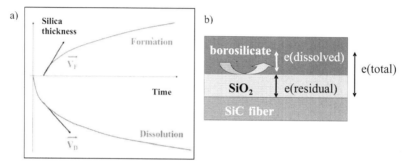

Figure 14. Representation of the dissolution "model" : a) competitive phenomena and b) multilayered oxide scale and different parts of silica involved

At a fixed temperature, the dissolution process can be considered as limited by the diffusion of silica in B_2O_3 far from the fiber surface. In a first approximation, a global average diffusion process can be considered across borosilicate, to understand and quantify the dissolution rate (the gradient of silica content and the associated one of SiO_2 diffusion coefficient are not yet considered since no satisfying data have been found).

In an established formation/dissolution silica phenomenon, a part of the global quantity of formed silica ($e(SiO_2$ total)) is dissolved ($e(SiO_2$ dissolved)) and the other one contributes to form the protective layer ($e(SiO_2$ residual)), (9).

$$e \ (SiO_2 \ total) = e(SiO_2 \ dissolved) + e(SiO_2 \ residual) \qquad (9)$$

$$\text{with } e^2(SiO_2 \ dissolved) = kd \times t \qquad (10)$$

(kd : parabolic kinetic constant, characterizing the SiO_2 diffusion in the surrounding borosilicate glassy phase, limiting the SiO_2 dissolution)

The global SiO_2 formation rate is only dependent on the thickness of the residual protective silica layer growing at the fiber surface (10) [18]. By introduction of expressions (9) and (10) in (11), a differential relation describing the total quantity of formed silica is obtained in function of the parabolic kinetic parameters, silica diffusion (k_d) and silica formation ($k_p(SiO_2)$). The logic solution of such an equation is the parabolic relation (13), with a constant parameter given by (14).

$$\frac{de(total)}{dt} = \frac{k_p(SiO_2)}{2 \times e(residual)} = \frac{k_p(SiO_2)}{2 \times (e(total) - e(dissolved))} = \frac{k_p(SiO_2)}{2 \times (e(total) - \sqrt{k_d}\sqrt{t})} \quad (11)$$

$$\frac{de(total)}{dt} \times \left(\frac{e(total)}{\sqrt{t}} - \sqrt{k_d} \right) = \frac{k_p(SiO_2)}{2 \times \sqrt{t}} \quad (12)$$

$$e(total) = a \times \sqrt{t} \quad (13)$$

$$\text{with } a = \frac{\sqrt{k_d} + \sqrt{k_d + 4k_p}}{2} \quad (14)$$

The final expression of the total formed silica, given by (15), has a parabolic shape, according to the evolution of weight gains during a well established oxidation process (characterized by a global parabolic constant, "K").

$$e(total)^2 = \left(\frac{\sqrt{k_d} + \sqrt{k_d + 4k_p}}{2} \right)^2 \times t = K \times t \quad (15)$$

The values of global parabolic kinetic parameters (K) are extracted from the experimental mass variation (Table 2). In a first step, if the boron oxide does not significantly modify the mechanism of silica formation and the efficiency of a protective silica layer, the parabolic kinetic parameters of silica formation ($k_p(SiO_2)$) remains similar to those measured in dry air (table 1). By imposing kp, values of parabolic kinetic parameters of silica diffusion, values of dissolution rate (k_d) are deduced: at different temperatures (Tables 2 and 3) and for different oxygen partial pressures (Table 4). The evolution of these values doesn't logically follow the variations of experimental parameters, according to the competition between complex phenomena (as illustrated through the relative position of weight gains curves in function of the environmental parameters (Figures 5-9)).

Table 2. Oxidation of SiC Hi-Nicalon fibers oxidized at 900°C, under atmospheric pressure with $P_{O_2} = 13$ kPa, under dry air and embedded in boron oxide (total flow 2 l.h^{-1}), in function of temperature.

	T (°C)	600	700	800	900	1000
fibres	kp (mg²/cm⁴.h)		7,93E-09	5,27E-08	2,07E-07	8,84E-07
	kp'(SiO₂) (m²/h)		2,12E-06	1,59E-05	6,49E-05	3,16E-04
	k'p(SiC) (m²/h)		8,57E-09	8,65E-08	4,12E-07	1,54E-06
fibres + B2O3	K (mg²/cm⁴.h)	2,93E-05	1,00E-04	3,76E-04	4,08E-04	
	K' (SiO₂) (m²/h)	5,52E-04	1,90E-03	7,10E-03	7,70E-03	
	k'p(SiC) (m²/h)	2,25E-04	7,70E-04	2,90E-03	3,10E-03	
	kd(SiO₂) (m²/h)	5,51E-04	1,88E-03	6,63E-03	7,45E-03	

Table 3. Values of oxidation kinetic rates of SiC Hi-Nicalon fibers oxidized, under atmospheric pressure with P_{O_2} = 1.2 kPa, under dry air and embedded in boron oxide (total flow 2 l.h^{-1}), in function of temperature.

	T (°C)	600	700	800	900
fibres + B_2O_3	K (mg²/cm⁴.h)	1,00E-05	8,16E-05	4,33E-04	3,32E-04
	k'p(SiO_2) (m²/h)	1,88E-04	1,54E-03	8,15E-03	6,25E-03
	k'p(SiC) (m²/h)	7,67E-05	6,26E-04	3,32E-03	2,55E-03
	kd(SiO_2) (m²/h)	1,88E-04	1,53E-03	8,11E-03	6,02E-03

Table 4. Values of oxidation kinetic rates of SiC Hi-Nicalon fibers oxidized at 900°C, under atmospheric pressure with different oxygen partial pressures and embedded in boron oxide (total flow 2 l.h^{-1}).

	T (°C)	13	9,8	6,4	3,1	1,2
fibres + B_2O_3	K (mg²/cm⁴.h)	4,08E-04	6,46E-04	1,04E-03	6,30E-03	3,32E-04
	k'p(SiO_2) (m²/h)	7,70E-03	1,22E-02	1,96E-02	1,19E-02	6,25E-03
	k'p(SiC) (m²/h)	7,45E-03	4,90E-03	7,97E-03	4,83E-03	2,55E-03
	kd(SiO_2) (m²/h)	7,45E-03	1,19E-02	1,94E-02	1,16E-02	6,02E-03

Table 5. Values of oxidation kinetic rates of SiC Hi-Nicalon fibers oxidized at 900°C, under atmospheric pressure with P_{O_2} = 13 kPa, under dry air and embedded in different borosilicate (total flow 2 l.h^{-1}).

	% massique B2O3	100	80	66	49
	% molaire B2O3	100	77,5	62,6	45,3
fibres + B_2O_3	K (mg²/cm⁴.h)	4,08E-04	2,28E-04	1,35E-04	2,52E-04
	k'p(SiO_2) (m²/h)	7,70E-03	4,29E-03	2,54E-03	4,74E-03
	k'p(SiC) (m²/h)	7,45E-03	1,75E-03	1,04E-03	1,93E-03
	kd(SiO_2) (m²/h)	7,45E-03	4,07E-03	2,32E-03	4,52E-03

In presence of B_2O_3 and oxygen, the consumption of SiC is often much more important than without B_2O_3, by a factor at least equal to 1000. This rate is higher for lower partial pressure since no protective SiO_2 layer is able to be formed (it is simultaneously dissolved). By increasing the silica content in borosilicate surrounding fibers, the oxidation rate is slowed since the SiO_2 dissolution rate is limited. However, a continuous dissolution (with respectively a sustained consumption of SiC) seems to always occur when borosilicate is saturated in silica. In this case, the dissolved SiO_2 at the fiber surface would precipitate far from this surface and this phenomenon may not stop.

CONCLUSION

The presence of boron oxide near SiC fibers leads to an increase of the fiber oxidation rate. The increase of the oxidation rate is due to the ability of B_2O_3 to dissolve SiO_2 depending on the wetting of SiO_2 by B_2O_3 This capability of boron oxide to dissolve silica is high. The thickness of the SiO_2 protective sale formed by oxidation of SiC becomes limited.

The kinetic laws of SiC fibers oxidation have been written for a specific configuration. On these kinetics, the effects of several parameters (P(O_2), the amount of B_2O_3 and the use of a borosilicate) have been highlighted and understood. The increase of the amount of B_2O_3 or its replacement by a borosilicate leads to a reduction of the SiC oxidation rate. On the other hand, the increase of the oxygen partial pressure results in a decrease of the oxidation rate as soon as the SiO_2 dissolution rate is lower than the formation one (for P(O_2)≥7kPa). This means that for very low levels of oxygen in a gaseous phase, the chemical degradation (respectively mechanical) of SiC material can become catastrophic : this behaviour may be compared to an active oxidation mechanism.

In this work, using the $\Delta m = k.t^n$ growth law allows providing a description of what going on during the oxidation of silica formers. The kinetic of SiC oxidation ceases to be linear when silica scale growth becomes significant. The kinetics is then best modelled with a combination of two parabolic terms, care being taken to ensure that the fitting coefficients agree with the physical processes underlying oxidation.

Now, in the modelling, improvements have to be mainly done on the definition of the coefficient diffusion of silica in borosilicate and its dependence on silica content in the generated borosilicate glassy phase (by taking into account the gradient of silica content in this glassy phase), in function of temperature.

REFERENCES
[1]F. Christin, Design, fabrication, and application of thermostructural composites (TCS) like C/C, C/SiC and SiC/SiC composites, Adv. Eng. Mat., **4(12)**, 903-912 (2002).
[2]R. Naslain, Fiber-matrix interphases and interfaces in ceramic matrix composites processed by CVI, Composite interface, **1(3)**, 253-286 (1993).
[3]L. Filipuzzi, Oxidation mechanisms and kinetics of 1D-SiC/C/SiC composite materials : I, An experimental approach, J. Am. Ceram. Soc., **77/2**, 459-466 (1994).
[4]W.G. Zhang, The effects of nanoparticulate SiC upon the oxidation behavior of C-SiC-B4C composites, Carbon, **36/11**, 1591-1595 (1998).
[5]L.U.J.T. Ogbuji, A pervasive mode of oxidative degradation in a SiC-SiC composites, J. Am. Ceram. Soc., **81/11**, 2777-2784 (1997).
[6]H. Hatta, High temperature crack sealant based on SiO2-B2O3 for SiC coating on carbon-carbon composites,Adv. Composite Mater, **12/2-3**, 93-106 (2003).
[7]L. Quemard, F. Rebillat, A. Guette, H. Tawil and C. Louchet-Pouillerie, Degradation Mechanisms of a SiC Fiber Reinforced Self-Sealing Matrix Composite in Simulated Combustor Environments, J. Eur. Ceram. Soc., **27**, Issue 1, 377-388 (2007).
[8]T.J. Rockett, Phase relations in the system boron oxide-silica, J. Am. Ceram. Soc. **82(7)**, 1817-1825 (1965)
[9]Takeda M. Sakamoto A. and Imai J., Microstructure and oxidation behavior of silicon carbide fibers derived from polycarbosilane. . J. Am. Ceram. Soc., **83(5)**, 1171-1176 (2000).
[10]T. Shimoo, F. Toyada, K. Okamura, Oxidation of oxidation of low-oxygen silicon carbide fiber. J. Mater. Sci, **35**, 3301-3306 (2000).
[11]K.G. Nickel, Corrosion of advanced ceramics: Measurements and modelling, Kluwer Academic Publishers, Boston, 59-71(1994).
[12]K.L. Luthra, A Mixed Interface Reaction/Diffusion Control Model for Oxidation of Si_3N_4, J. Electrochem. Soc. **138**, 3001-3007 (1991).
[13]J. Schlichting, Oxygen transport through silica surface layers on silicon-containing ceramic materials, High Temperatures-High Pressures, **14**, 717-724 (1982).
[14]J. Schlichting, Oxygen transport through glass layers formed by a gel process, J. of Non-Crystalline Solids, **63**, 173-181 (1984).
[15]D.W. McKee, Oxidation protection of carbon materials, Science of Carbon Materials Chap. 12 (2000)
[16]S. WERY, Etude de la réactivité de composites à matrice céramique à haute température, Phd thesis, University of perpignan via domitia,september (2008).
[17]Martin, X., Oxydation/corrosion de matériaux composites (SiCf / SiBCm) à matrice auto-cicatrisante, PhD thesis, Bordeaux I University, n°2749, (2003).
[18]B. E. Deal et A.S. Grove, General relationship for the thermal oxidation of silicon, *Journal of Applied Physic*, **36** [12] ,3770-78 (1965).

OVERVIEW ON THE SELF-SEALING PROCESS IN THE SIC$_F$/[SI,C,B]$_M$ COMPOSITES UNDER WET ATMOSPHERE AT HIGH TEMPERATURE

F. Rebillat, X. Martin, E. Garitte and A. Guette
University of Bordeaux, Laboratoire des Composites Thermostructuraux, UMR 5801 (CNRS-SAFRAN-CEA-Université Bordeaux 1),
3 Allée de la Boétie, 33600 Pessac, France,

ABSTRACT
    Ceramic Matrix Composites such as SiC/SiC are proposed for applications in the aeronautical and space domains because of their thermomechanical properties and their resistance to corrosive atmosphere. The matrix cracks represent pathway for oxygen diffusion up to fibers. The incorporation of boron bearing species such as B$_4$C in the matrix allows a self-healing process of these cracks from temperatures around 500°C by formation of B$_2$O$_3$, a liquid oxide. However, in presence of water vapour, boron oxide forms gaseous hydroxide species leading to a reduction of the self-healing capability.
Our objective was to understand the role of the borosilicate glassy phase, formed by simultaneous oxidation of the different matrix constituents, in the self-healing process of cracks across this multi-layered matrix. Finally, under moist air, the sealing appears as a compromise between a quick poorly stable boron-enriched oxide formation and a slower generation rate of a much more chemically stable boria-containing phase.

INTRODUCTION
    Ceramic Matrix Composites such as SiC/SiC or C/SiC composites are constituted by a SiC matrix reinforced by carbon or SiC fibres. They are currently used in the aeronautic and space fields because of their thermomechanical properties and resistance to corrosive atmosphere [1,2]. However, they can exhibit cracks due to their fabrication process (especially for C/SiC composites because of the mismatch of their thermal expansion coefficient [3,4] and/or to mechanical loadings. These matrix cracks represent pathways for oxygen diffusion to fibres and the interphase. Carbon fibers or the carbon interphase (PyC) undergo an active oxidation when exposed to oxygen or water vapour at temperature as low as 400°C [5,6]. Moreover, the formation of protective silica layer in a sufficient quantity due to the oxidation of the SiC matrix occurs only at temperature higher than 1000°C [7]. Therefore, this silica layer will be able to seal matrix crack only at temperature higher than 1000°C. It is consequently necessary to introduce in the matrix a ceramic material such as B$_4$C which is able to generate an oxide at lower temperature [8,9]. Thus, some studies have been made on SiC and B$_4$C mixtures in order to determine their resistance to oxidizing environment [10,11]. The oxidation resistance is ensured for a mixture of 10 wt% of B$_4$C and 8 wt% of SiC for temperature ranging between 800 and 1200°C under dry air by formation of a glassy borosilicate coating [10]. Guo et al. had pointed out that the composition of the mixtures (B/Si ratio) has to be adjusted depending on the exposure temperature [11]. In the literature, boron carbide is also currently used in order to protect carbon/carbon composites from oxidation in particular in multilayer systems [12-14]. Further, the air oxidation resistance of carbon/carbon composites is enhanced in the presence of boron oxide [15-17].
More recently, a multilayer Si-B-C ceramic matrix, constituted by a succession of B$_4$C and SiC layers, has been developed so as to provide self-healing properties to the new composite onto a wide range of temperature [8,18]. However, no detailed oxidation kinetic law has been written yet for such materials [19-21]. It clearly appears that B$_4$C exhibits interesting properties leading to the healing of matrix cracks.
The aim of this study is to understand the self-healing behaviour and predict the lifetime of this new multilayer SiBC ceramic matrix composite. Thus, it is necessary to progressively study : (i) the

oxidation behaviour of the different B$_4$C-contening components and (ii) the oxidation behaviour of this multi-layered matrix, in dry and wet atmospheres.

EXPERIMENTAL APPROACH
Monolithic materials
    In this work, the monolithic samples used are constituted by boron carbide coatings, noted B$_4$C, or SiBC phase deposited by a chemical vapor deposition (CVD) technique on silicon carbide pellets (diameter ≈ 8 mm and thickness ≈ 2 mm, BOOSTEC, purity ≈ 99.7%). The average thickness of the deposit is about 33 μm.
The composition of the boron carbide deposit has been determined from electron probe micro analysis (SX 100 from CAMECA France). The atomic percentage of carbon is higher than 0.2 (value of the highest atomic percentage of carbon of the "B$_4$C" existence domain). Thus the studied deposit exhibits a composition in the B$_4$C/C biphasic domain of the B-C binary diagram (Figure 1)[22].
The Si-B-C matrix layer can be described as a mixture of SiC nanocrystals in an amorphous B$_4$C phase. The B/Si ratio is much higher than 1.
Further, volatilization kinetic of borosilicate glasses can be separately measured from glassy samples. Vitreous borosilicate oxide is prepared from commercial boron oxide powder (Chempur, 99%, amorphous) and silica powder (Chempur, 99,9%, quartz α). After low heating at 300°C for 3 hours in order to dehydrate powders, the mixture is heated at 900°C for one hour, and then undergoes a quenching. The volatilization tests on borosilicate glassy phases allow pointing out the impact of the boron oxide volatilization on the boron carbide oxidation rate.

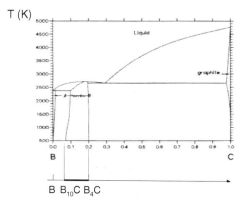

Figure 1. B-C binary diagram[22].

Composite materials
    Moreover, in order to study the role of each boron carbide-containing constituent in the self-healing mechanism of a composite, a planar fissure is made in the SiBC ceramic matrix composite (Figure 2). To get a controlled geometry of this single crack, the composite is cut perpendicularly to one principal fiber direction of the fabric and the two machined faces are polished. Finally, they are maintained in front of each other by a platinum thread. A planar model crack, in a real material, with a wideness of around (9±3) μm is obtained. Later, the assembling is placed in a furnace under oxidizing atmosphere and the weight changes are measured punctually.

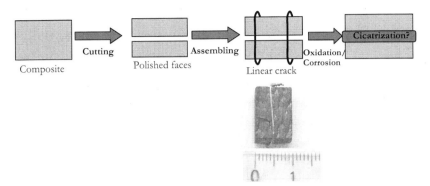

Figure 2. Preparation of a model linear matrix crack in a real composite.

Oxidation chemical reactions

In the SiBC ceramic matrix composite, boron carbide is very useful as it gives a liquid oxide (B_2O_3) when exposed at temperature upper than around 500°C according to the reactions 1, 2 and 3 [19,23,24]. The reaction 3 is dominating as soon as the oxygen partial pressure becomes too low.

$$\tfrac{1}{2}\, B_4C + 2\, O_2\, (g) \rightarrow B_2O_3\, (l) + \tfrac{1}{2}\, CO_2\, (g) \tag{1}$$
$$\tfrac{1}{2}\, B_4C + 7/4\, O_2\, (g) \rightarrow B_2O_3\, (l) + \tfrac{1}{2}\, CO\, (g) \tag{2}$$
$$\tfrac{1}{2}\, B_4C + 3/2\, O_2\, (g) \rightarrow B_2O_3\, (l) + \tfrac{1}{2}\, C\, (s) \tag{3}$$

This oxide exhibits interesting properties such as a low melting point (about 460°C) and a low viscosity so as to fill up a matrix microcrack [16,25]. However, the boron oxide will react with H_2O in order to form boric acid (noted $H_xB_yO_z$) according to the following equations, depending on the temperature [26,27].

$$3/2\, H_2O\, (g) + 3/2\, B_2O_3\, (l,\, g) \rightarrow (HBO_2)_3\, (g) \tag{4}$$
$$3/2\, H_2O\, (g) + 1/2\, B_2O_3\, (l,\, g) \rightarrow H_3BO_3\, (g) \tag{5}$$
$$1/2\, H_2O\, (g) + 1/2\, B_2O_3\, (l,\, g) \rightarrow HBO_2\, (g) \tag{6}$$

At low temperatures (T < 900°C), the predominant species are $(HBO_2)_3$ and H_3BO_3 whereas at high temperature (T > 900°C), HBO_2 is the only predominant species. This volatilization process leads to a reduction of the self-healing ability of this oxide

In high temperature environments, the SiC matrix layers can oxidize significantly under dry air above 1000°C to form a protective SiO_2 scale (equations 7-8).

$$SiC + {}^3\!/_2\, O_2 \rightarrow SiO_2 + CO_{(g)} \tag{7}$$
$$SiC + 2\, O_2 \rightarrow SiO_2 + CO_{2(g)} \tag{8}$$

In environments containing O_2 and H_2O, the formation of silica is dramatically enhanced. Under steam-containing environments, the SiO_2 scale may also volatilize, by formation of $Si(OH)_{4\,(g)}$ species[28] (equation 9).

$$SiO_{2\,(s)} + 2\, H_2O\,_{(g)} \rightarrow Si(OH)_{4\,(g)} \tag{9}$$

Finally, the silica content increases in a borosilicate, when exposed to moisture environment, due to a favoured B_2O_3 volatilization.

Oxidation apparatus

Several oxidation equipments are used in order to highlight the effect of the gas velocity, the oxygen partial pressure or the furnace geometry on the boron carbide oxidation rate.

On the one hand, the thermogravimetric analysis (TGA) is performed at atmospheric pressure using an apparatus (Setsys 16/18 provided by Setaram, France) equipped with a vertical alumina tube (inner tube diameter = 17.8 mm) allowing heating of the specimen up to 1600°C. The atmosphere consists of dry gas mixtures, with O_2 and N_2 flows controlled independently by mass flow meters. The oxygen partial pressure ($P(O_2)$) is regulated from 0 to 100 kPa and the nitrogen one is fixed by $100 - P(O_2)$. The total gas flow is fixed to around 2 l/h and the gas velocity is then 0.2±0.05 cm/s, in the cold furnace zone.

On the other hand, some oxidation tests are performed in a horizontal tubular furnace at atmospheric pressure under dry air (20 kPa of oxygen). The furnace is equipped with a silica tube with an inner diameter more than 1.5 times higher than the TGA equipment (29.7 mm against 17.8 mm). The samples were only weighed punctually with an accurate balance (0.01 mg). At fixed temperature, the oxidation tests are made in order to highlight the effect of : (i) the furnace geometry on the constituent oxidation rate (to verify that the work is done in a reactive domain and not in a diffusive one) and (ii) the gas velocity on the volatilization rate of borosilicate phase. Composites materials are mainly tested in this apparatus.

Oxidation tests

In order to establish the oxidation kinetic laws of these materials, the oxidation tests were performed by heating the sample (in a TGA or in a tubular furnace) under controlled gaseous environment. The pellet is placed in a silica crucible. All the tests were done at fixed temperatures ranging between 460 and 900°C under dry or wet oxidizing atmosphere (oxygen partial pressure ranging between 1.2 and 80 kPa and water between 0 to 20 kPa) while the heating and the cooling are made under argon. The mass variation is measured with time and expressed per unit of reactive surface area. The test duration depends on the temperature, but it does not exceed 72 hours at low temperatures and 20 hours at higher temperatures. These durations are fixed according to the oxidation progress in order to avoid any overflow of the liquid oxide out of the reactive surface.

Moreover, the borosilicate volatilization rate as is measured on a pure sample of glass. For the volatilization tests, the glassy phase is placed in a platinum crucible containing 5% of gold so as to decrease the wetting of the oxide (to limit the capillary rise profile). So, the exchange surface between boron oxide and atmosphere is close to the surface area of the crucible section. The boron oxide is initially dehydrated at about 300°C during one hour and then heated at temperature ranging between 600 and 1200°C. During an isothermal volatilisation test, the mass lost is measured and expressed per unit of reactive surface area of the glass in contact with the gas flow.

Material characterization

After oxidation tests, the oxide scales, formed at sample surfaces, are characterized by Raman Microspectrometry (Labram 10 spectrometer from Jobin Yvon, France, with a laser He-Ne ($\lambda = 632.8$ nm)), Auger Electron Spectroscopy (AES, MICROLAB VG 310-F). The Raman Microspectrometry, equipped with a confocal microscope, allows making analyses below the surface. This apparatus is used to point out the presence of precipitation at the B_4C/B_2O_3 interface (as free carbon). By AES, profiles of atomic composition in depth can be made using ionic etching, to highlight a composition gradient.

Further, X-Ray Photoelectron Spectroscopy (XPS, ESCALAB VG 220i-XL) is used to evaluate the composition of the oxide scales and mainly to identify the chemical bonds. Its main interest is to correctly detect light elements up to boron.

Moreover, the morphology of the oxidized surfaces is observed by Scanning Electron Microscopy (SEM, Hitachi S4500) in order to measure the thicknesses of each consumed constituent and formed oxide scale and the homogeneity (uniform repartition) of the oxide scale. These thicknesses are also measured from examination of polished cross sections using an optical microscope (Nikon Eclipse M600L).

SELF SEALING IN A DRY OXIDIZING ENVIRONMENT

Thermal stability of B$_2$O$_3$ under dry oxidizing environments

The weight changes due to the boron oxide volatilization under 20kPa of oxygen at several temperatures are measured. The linear weight losses are representative of an interfacial exchange between the atmosphere and B$_2$O$_3$. The linear volatilization rate constants k$_l$ (mg/cm².h) is obtained from the slopes of the straight lines. The values of volatilization rates are in agreement with the low B$_2$O$_3$ equilibrium partial pressures expected from thermodynamic calculations[19] and with other experimental works [23,29,30]. Thus, the boron oxide volatilization can be considered as insignificant at temperature lower than 1000°C. The boron oxide volatilization kinetic law (mg/cm².h) under dry atmosphere can be written for temperatures ranging from 800 to 1200°C, through equation (10).

$$k_l = 2,34.10^6.\exp(-\frac{178000}{R.T})$$
(10)

So, the further experimental weight changes measured during the oxidation tests on boron carbide are reasonably assumed to be exclusively attributed to the formation of boron oxide.

Oxidation behaviour of boron carbide under dry oxidizing environments

The oxidation is a thermally activated phenomenon. In order to write the oxidation kinetics laws, the activation energy (Ea, in kJ/mol) has to be determined from the slope of the logarithm of the B$_4$C oxidation rate constant, k$_p$, as a function of the reverse of the temperature for a constant oxygen partial pressure (Figure 3).

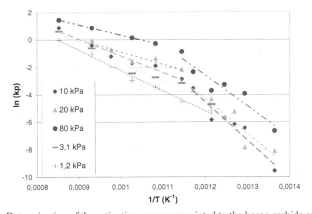

Figure 3. Determination of the activation energy associated to the boron carbide oxidation under dry atmosphere for several oxygen partial pressures from 2 to 80 kPa.

The change of slopes between 550 and 600°C depending on the oxygen partial pressure pointed out that two reactive domains have to be considered according to the temperature. The distinction between these two domains is in agreement with results from the chemical analyses. At temperature lower than 600°C, the B$_2$O$_3$ formation is accompanied by a total release of carbon, as no carbon is highlighted in the oxide scale by Raman micro-spectrometry (Figure 4) or Auger Electron Spectroscopy Analyses. On the contrary, at temperature higher or equal to 600°C, a thin carbon layer is formed between the oxide and the boron carbide and a gradient of carbon concentration is noticed in the oxide scale (with no carbon at the oxide surface) (Figure 4b).

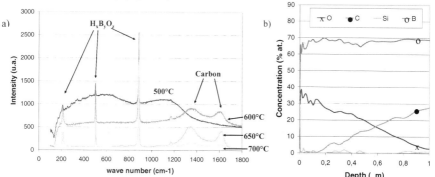

Figure 4. (a) Raman spectra (λ=632.8nm) of the boron oxide layers
formed by oxidation of B$_4$C coatings under 10 kPa of oxygen at several temperatures
and (b) Auger Electron Spectroscopy profiles in-depth of boron oxide coating formed by oxidation of
B$_4$C at 750°C under 10 kPa of oxygen during 30 minutes.

At temperatures lower than 600°C, boron oxide formation rate is very slow and all the oxygen reaching the surface of the deposit is not totally consumed; the oxidation of free carbon in the deposit is favoured (1 and. 2). The boron carbide oxidation kinetic law expressed in global weight change (mg^2/cm^4.h) (11) can be written for oxygen partial pressure ranging between 10 and 80kPa.

$$k_p = 1.4.10^7.\exp(-\frac{204000}{R.T}).P_{O2}^{0.9} \tag{11}$$

At temperatures higher or equal to 600°C (550°C for the lowest P(O2), the boron oxide formation rate is higher and allows to consume a large part of the diffusing oxygen. Comparatively, the carbon should be less reactive since its activity is decreased once surrounded by B$_2$O$_3$ in this temperature domain [31]. These experimental observations are in agreement with thermodynamic calculations on the (B, C, O) system [19]: for low oxygen partial pressures, free carbon is attempted to be in equilibrium with B$_2$O$_3$ and O$_2$. So, the chemical reaction operating under these experimental conditions is (12).

$$B_4C(s)+ y. C(s) + (3+(1+y-a)) O_2 (g) \rightarrow 2 B_2O_3(l) + a C(s) + (1+y-a) CO_2 (g) \tag{12}$$

with y (the initial amount of carbon in excess) and a (the amount of carbon remaining at the solid state), able to be higher than the amount of initial carbon y. For instance, a ≈ 0.34 on a B$_4$C deposit oxidized at 750°C under 10kPa of oxygen.

At temperature higher or equal to 600°C, the boron carbide oxidation kinetic law expressed in global weight change (mg^2/cm^4.h) (13) can be written for oxygen partial pressure ranging between 2 and 80kPa.

$$k_p = 58.\exp(-\frac{98000}{R.T}).P_{O2}^{0.67} \tag{13}$$

Lastly, these laws, whatever the domain of temperatures, allow estimating the boron carbide oxidation rate under dry atmosphere with less than 15% of dispersion (among all experimental data) in the large oxidizing domain explored.

Oxidation behaviour of SiBC phase under dry oxidising environments

Compared to B$_4$C, the oxidation of SiBC phase allows forming a borosilicate glassy phase, more thermally stable and viscous than B$_2$O$_3$. As the oxygen diffusion through this oxide scale is comparatively limited, the lower oxidation rate of SiBC phase is neatly slowed (Figure 5a). The oxidation behaviour is much more complex and the weight gain rate can be described in a first step by a parabolic evolution, followed by a linear one.
The composition of the formed borosilicate oxide scale is not constant during oxidation (Figure 5b). In the first step, the amount of boria in this borosilicate is high and progressively the content of SiO$_2$ increases up to the limit of solubility of silica in B$_2$O$_3$ (according to the liquidus curve in the B$_2$O$_3$-SiO$_2$ binary diagram[22]). Higher is the oxidation temperature, higher is the silica content in the borosilicate. Further, the silica content in the oxide becoming higher than the silicon content (proportionally to the boron one in the SiBC material), a depletion zone in silicon carbide is created between the oxide scale and the bulk. This feature is possible due to the high reactivity of SiC nanocrystals against oxygen, enhanced in presence of B$_2$O$_3$.

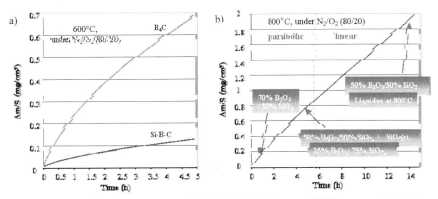

Figure 5. Weight changes measured by TAG : a) oxidation rate of SiBC compared to B$_4$C and b) following by AES of the borosilicate composition with time of oxidation.

A similar complex behaviour can be met during a selective oxidation of one of the elements in an alloy [33]. Further, this shape of thermo-gravimetric curve is also seen when the oxide scale is formed by two layers whom one is much less protective than the other (porous) [34]. A parabolic-linear model can help to explain this behaviour. Thus, a complex model has been developed to take into account this kind of parabolic-linear behaviour in material containing many constituents [35] (Figure 6). From the oxidation rates of each constituent (extracted from an experimental thermo-gravimetric curve), the consumed thickness of each constituent, in the SiBC phase, can be calculated and the depletion zone in silicon carbide highlighted. These values are in agreement with those measured by microscopy. Finally, the simulated curve is very similar to the experimental one.

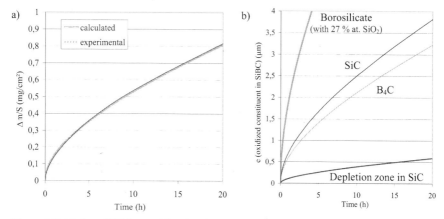

Figure 6. Oxidation of SiBC material at 650°C, at atmospheric pressure, with P$_{O2}$ = 10 kPa (v(gas) = 0.2 cm/s) : a) experimental thermo-gravimetric curve and calculated curve by a parabolic-linear model and b) calculated consumed thickness of each constituent in the SiBC phase.

Oxidation of composites under dry oxidising environments

In a SiC/SiBCcomposites, it is assumed that for temperature lower than 1000°C, the filling up of the crack is mainly ensured by the oxidation of the B$_4$C coating. Under dry atmosphere, the boron oxide volatilization is negligible for temperatures lower than 1000°C. So, the delay to fill up a matrix, crack of width "e", depends only on the B$_4$C oxidation rate. The precedent kinetic data may be used to predict the delay necessary to fill up a matrix crack by the oxide formed in a composite.

The time necessary to fill up a matrix crack decreases when the temperature is increased and the oxygen partial pressure is raised (figure 6). Thinner matrix cracks are sealed faster. For higher oxygen partial pressures or temperature higher than 600°C, wide cracks can be quickly sealed.

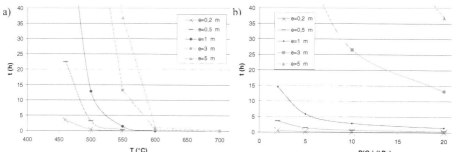

Figure 6. Effect of temperature (a) and of oxygen partial pressure (b) on the cicatrisation time estimated at 550°C under 20 kPa of oxygen (v(gas) = 0.2 cm/s).

Experimentally, oxidation tests are done under 20kPa of oxygen on a linear model crack in a real composite in order to check the validity of our simulations. The weight changes measured at 550 and 700°C with a gas velocity fixed to 0.2cm/s are reported in Figure 7. The weight losses observed at the beginning of the test are representative of a consumption of C interphase before the formation of an enough amount of B$_2$O$_3$ to seal the matrix crack. The healing occurs faster at 700°C as predicted by our simulations. These experimental required times to get a weight gain are in agreement with the calculated delay necessary to fill up a crack of 13 μm : about 2 hours at 700°C and more than 100 hours at 550°C (figure 6).

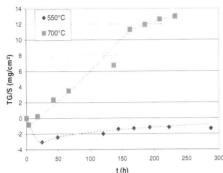

Figure 7. Effect of temperature on the global weight changes
measured under 20 kPa of oxygen (v(gas) = 0.2 cm/s), e = (13±5) μm)
on a planar model crack in a real composite.

SELF SEALING IN A WET OXIDIZING ENVIRONMENT
Thermal stability of borosilicate under moist environments

Important weight changes are measurable on boron oxide in a moisture environment for temperature lower than 500°C. The linear volatilization rate constants, k_l, (mg/cm^2.h) are obtained from the slopes of these straight lines. Few domains of volatilization mechanisms can to be considered depending on the main hydroxide species formed in function of the temperature[26,27] (4-6). The average boron oxide volatilization kinetic law (mg/cm^2.h) under different wet atmospheres (P$_{H_2O}$ ranging between 1.5 and 20kPa) can be written for temperatures ranging from 700 to 1200°C (14).

$$k_l = 0,05. \exp(-\frac{35000}{R \times T}) \times P(H_2O)^{0,9} \qquad (14)$$

With the dissolution of SiO$_2$ in B$_2$O$_3$, chemical interactions allow making B$_2$O$_3$ less reactive against H$_2$O. The increase of chemical stability of B$_2$O$_3$ is proportional to the SiO$_2$ content in borosilicate (Figure 8).

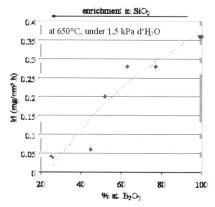

Figure 8. Volatilization rates of borosilicate in a moist environment
(v(gas) = 0.2 cm/s, in the cold zone of the furnace).

Oxidation behaviour of boron carbide under moist environments

Under wet atmosphere, two competitive phenomena have to be considered: (1) the coating undergoes an oxidation that leads to a weight gain and (2) the volatilization of this formed oxide by reaction with H_2O (g) leads to a weight loss. As these two phenomena occur simultaneously, a parabolic-linear model describes the mass variation [36]. Moreover, if the volatilization rate is initially higher, the oxidation regime is active [19,35].

On figure 9, the measured global weight changes are reported for the oxidation of B_4C under 1.5 kPa of water and for a gas velocity equal to 0.2 cm/s.

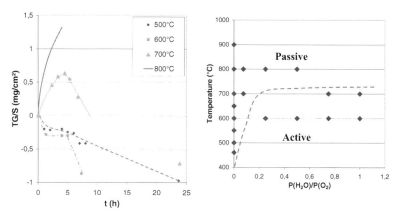

Figure 9. (a) Effect of the temperature on the global weight change measured during the B_4C oxidation under continuous flow (P(H_2O) = 1.5 kPa, P_{O_2} = 20 kPa, v(gaz) = 0.2 cm/s) and (b) Effect of the composition of the gaseous phase on the transition temperature between an active and a passive regime of B_4C oxidation (v(gas) = 0.2 cm/s, P_{O_2} = 20 kPa).

At temperatures lower than 650°C, the oxidation regime of B$_4$C coatings is active with a weight loss as soon as the beginning of the oxidation test. A passive regime is highlighted at temperatures higher than 700°C as in a first time a weight gain is observed (Figure 9a). The presence of an oxide layer at the surface of the coating contributes to a decrease of the oxidation rate up to equalize the volatilization rate of the oxide. Then, a linear weight loss is obtained and the coating recession rate is calculated from this part of the curve.

These results enable to determine the transition temperature between active and passive oxidation domains in function of the composition of the gaseous atmosphere and temperature (Figure 9b), for a gas velocity fixed to 0.2 cm/s. For water partial pressures lower or equal to 1.5 kPa, a passive oxidation is observed at temperatures at least higher than 700°C. On the contrary, for higher water partial pressures, a passive oxidation will be reached at temperatures higher than 800°C.

For a water partial pressure about 1.5kPa with the increase of the gas velocity by a factor of ten, the passive oxidation domain is shifted toward the upper temperature (T ≥ 700°C). On the contrary, under a similar water partial pressure, by decreasing the gas velocity (free convection), a passive oxidation can be obtained at temperatures as low as 600°C.

Under wet atmosphere, two oxidizing species may be considered: O$_2$ and H$_2$O. It is assumed that the oxidation phenomenon is mainly due to the action of O$_2$ whereas the oxide volatilization is related to H$_2$O. In order to verify this assumption, the mass variation in a moist oxidizing atmosphere can be respectively calculated by summing: (i) the weight gain, due to the oxide formation, obtained from the parabolic oxidation rate constant k$_p$ (determined under dry atmosphere) and (ii) the weight lost obtained from the linear volatilization rate constant k$_l$ (measured under wet atmosphere). The simulated curves are similar to the experimental results (Figure 10). So, oxygen is mainly responsible for the oxidation of the boron carbide whereas the oxide volatilization is due to water.

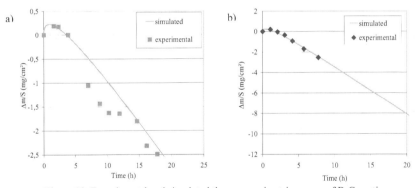

Figure 10. Experimental and simulated thermo gravimetric curves of B$_4$C coatings oxidized at 700°C, under P$_{H2O}$ = 1,5 kPa and P$_{O2}$ = 20 kPa : a) v(gas)= 0.2 cm/s and b) v(gas) = 2cm/s (in the cold zone of the furnace).

Oxidation behaviour of SiBC phase under moist environments

During the oxidation of SiBC, in presence of moisture, the SiO$_2$ content in the borosilicate oxide scale formed rises with the B$_2$O$_3$volatilization.

At low temperature and gas velocity, this quantity of silica being initially small, a large part of B$_2$O$_3$ is volatized and a higher weight lost rate is measured (Figure 11). With the increasing temperature, the initial quantity of SiO$_2$ present in the borosilicate is more important, and its enrichment is significantly favoured by the volatilization of B$_2$O$_3$. This enrichment leads to the formation a more protective oxide scale against oxidation and more stable against moisture.

With high gas velocities and increased volatilization rates, these behaviours are reversed with : (i) a rapid protective oxide scale formation at low temperature (the oxide formation rate being limited) and (ii) a higher weight lost rate at higher temperature (the quantity of oxide available to volatize being important).

Analyses of oxide compositions and observations on cross section of formed oxide scales revealed that a silica layer grows at the interface borosilicate/SiBC coating (Figure 12). With the B$_2$O$_3$ volatilization, the SiO$_2$ content in the borosilicate, close to surface, may exceed the limit of SiO$_2$ solubility in B$_2$O$_3$, and provokes a SiO$_2$precipitation.

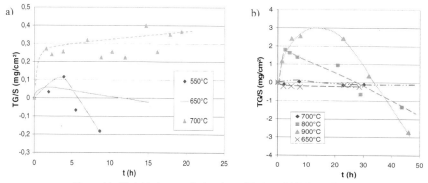

Figure 11. Weight changes during the oxidation of SiBC coatings
under P$_{H2O}$ = 1,5 kPa and P$_{O2}$ = 20 kPa : a) v(gas) = 0.2 cm/s and b) v(gas)= 2 cm/s
(in the cold zone of the furnace).

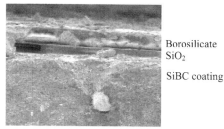

Borosilicate
SiO$_2$

SiBC coating

Figure 12. Observation of a fractured cross section of Si-B-C coating oxidized at 650°C at atmospheric pressure under P$_{H2O}$ = 1,5 kPa and P$_{O2}$ =20 kPa (v(gas) = 0.2 cm/s, in the cold zone of the furnace).

Oxidation of composites under moist environments
    In a composite with a self-sealing matrix, mainly B$_2$O$_3$ formed by oxidation of B$_4$C fills up a matrix crack in a first time. Then, the silica content progressively increases, since the oxidation rate of SiBC is much slower.

It is noticeable that in matrix microcracks, the gas atmosphere is confined and it has been verified that the B$_2$O$_3$ evacuation outside of this crack is limited by the convection around the sample. The oxide formation flow rate depends on the surface are of B$_4$C coating (P$_{O2}$ is considered similar to the outside).

Respectively, the volatilization flow rate depends on the width of the matrix crack, "e", and the gas velocity (this evacuation rate of H$_x$B$_y$O$_z$ outside of the crack is lower the volatilization rate measured directly on a glassy sample, k$_1$). By the comparison of these both flows rates, it is possible to determine the environmental conditions and temperature under which the composite is able to self-heal[35].

The calculated required delay to fill up matrix cracks with variable widths ranging between 0.2 and 5µm under 1.5kPa of H$_2$O (v(gas)=0.2cm/s) are reported in Figure 13 for several temperatures. The required time to fill up a matrix crack decreases as the temperature is increased: the thinner the matrix cracks, the faster the cicatrisation.

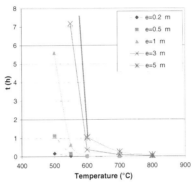

Figure 13. Effect of temperature on the cicatrisation time
estimated under 1.5 kPa of water (P$_{O_2}$= 20kPa, v(gas) = 0.2cm/s).

Experimental tests on linear model cracks

Oxidation tests are done on a linear model crack in a real composite in order to check the validity of our simulations under 1.5kPa of water and 20kPa of oxygen. Depending on the preparation, the crack width is about (9±3)µm. The weight changes measured at 550 and 700°C with a gas velocity fixed to 0.2cm/s are reported in Figure 14.

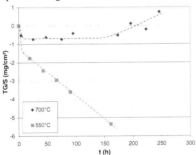

Figure 14: Effect of temperature on the global weight changes measured
under 1.5 kPa of water (v(gas) = 0.2 cm/s, e = (9±3) µm) on a linear model crack

From the previous simulation, for a crack of 8μm at 700°C, the healing should take place in less than one hour and is never reached at 550°C. These results are in agreement with the increase of the mass variations at 700°C and the linear weight loss at 550°C on figure 14.

The weight losses observed at the beginning of the test are representative of a consumption of C interphase and the volatilization of a large amount of B_2O_3 before the complete sealing of the crack entrance. The increase of the weight gain observed at 700°C is linked to a decrease of the B_2O_3 volatilization rate by the sealing of the matrix crack and the higher quantity of silica in the oxide (more stable). Moreover, the matrix crack sealing is delayed by the increase of the gas velocity (increase of the B_2O_3 volatilization rate).

CONCLUSION

This work leads to a better understanding of the boron carbide oxidation mechanisms under dry atmosphere (oxygen partial pressure ranging between 2 and 80kPa and with different ratio of moisture) at temperature between 460 and at least 900°C. The oxidation mechanisms of different matrix constituents could be described.

In a dry environment, the boron oxide volatilization remains negligible as temperature remains lower than 1000°C. For B_4C, two oxidation mechanisms have to be considered depending on temperature. At temperature lower than 600°C, the oxidation of boron carbide is accompanied by a total release of carbon initially present in the deposit. However, at temperature higher or equal to 600°C, a part of this carbon remains at the solid state during the boron carbide oxidation under dry atmosphere. The previous estimated oxidation kinetic laws allow to predict the delay necessary to fill up a matrix crack under dry atmosphere. The reliability of our simulation has been verified through the good agreement between the experimental results on linear model cracks in a real composite and the estimated cicatrisation times, at 550 and 700°C.

In a wet atmosphere, an important volatilization of boron oxide occurs. An active oxidation of B_4C can be observed for temperature lower than 700°C, depending on P_{H2O} and the gas velocity. The passive oxidation domain is shifted at temperature higher or equal to 800°C for a high gas velocity. Moreover, in the global oxidation mechanism, it has been proved that the oxygen is mainly responsible of the B_4C oxidation while the oxide volatilization is due to reaction with moisture. In a matrix microcrack, the boron oxide volatilization will be limited and the transition between an active and a passive oxidation of B_4C will be observed at a lower temperature. The reliability of the oxidation kinetic laws determined on monolithic ceramic has been verified with oxidation of planar model cracks in a real composite at 550 and 700°C under 1.5kPa of water and a gas velocity equal to 0.2cm/s. There is a good agreement between our experimental results and the estimated sealing times. Boron carbide is able to ensure the healing of cracks in a SiC/SiC composite at temperature higher or equal to 700°C. At these temperatures, the progressive increase of the silica content in the borosilicate allows improving the chemical stability of the healing glassy phase against moisture.

REFERENCES
[1]G. Boitier and J.L. Chermant, Les composites à matrices céramiques, Lettres des Sciences Chimiques du CNRS, 23-28 (march 2002).
[2]R. Naslain and F. Christin, MRS bulletin, 654-658 (September 2003).
[3]L. Cheng, Y. Xu, L. Zhang, and D. Wang, Thermal expansion and oxidation of 3D C/SiC composites with different coatings from room temperature to 1400°C, Sc. And Eng. of Composite Materials, **10** [5], 377-382 (2002).
[4]Q. Zhang, L. Cheng, L. Zhang and Y. Xu, Thermal expansion behavior of carbon fiber reinforced chemical-vapor-infiltrated silicon carbide composites from room temperature to 1400°C, Materials Letters, **60** [27], 3245-3247 (2006).

[5] S.M. Gee and J.A. Little, Oxidation behaviour and protection of carbon/carbon composites, J. Mat. Sci., **26**, 1093-1100 (1991).

[6] D.W. McKee, Oxidation behaviour and protection of carbon/carbon composites, Carbon, **25** [4], 551-557 (1987).

[7] L. Filipuzzi, G. Camus and R. Naslain, Oxidation mechanisms and kinetics of 1D-SiC/C/SiC composite materials: I, An experimental approach, J. Am. Ceram. Soc., **77** [2], 459-466 (1994).

[8] L. Vandenbulcke & S. Goujard, Multilayer systems based on B, B$_4$C, SiC and SiBC for environmental composite protection, Progress in Advanced Materials and Mechanics, 1198-1205 (1996).

[9] F. Lamouroux, S. Bertrand, R. Pailler, R. Naslain, & M. Cataldi, Oxidation-resistant carbon-fiber-reinforced ceramic-matrix composites, Composites Science and Technology, **59**, 1073-1085 (1999).

[10] Z. Fan, T. Wei, J. Shi, G. Zai, J. Song, L. Liu, J. Li and J. Chen, New route for preparation of SiC-B4C/C composite with excellent oxidation resistance up to 1400°C, J. Materials Science Letters, **22**, 213-215 (2003).

[11] Q. Guo, J. Song, L. Liu and B. Zhang, Relationship between oxidation resistance and structure of B4C-SiC/C composites with self-healing properties, Carbon, **37**, 33-40 (1999).

[12] F.J. Buchanan, and J.A. Little, Glass sealants for carbon-carbon composites, J. Materials Science, **28**, 2324-2330 (1993).

[13] C.A.A. Cairo, M.L.A. Graça, C.R.M. Silva and J.C. Bressiani, Functionally gradient ceramic coating for carbon-carbon antioxidation protection, J. European Ceramic Society, **21**, 325-329 (2001).

[14] J.W. Fergus, and W.L., Worrell, Silicon-carbide/boron containing coatings for the oxidation protection of graphite, Carbon, **33** [4], 537-543 (1995).

[15] D.W. McKee, Borate treatment of carbon fibers and carbon/carbon composites for improved oxidation resistance, Carbon, **24** [6], 737-741 (1986).

[16] T. Sogabe, O. Okada, K. Kuroda, and M. Inagaki, Improvement in properties and air oxidation resistance of carbon materials by boron oxide impregnation, Carbon, **35** [1], 67-72 (1997).

[17] D.W. McKee, C.L. Spiro and E.J. Lamby, The effects of boron additives on the oxidation behavior of carbons, Carbon, **22** [6], 507-511 (1984).

[18] S. Goujard, L. Vandenbulcke & H. Tawil, Oxidation behaviour of 2D and 3D carbon/carbon thermostructural materials protected by CVD polylayers coatings, Thin Solids Films, **252**, 120-130 (1994).

[19] X. Martin, Oxydation/corrosion de matériaux composites (SiCf / SiBCm) à matrice auto-cicatrisante, PhD thesis, Bordeaux I University, n°2749, 2003

[20] L.W. Litz, and R.A. Mercuri, Oxidation of boron carbide by air, water, and air-water mixtures at elevated temperatures, J. of Electrochem. Soc., **110** [8], 921-925 (1963).

[21] J.P. Viricelle, P. Goursat, and D. Bahloul-Hourlier, Oxidation behaviour of a boron carbide based material in dry and wet oxygen, J. of Thermal Analysis and Calorimetry, **63**, 507-515 (2001).

[22] H. Okamoto, B-C (boron-carbon), J. oh Phase Equilibria, **13** [4], 436 (1992).

[23] T. Piquero, H. Vincent, C. Vincent and J. Bouix, Influence of carbide coatings on the oxidation behavior of carbon fibers, Carbon, **33** [4], 455-467 (1995).

[24] R. Naslain, A. Guette, F. Rebillat, R. Pailler, F. Langlais and X. Bourrat, Boron-bearing species in ceramic matrix composites for long-term aerospace applications, J. of Solid State Chemistry, **177** [2], 449-456 (2004).

[25] R. Naslain, A. Guette, F. Rebillat, S. Le Gallet, F. Lamouroux, L. Filipuzzi and C. Louchet, Oxidation mechanisms and kinetics of SiC-matrix composites and their constituents, J. of Materials Science, **39**, 7303-7316 (2004).

[26] K. Kobayashi, Formation and oxidation resistance of the coating formed on carbon material composed of B4C-SiC powders, Carbon, **33** [4], 397-403 (1995).

[27] N.S. Jacobson, High-temperature oxidation of boron nitride : II, Boron nitride layers in composites, J. Am. Ceram. Soc., **82** [6], 1473-1482 (1999).

[28]E.J. Opila, Oxidation and volatilisation of silica formers in water vapors, J. Am. Ceram. Soc., **86** [8], 1238-1248 (2003).

[29]V.A. Lavrenko, A.P. Pomytkin, P.S. Kislyj and B.L. Grabchuk, Kinetics of high-temperature oxidation of boron carbide in oxygen, Oxidation of Metals, **10** [2], 85-95 (1976).

[30]W.G. Fahrenholtz, The ZrB2 volatility diagram, J. Am. Ceram. Soc., **88** [12], 3509-3512 (2005).

[31]P. Ehrburger, P. Baranne and J. Lahaye, Inhibition of the oxidation of carbon-carbon composite by boron oxide, Carbon, **24** [4], 495-499 (1986).

[32]T.J. Rockett, Phase relations in the system boron oxide-silica, J. Am. Ceram. Soc., **82** [7], 1817-1825 (1999).

[33]D. Landolt, Traité des Matériaux 12 : Corrosion et chimie de surface des métaux, Alden Press Edité par Presses Polytechniques et Universitaires Romandes (1997).

[34]P. Sarrazin, A. Galerie & J. Fouletier, Les mécanismes de corrosion sèche, Edité par EDP SCIENCES, part 5, (2000).

[35]E. Garitte, Etude de l'oxydation/corrosion des composites céramiques, PhD thesis, Bordeaux I University, n°3484, (2007).

[36]C. S., Tedmon, The Effect of Oxide Volatilisation on the Oxidation Kinetics of Cr and fe-Cr Alloys, J. electrochem. Soc., **113** [8], 766-768 (1966).

NDE FOR CHARACTERIZING OXIDATION DAMAGE IN REINFORCED CARBON-CARBON

Don J. Roth[*]
National Aeronautics and Space Administration
Glenn Research Center
Cleveland, Ohio, United States

Richard W. Rauser
University of Toledo
Toledo, Ohio, United States

Nathan S. Jacobson
National Aeronautics and Space Administration
Glenn Research Center
Cleveland, Ohio, United States

Russell A. Wincheski
National Aeronautics and Space Administration
Langley Research Center
Hampton, Virginia, United States

James L. Walker
National Aeronautics and Space Administration
Marshall Space Flight Center
Marshall Space Flight Center, Alabama, United States

Laura A. Cosgriff
Cleveland State University
Cleveland, Ohio, United States

ABSTRACT

In this study, coated reinforced carbon-carbon (RCC) samples of similar structure and composition as that from the NASA space shuttle orbiter's thermal protection system were fabricated with slots in their coating simulating craze cracks. These specimens were used to study oxidation damage detection and characterization using NDE methods. These specimens were heat treated in air at 1143 and 1200 °C to create cavities in the carbon substrate underneath the coating as oxygen reacted with the carbon and resulted in its consumption. The cavities varied in diameter from approximately 1 to 3 mm. Single-sided NDE methods were used since they might be practical for on-wing inspection, while X-ray micro-computed tomography (CT) was used to measure cavity sizes in order to validate oxidation models under development for carbon-carbon materials. An RCC sample having a naturally-cracked coating and subsequent oxidation damage was also studied with X-ray micro-CT. This effort is a follow-on study to one that characterized NDE methods for assessing oxidation damage in an RCC sample with drilled holes in the coating. The results of that study are briefly reviewed in this article as well. Additionally, a short discussion on the future role of simulation to aid in these studies is provided.

[*]Corresponding author: Phone: 216–433–6017, Fax: 216–977–7150, E-mail: donald.j.roth@nasa.gov

INTRODUCTION

Reinforced carbon-carbon (RCC) with a silicon carbide coating for oxidation resistance is used on the NASA Space Shuttle Orbiter's wing leading edge, nose cap, and arrowhead attachment point to the external tank for thermal protection during re-entry. The strength and light weight of RCC make it an ideal aerospace material; however, oxidation is a major concern. Oxidation damage to RCC can occur if the silicon carbide coating is itself damaged but still intact such that hot gases have access to the carbon beneath the coating.[1] In such cases, it is critical to evaluate the extent of the oxidation damage underneath the intact SiC coating. Even small breaches in the RCC coating system have recently been identified as potentially serious. In a prior study, small breaches in the coating of an RCC sample were created by drilling holes followed by oxidation of the sample and subsequent nondestructive (NDE) to characterize oxidation damage.[2] In that study, RCC samples were oxidized to create approximately hemi-spherical holes underneath the silicon carbide coating and subsequently inspected using various nondestructive evaluation methods. In the current study, small breaches were created by machining slots of various widths to simulate cracks of various sizes in the coating, and NDE was again subsequently used to characterize the oxidation damage.

The NDE techniques employed in this study included state-of-the-art backscatter x-ray (BSX), ultrasonic guided waves, eddy current (EC), and thermographic methods. All of these methods are single-sided techniques thereby lending themselves to practical inspections of components only accessible from one side. Samples were also inspected with x-ray micro-computed tomography (CT) to evaluate the true dimensions and morphology of the holes, as well as natural crack formations.[3] The controlled oxidation damage provides standards for investigating the effectiveness of various nondestructive evaluation techniques for detecting and sizing oxidation damage in this material. NASA Glenn Research Center led this investigation that had some of the top NDE specialists/facilities from NASA Glenn, NASA Langley, and NASA Marshall inspect these samples with the various NDE methods. The results of this study are also discussed in reference 4.

EXPERIMENTAL

RCC Material

Figure 1 is a schematic of reinforced carbon/carbon (RCC) with a Silicon Carbide (SiC) conversion coating. Briefly, this material is made with a two-dimensional layup of carbon-carbon fabric with repeated applications of a liquid carbon precursor to fill voids. An oxidation protection system is based on a SiC conversion coating. Because of the difference in coefficient of thermal expansion (CTE) of the SiC coating and carbon/carbon substrate, the SiC coating shrinks more than the underlying carbon/carbon on cooldown from the coating application temperature. This leads to vertical cracks in the coating, and these cracks are pathways for oxygen to reach the carbon/carbon substrate. Actual RCC used on the Space Shuttle Orbiter is infiltrated with tetraethyl orthosilicate (TEOS) which decomposes on a mild heat treatment to silica. Then the RCC surfaces are painted with a sodium silicate glass, which melts and seals cracks on re-entry. All the samples used in this study had the TEOS treatment. In addition, one of the samples studied had the sodium silicate glass. The samples were all approximately 5 mm thick.

The sample with SiC plus glass coating (RCC1) was a flat plate with an approximately 1 mm thick SiC plus glass coating and coated on all sides. The plate had an artificial craze crack of linear geometry made with a diamond blade (Keen Kut Products, Hayward, CA) of 0.25 mm thickness. This plate was used for ultrasonic studies. The slot was cut to the SiC plus glass coating/carbon-carbon interface on one side of the sample.

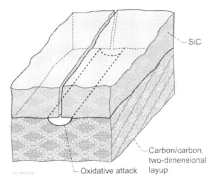

Figure 1. Schematic of SiC-protected carbon/carbon used in this study. Reprinted with permission from Elsevier.

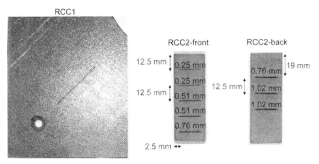

Figure 2. Photographs of machined RCC samples with machined slots. Nominal widths of slots are shown above the slot and spacing between slots and from edges is also shown.

The samples with only the SiC coating had an approximately 1 mm thick coating and were coated on all sides. Of these, one was a plate. The plate (RCC2) had machined slots made with diamond blades of 0.25, 0.51, 0.76, and 1.02 mm thicknesses. These slots were cut to the SiC/carbon-carbon interface on both sides of the plate ample. These slotted specimens are shown in figure 2.

Other SiC-only coated samples included several flat 1.91 cm diameter and 1.52 cm thick disks.[3] Some of these had slots machined in them; others were used in their as-fabricated form with the naturally-occurring craze cracks acting as paths for oxidation. Polishing ~300 μm of SiC off the surface revealed the cellular crack pattern as shown in figure 3(a) together with a 'skeleton' trace of the cracks in figure 3(b).

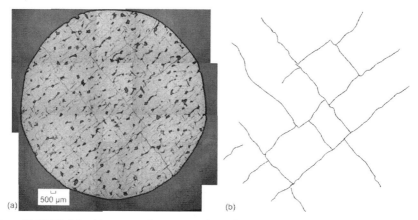

Figure 3(a) RCC disk with craze crack pattern on the surface. (b) 'Skeleton' trace of crack pattern. Reprinted with permission form Elsevier.

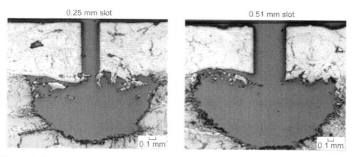

Figure 4. Cross-sections of cavity resulting from oxidation treatment for the slotted disks with an oxidation treatment for 0.5 hr in 1200 °C static laboratory air.

Controlled laboratory oxidation exposures were performed as follows: (a) Plates and disks with machined slots: 0.5 hr at 1200 °C in static laboratory air; (b) Disk with craze cracks only: 2.5 hr in bottle flowing air at 1143 °C. The carbon/carbon substrate below the machined slot oxidized to form an approximate hemispherical void (fig. 4). This uniform attack pattern indicates diffusion control; and reference 3 describes an oxidation model for given slot and crack breaches in the coating. The resulting size of the void would be expected to increase with increasing slot width. Depth and diameter of voids in this study tended to be on the order of 1 to 3 mm. This controlled oxidation damage provides standards for investigating the effectiveness of various nondestructive evaluation techniques for detecting and sizing this type of oxidation damage in this material.

Nondestructive Evaluation (NDE)

Various single-sided state-of-the-art NDE methods were used to characterize the RCC samples prior to (baseline condition), and after, oxidation. These methods included backscatter x-ray (BSX), eddy current (EC), thermography, and guided wave ultrasonics. X-ray micro-computed tomography (CT) was used to size oxidation damage and cracking, and develop three-dimensional volumetric visualizations. References 5 to 9 provide basic principles for the various methods, and experimental parameters for the methods are given here. Specialized image processing was employed as needed to highlight indications from the NDE methods using software developed at NASA and available in the public domain.[10] The software processing generally consisted of the following operations (in the following sequence): image crop, contrast expansion, outlier (bad value) removal, wavelet denoise. If the processing was applied, it was applied identically to the images of pre- and post-oxidation NDE images.

Backscatter X-ray

A scanning backscatter x-ray (BSX) system was used to image the RCC samples.[5] Several parameter settings were employed to determine optimized conditions but for the purposes of comparing prior and post-oxidation images, the following settings were used in those cases. The settings were aperture of 2 mm, voltage of 55 kV, current of 11.6 mA, focal spot size of 1 mm, finned collimator angle of 90° and exposure time per scan position of 0.2 sec. The scan parameters included scan (X) and step (Y) increments of 0.5 mm and scan velocity of 2.5 mm/sec. Scans were performed from both sides of the RCC2 sample.

Eddy Current

Scanning eddy current (EC) inspection of the RCC sample was accomplished using a high frequency eddy current surface probe connected through a spring loaded z-axis gimbal to an x/y scanning system. In the present work a probe drive frequency of 5 MHz and scan spacing of 0.025 in. were used. Calibration of the system is performed using an uncoated RCC test article of nominally matching conductivity to the part under test. The probe response is rotated such that lift-off is in the horizontal direction. Nonconductive plastic shims are then used to measure the nominal SiC coating thickness (lift-off) and lift-off sensitivity. Oxidation damage under the SiC coating is measured as a localized increase in lift-off due to the increased spacing between the sensor and the conducting carbon-carbon substrate in the oxidized areas. Scans were performed from both sides of the sample. The differences in conductivity/lift-off values are displayed in various shades of gray in the c-scan image. The system used in this study minimized edge effects as compared to the one employed in reference 2.

Thermography

A pulsed full-field thermographic NDE method utilizing flash lamps and a high speed camera was used to obtain images of the RCC sample.[7] The system consists of two high energy xenon flash lamps, each capable of producing a 1.8 kJ flash with a 5 msec duration. The flash lamps were placed at locations that provide a relatively uniform distribution of heat across the surface of the specimen. The transient thermal response of the specimen after flashing was captured using high-speed infrared camera. The camera used in the study is a 640 by 512 InSb focal plane array type with a 14 bit dynamic range. The camera operates in the 3 to 5 μm wavelength range and is capable of capturing thermal data at rates of 30 Hz for the full array size. For this study, a 320 by 256 portion of the full array was utilized in order to increase the frame rate to approximately 60 Hz. Flash initiation, data collection, storage and processing were all performed using software on the acquisition computer. Experimental data was collected using the following procedure. The specimen was placed in front of the IR camera at a distance that allowed the sample to fill most of the active focal plane and then focused. Flash lamps were set at a distance of approximately 300 mm. from the sample at an angle of 45°. Along with the images captured after the flash, 6 preflash images were collected. Instantaneous and derivative images (from relative temperature vs. time) were obtained, and the operator normally selected the best images for analysis using a subjective process of selecting frames of maximum contrast. Thermography was performed from both sides of the sample. The differences in surface temperature are displayed in various shades of gray in the image.

Ultrasonic Guided Waves

An ultrasonic guided wave measurement system[8] was employed to determine whether total ultrasonic energy (M_0) of the time domain guided waveform was altered by the addition of the slot (artificial crack) and oxidation of the RCC1 sample containing the single slot. M_0 is calculated from the area under the curve of the power spectral density $S(f)$ of the time domain waveform according to:

$$M_0 = \int_{f_{low}}^{f_{high}} S(f)\,df \qquad (1)$$

where f_{low} and f_{high} are the lower and upper frequency (f) bounds of the integration range, respectively. Total energy is a physically understandable parameter that would likely be altered due to ultrasonic scattering both by the addition of the slot into the ultrasonic path and after further alterations of the slot due to oxidation and glass filling. Broadband ultrasonic transducers were used with center frequencies of approximately 1 MHz (both sender and receiver were of the same frequency). Multiple mode excitation is likely due to the use of broadband transducers and the existence of multiple plate wave modes is confirmed by the complicated nature of the signal.[8] Ultrasound was coupled to the material via elastic coupling pads. The distance between sending and receiving transducers was 2.5 cm. For the baseline (prior to slotting) condition of the RCC1 sample (fig. 2), the transducers were positioned so that they would straddle the future position of the slot, and then after slotting, they were positioned identically such that they straddled the slot.

Analog-to-digital sampling rate for the ultrasonic testing was 10 MHz. A measurement was made (contact load = 3.63 + 0.23 kg [8 + 0.5 lb]), the sender-receiver pair was lifted, moved to the next location, lowered to be in contact with the sample, and another measurement made. 20 measurements were obtained (2 columns by 10 rows) with measurements separated by 1 mm, and mean and standard deviation of (M_0) was calculated. The identical pattern of measurements was made prior to slotting, after slotting, after oxidation, and after removing glass sealant from the crack followed by a second oxidation. Additionally, the final scan was done 5 times to measure reproducibility of the technique. In a future investigation, the effect of the slot and oxidation on other ultrasonic parameters that can be derived from broadband ultrasonic guided wave signals will be considered.[8,10]

X-ray Computed Tomography

X-ray computed tomography (CT) was used to provide additional images of the oxidation damage and study oxidation in the naturally-cracked RCC sample, without destructive sectioning.[9] This SmartScan Model 100 (CITA Systems, Inc., Pueblo, CO) system utilizes a Feinfocus FXE–160 (COMET AG, Flamatt, Switzerland) microfocus x-ray source to produce very high resolution imaging of samples, approaching 0.025 mm, in the CT mode of operation. The major hardware components of this system included the x-ray source, an area detector system, a five-axis object positioning subassembly, and a lead-lined radiation cabinet. A dual-processor computer system controlled the data acquisition and image processing. The slice plane thickness was 0.120 mm per slice for these samples. Putting together slices electronically gave a three-dimensional view of oxidation damage. The sample was placed on a micropositioner between the source and detector which allowed positioning and rotation to obtain the slice images. The differences in x-ray density are displayed in various shades of gray.

RESULTS AND DISCUSSION

Figures 5 to 9 show NDE results including images of the samples prior to and after oxidation. Figure 5 shows backscatter x-ray results. Prior to oxidation, indications of the slots were difficult to discriminate when the x-ray source faced the RCC2 sample face with five slots (front) except for the bottommost slot which was approximately 0.76 mm wide (the widest of the five slots on that face). When the x-ray source faced the sample face having with three slots (back), those three slots could be discriminated fairly easily due to their large width (0.76 and 1.02 mm) prior to oxidation. Post oxidation, all eight slots were easily discriminated with the x-ray source facing the front face indicating the extensive depth of damage created by the oxidation. A post-oxidation bsx image with x-ray source facing the back face is not shown (nor necessary) since all slots were revealed with x-ray source facing the front face post-oxidation.

Figure 6 shows eddy current results. Slots were impossible to discriminate prior to oxidation but were easily discriminated after oxidation. Eddy current required scanning from both sides to allow visualization of each face's slots. Qualitatively, the larger slots appeared to give more pronounced indications.

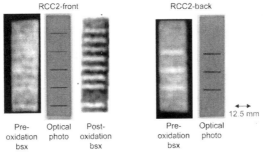

Figure 5. Backscatter x-ray results.

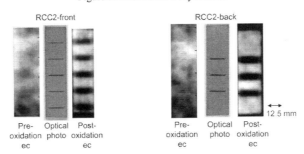

Figure 6. Eddy Current results.

Figure 7 shows thermography (derivative image) results. Slots were modestly discriminated prior to oxidation and were easily discriminated after oxidation. The heat source and thermography camera were facing the slots in order for oxidation damage detection to occur.

Figure 8 shows ultrasonic guided wave (mean total energy M_0) results for RCC1 with the single machined slot. The guided wave technique is extremely sensitive to surface condition and the optical appearance of the surface appeared the same both before (baseline) and after oxidation. Controlled, identical force was present on the ultrasonic transducers for the ultrasonic measurements prior to and after oxidation. As compared to mean M_0 of the time domain wave for the baseline condition (prior to machining slot), mean M_0 had decreased after machining, after first oxidation, and after machining out glass in the slot and performing a second oxidation. However, the decrease was not monotonic for these steps. The general decrease in M_0 after slotting and subsequent oxidation treatments is indicative of the slot structure scattering the ultrasound such that less ultrasonic energy reaches than receiving transducer as compared to the baseline (no slot) condition. Note that the repeatability of the measurement method was good. The five scans that were repeated after the final step show variability of mean M_0 to be about 2 percent which is significantly less than the 8 percent decrease in M_0 noted from baseline to after slot machined condition. This variability is dependent upon surface condition and measurement load remaining very similar.

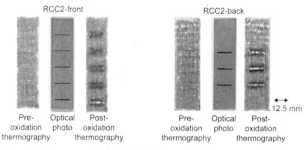

Figure 7. Thermographic (derivative) image results. Pre-oxidation images are frames at 0.18 sec in the temperature versus time cooldown stream of frames. Post-oxidation images are frames at 0.164 sec in the temperature versus time cooldown stream of frames.

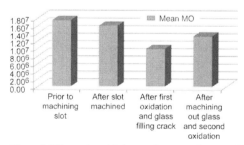

Figure 8. Ultrasonic guided wave (mean total energy M_0) results on sample RCC1 (fig. 2). Measurement variability is about 2 percent based on repeated trials.

Figure 9. (a) Contrast-enhanced x-ray CT slice of portion of RCC2 sample. Ring pattern is an artifact of the CT processing. (b) Solid three-dimensional visualization constructed from CT slices.

Figure 9(a) shows an x-ray CT slice of a portion of the RCC2 sample showing the 0.25, 0.51, and 1.02 mm slots and the hemi-spherical oxidation damage beneath the SiC coating and slots. CT has been used to size the hemispherical regions in order to help validate and/or reveal deviations from the oxidation model discussed in references 1 and 3. The size of the voids revealed by CT was on the order of 1 to 3 mm for depth and diameter, depending on the slot width. CT sizing was generally within 5 to 10 percent of actual values from destructive sectioning of several of the sample portions. X-ray CT allows the ability to sweep through various cross-sections and find the maximum diameter nondestructively and thus can result in a more accurate value for void diameter than destructive sectioning. Three-dimensional visualizations and animations composed of 11 consecutive CT slices were prepared on different sections of test samples, with one solid visualization shown in figure 9(b). These help reveal additional morphological features of the damage due to the three-dimensional nature.

The high resolution of the x-ray CT technique makes it suitable for probing the oxidation damage below craze cracks, as this damage tends to be smaller and much more irregular than the damage below the machined slots. Figure 10 shows two x-ray CT slices for a naturally-cracked RCC sample with the outlined area indicating the oxidation damage below the natural crack paths. The oxidation cavity shapes are highly irregular, as expected from the varying coating thickness and irregular nature of the cracks.[3] Figure 11 shows a translucent three-dimensional visualization constructed from 11 consecutive CT slices of this sample.

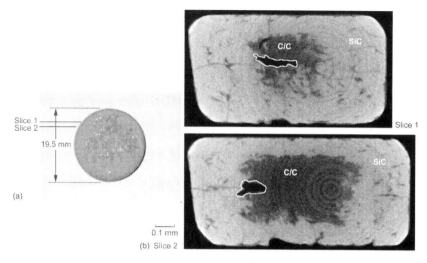

Figure 10. X-ray CT of SiC-coated RCC oxidized for 2.5 hr. in air at 1143 °C. (a) Location of CT slices. (b) Two CT slices. Reprinted with permission from Elsevier.

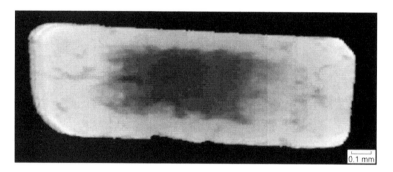

Figure 11. Translucent three-dimensional visualization constructed from X-ray CT slices of sample shown in figure 10.

PRIOR STUDIES

It is pertinent here to review the results of a prior related study in which a more quantitative approach to NDE capability was evaluated.[2] In that study, RCC material that was the same as that used for the wing-leading-edge of the shuttle had holes rather than cracks incorporated to the depth of the coating in order to study oxidation damage with NDE methods. Seventeen 0.5 mm diameter holes, spaced 13 mm apart, were ultrasonically drilled through the coating system to the depth of the SiC coating, followed by oxidation treatments. The oxidation treatments were conducted in static (nonflowing) air at 1 atm. and 1200 °C in a box furnace. A series of drillings followed by oxidation treatments was performed; first Batch 1 was drilled, then the disk was oxidized for 10 min, then Batch 2 was drilled, then the disk was oxidized for 10 min (giving 20 min on Batch 1 and 10 min on Batch 2), and so forth. The sample underwent seven drillings, and oxidation treatments varying from 10 to 70 min. The ability to measure the diameter and depth of resulting oxidation damage using NDE methods was evaluated. Figure 12 shows apparent diameter results from the various NDE methods and comparison to optically-measured diameters. Xray CT and thermography revealed the trend of void size increasing with increasing oxidation treatment time. Digital and Film Xray were not able to differentiate the void sizes. This is most likely due to the fact that the drilled hole and its extension into the cavity provided the dominant x-ray contrast over the oxidation damage. Figure 13 shows apparent depth of the holes beneath the coating from Xray CT and Eddy Current and their comparison to depth measured from optical measurements. Eddy current and CT measurements revealed the trend of void depth increasing with increasing oxidation treatment (with Xray CT providing the best results). Eddy current measurements will (and did) underestimate the actual void depth for small void diameters as unflawed material near the void will be averaged into the impedance measurement of the relatively large (0.1 by 0.05 in.) coils. Xray CT, the two-sided method, is the most accurate NDE method with regards to void sizing and measuring void depth.

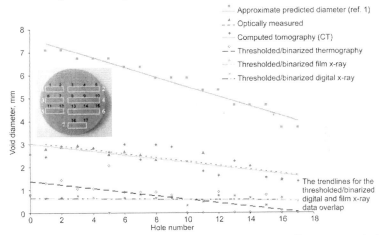

Figure 12. Apparent Void Diameter versus Hole Number from the different NDE methods and from optical measurement after sectioning. Disk schematic shows hole numbers (black) and oxidation batch numbers (white). Linear trend lines are shown.

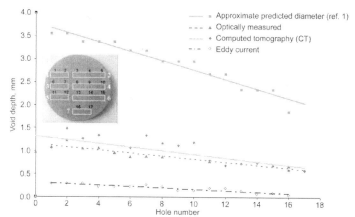

Figure 13. Apparent Void Depth versus Hole Number from different NDE methods and from optical measurement after sectioning. Disk schematic shows hole numbers (black) and oxidation batch numbers (white). Linear trend lines are shown.

THE ROLE OF NDE SIMULATION STUDIES

The ability to simulate NDE inspection scenarios is of great interest to NASA in order to quickly determine feasibility of specific inspection scenarios, design future inspections, and perhaps aid in the determination of the smallest detectable flaw size for a given inspection scenario (technique, material, geometry, access). Recently a NASA Engineering and Safety Center team of NDE experts engaged in an activity to evaluate existing inspection simulation tools and provide seed funds for development of usable simulation tools where physical models have been developed and at least partially validated. The team found that as with all simulation tools for physical phenomena, computational NDE tools represent an idealization of the true physical situation. Approximations are made with respect to geometry, material properties, measurement systems, underlying physical laws, and solution methods. A thorough understanding of these approximations, along with knowledge of the specific application requirements, is needed to develop appropriate and useful simulation results. Of pertinence to this particular study and that of reference 2 are nondestructive evaluation simulation methods for x-ray and x-ray computed tomography, thermography, eddy current and possibly ultrasonic guided wave inspection. In future studies, we will report on the use of some of these tools as applied to ceramic and composite structure inspection scenarios. Reference 11 provides an excellent introduction to the topic of NDE simulation.

SUMMARY

In this study, RCC samples having slots in their coating machined to the depth of the SiC coating, underwent oxidation treatments to create void-like damage in the carbon-carbon substrate underneath the coating. The size of the voids revealed by x-ray micro-CT was on the order of 1 to 3 mm for depth and diameter, depending on the slot width. State-of-the-art single-sided NDE methods, practical where access to only one side of the structure is available such as for on-wing inspections, were used to detect the damage. These methods included backscatter x-ray, eddy current, thermography, and ultrasonic guided waves. All of the methods were successful at detecting the oxidation damage, whereas only thermography unambiguously revealed the slots prior to oxidation. The higher resolution of the CT technique made it suitable to quantitatively assess oxidization damage both below machined slots and craze cracks. Further this technique revealed the patterns of oxidation attack, which tended to be smaller and much more irregular below the craze cracks than under the machined slots. Changes in ultrasonic guided wave total energy show this parameter is sensitive to structural changes in RCC. In prior studies, thermography revealed the trend of void size increasing with increasing oxidation treatment time, and eddy current revealed the trend of void depth increasing with increasing oxidation treatment. NDE simulation methods are expected to play a greater role in the future in assessing the feasibility of specific inspection scenarios and perhaps aid in the determination of the smallest detectable flaw size for a given inspection scenario (technique, material, geometry, access).

REFERENCES
[1]Jacobson, N.S., and Curry, D.M. , "Oxidation microstructure studies of reinforced carbon/carbon," Carbon 44 (2006) 1142-1150.
[2]Roth, D.J., Jacobson, N.S, Gray, J.N., Cosgriff, L.M., Bodis, J.R., Wincheski, R.A., Rauser, R.W., Burns, E.A., and McQuater, M.S., "NDE for Characterizing Oxidation Damage in Reinforced Carbon-Carbon Used on The NASA Space Shuttle Thermal Protection System," Ceram. Eng. Sci. Proc., vol. 26, no. 2, 2006, pp. 133-141.
[3]Jacobson, N.S., Roth, D.J., Rauser, R.W., Cawley, J. D., and Curry, D.M., "Oxidation through coating cracks of SiC-protected carbon/carbon," Surf. & Coatings Tech. 203 (2008) 372-383.
[4]Roth, D.J., Rauser, R.W., Jacobson, N.S., Wincheski, R.A., Walker, J.L., Cosgriff, L.M., "NDE for Characterizing Oxidation Damage in Cracked Reinforced Carbon-Carbon," NASA TM-2009-215489 and accepted for publication in International Journal of Applied Ceramic Technology.
[5]Dugan, E., Jacobs, A., Shedlock, D., and Ekdahl, D., "Detection of Defects in Foam Thermal Insulation Using Lateral Migration Backscatter X-ray Radiography," Proceedings of SPIE 49th Annual Meeting, Symposium on Optical Science and Technology, Penetrating Radiation Systems and Applications VI, vol. 5541, Denver, August, 2004.
[6]Wincheski, B.A. and Simpson, J.W., "Application of Eddy Current Techniques for Orbiter Reinforced Carbon-Carbon Structural Health Monitoring," Contract# 23-376-70-30-05, 2005.
[7]Shepard, S.M., Lhota, J.R., Rubadeux, B.A. and Ahmed, T., "Onward and Inward: Extending the Limits of Thermographic NDE," Proc. SPIE Thermosense XXII, vol. 4020, 2000, p. 194.
[8]Roth, D.J., Cosgriff, L.M., Verilli, M.J., and Bhatt, R.T., "Microstructural and Discontinuity Characterization in Ceramic Composites Using an Ultrasonic Guided Wave Scan System," Materials Evaluation, vol. 62, no. 9, 2004, pp. 948-953.

[9]NDE of advanced turbine engine components and materials by computed tomography, Yancey, R.N.; Baaklini, George Y.; Klima, Stanley J., ASME, International Gas Turbine and Aeroengine Congress and Exposition, 36th, Orlando, FL, June 3-6, 1991.

[10]Roth, D.J., Martin, R.E., Seebo, J.P., Trinh, L.B., Walker, J.L., and Winfree, W.P., "A Software Platform for Post-Processing Waveform-Based NDE," SAMPE Proceedings and presentation, June 2007, Baltimore, MD.

[11]Aldrin, J.C. and Knopp, J.S., "Modeling and Simulation for Nondestructive Testing With Applications to Aerospace Structures," Materials Evaluation, vol. 66, no. 1, 2008, pp. 53-59.

SILICON NITRIDE AND SILICON CARBIDE COMPONENTS AS ENABLING TOOLS IN AVIONICS, SPACE AND DISPERSING TECHNOLOGY

Dr. Karl E. Berroth
FCT Ingenieurkeramik GmbH, Rauenstein, Germany

INTRODUCTION

During the last years a large effort was made, to be able to supply even very large and complex shaped components made of sintered silicon carbide (SSiC) and of gas pressure sintered silicon nitride (GPSN) ceramics. This approach has opened new applications for such ceramic materials and also new markets for ceramic producers. On the other side, designers and engineers are now allowed to think much more complex in designing of ceramic components. In this paper, a new rapid prototyping routine for very complex components as well as the properties of the corresponding materials will be presented. Components for innovative optical aviation and space applications, for dynamic materials testing, for liquid metal processing, metal forming and chemical engineering are shown. Not only their unique properties, but also newly developed and adopted shaping and machining technologies for this specific ceramics have let to highly valued products.

By producing the housing structure of an infrared camera for aerospace with silicon nitride, higher resolution was reached due to a very low coefficient of thermal expansion, higher stiffness and less weight. Using silicon nitride, highly stiff, lightweight large scale structures for satellites can be realized with a material whit a very low CTE and improved outgasing behaviour. Also dynamic testing equipment mainly for avionic turbine blades can reach much higher frequencies and so reduce fatigue testing time and costs. With silicon nitride rollers for various functions, the lifetime of components in stainless steel and titanium rolling operations was extended by more than a magnitude, additionally improving the quality of the rolled wires, sheets, foils and thin walled structures. Improved efficiency of attrition mill without any contamination through to no metal contact was reached with silicon carbide liners and silicon nitride agitators. Setters of highly strong, lightweight and corrosion resistance SSiC led to triple the capacity of a sintering furnace for high volume technical ceramic parts and also reducing the specific energy consumption. As a result it can be stated that the economic availability of new advanced non oxide ceramic materials can generate new technical solutions for avionics, space and industrial equipment and processes.

Due to their very specific set of material properties, silicon nitride and silicon carbide have gained a lot of interest in the past 20 years. Moreover, many new approaches in technical equipment like gas turbine and motor engines, cooking systems, seals and bearings, cutting tools, metal pumps, electronic substrates and many others were put in place with corresponding research activities.

On the other side, a lot of less spectacular applications are state of the art in industry today. They have opened a wide field for niche products and lead to new technical solutions with higher performance, extended service time, less wear and corrosion and improved product properties.

1. MATERIAL AND FABRICATION

FCT Ingenieurkeramik has established material grades for GPSN and SSiC as well as for some composite Materials like $C_f/CSiC$ and the corresponding processing routes for the commercial fabrication of a broad range of components with mainly large and complex sizes and shapes and i. g. very narrow tolerances.

The standard grade for GPSN is a composition of 90% Si_3N_4, 6% Al_2O_3, 4% Y_2O_3 which is sintered at 1 MPa of nitrogen gas pressure. It can be brought to improved properties using our HIP process up to 1850°C and 200 MPa. For SSiC the standard is 99,5% SiC, with B and C as sintering aids.

As major fabrication routines we use:

1. slip casting for complex, thin walled components
2. cold isostatic or uniaxial pressing of preforms
   with
3. subsequent green machining,
4. sintering, gas pressure sintering, HIPing
   and if necessary but most likely for narrow tolerances
5. final machining by laser cutting, grinding, honing, drilling, lapping and polishing

Details of our production process are described elsewhere.
Rather large components with one dimension up to 1,25 m in length and/or 0.5 m in diameter, but also tiny things with only some mms feed into our product range. We typically start to produce prototypes according to customers design requirements but also do small and intermediate series up to 10,000 pieces per year or fabrication lot, whenever prototypes were successfully tested.

2. APPLICATIONS

2.1. Light metal cast house applications

One main application area has been located and developed in light metal casting.
Here mainly two of the materials properties are required:

1. excellent corrosion resistance against metal melt
2. excellent thermal shock resistance
   additionally helpful is
3. high strength and fracture toughness

Components like thermocouple sheaths, heater sheaths, riser tubes, valve seats and plungers, degassing agitators and a lot of other very specific parts are in use or are tested in new advanced casting systems. The size is medium to large, the complexity is simple to medium. Rather rigid components are required for the rough working conditions in foundries

Figure 1 Example for foundries: degassing impeller and immersion heater sheath for transport Aluminium tank

2.2. METAL WORKING

2.2.1. Rolling

In metal industry, new shaping approaches for even complex part are rolling operations. For such shaping processes steel, WC and ceramic rollers are used. Silicon nitride due to its excellent thermal shock behaviour, high strength, hardness and toughness shows improved service time and product quality in both, at cold and even at hot rolling with temperatures up to 1050°C in rolling of stainless steel, titanium and aluminium alloys.

Further advantages are the possibility of cooling reduction and the lack of material transfer by friction welding to the rolled metal.

Even more successful are wear parts in metal forming facilities, like guiding, drive and break rollers where we could extend the service time by more than a magnitude.

Figure 2 press rollers for wheels and strip rollers with silicon nitride barrel for Al-foil

2.2.2. Welding

For the production of welded tubes and profiles, highly wear resistant and tough calibration rollers have become state of the art because of their very much extended service time and accuracy.

Also for the fixation and precise, reproducible alignment of steel sheets, fixation pins are widely used in automotive industry. The advantage of silicon nitride in welding is the improved strength, thermal shock capability and the non wetting for sparks created through the welding process.

Figure 3 guiding and welding rolls and welding caliber rolls in service

2.3. MECHANICAL AND CHEMICAL ENGINEERING

In mechanical engineering, many applications need a wear resistant material which also can withstand a certain impact stress or mechanical load. The material preferably has to have a high thermal conductivity, corrosion and erosion resistance. This is met with SSiC.

Highly precise sleeves for calandering equipment with slits down into the µm range are state of the art for high efficiency dispersion in colour industry. Agitator arms and rotors as well as a broad variety of shapes and dimensions of lining inserts in mixers and attrition mills are other typical examples for this industry. Liners can be made monolithic with about 500 mm in diameter and up to 1 m in length with SSiC and GPSN.

Figure 4   SSiC roller sleeves for calendaring, SSiC cylindrical mill barrels inserts and GPSN agitators for attrition mill

2.4. OPTICAL ENGINEERING

For optical applications, mainly for instruments which are used in airplanes or space, the density is very important beside stiffness, coefficient of thermal expansion and long term durability. Also the mechanical strength and fracture toughness is important, because on starting and landing, very high mechanical loads can appear. Another point is the thermal conductivity, which sometimes should be as high as possible and sometimes as low as possible.

Here with a sophisticated green machining and sintering technology, very large and complex structures for optical equipment like space telescopes and avionic cameras have opened a new promising application. Also long term stable guiding beams for wafer steppers and optical measuring systems as well as supporting structures for lenses, can be either made from silicon nitride or silicon carbide, depending on the best fitting property set.

One of the most sensitive parameters for large optical components is the CTE. At room temperature, for silicon nitride a value of $1x\ 10^{-6}$ /K is the lowest available value in combination with low density, high stiffness, strength and durability. The CTE of corresponding materials for optical use is shown for the temperature range between -100°C and +100°C.

Table 1 Coefficient of thermal expansion CTE for structural "optical" materials

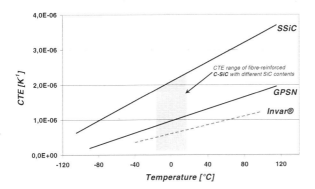

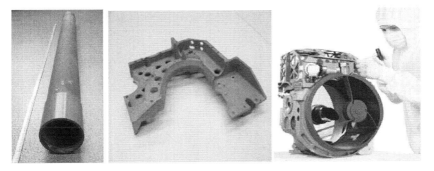

Figure 5 lightweight beam, instrument carrier and ceramic camera structure at final set up in clean room environment

2.5. ELECTRICAL ENGINEERING AND ELECTRONICS

Also in electrical engineering, silicon nitride components have gained specific application niches. Highly strong, thermo mechanical and thermal shock resistant insulator components show improved service behaviour compared to alumina or porcelain. Coil supports for inductive heating equipment for metal heat treatment have tripled the service time. Heater separation and shielding plates as well as drive rollers which withstand temperature and corrosive atmosphere e.g. for thin film solar cell production have improved the performance of CVD and PVD coating equipment. Power feed troughs for vacuum and high pressure furnaces can better withstand the thermal gradient which occurs in this application.

In electronics, high stiffness and low wear is reached for guiding beams and supporting plates for wafer machining and handling. Also probe cards in chip manufacturing and testing equipment has gained some importance and reasonable market volume.

Figure 6 supporting plate for wafer handling, thickness 1 mm, diameter up to 380 mm

Another approach can be made with composites, based on silicon nitride having a rather high electrical conductivity. Wear resistant EDM tools and Sensors can be fabricated as well as non metallic heating elements for temperatures up to 1400°C. One advantage of such heater is the fact, that they keep very flat and do not deform or build up stresses due to their low CTE.

2.6. MEDICAL APPLICATION

screws and bone plates are tested as replacement for alumina and zirconia in order to provide higher strength lower density
for lower Xray shadowing in post operation inspection and improved bio combatibility in implants.

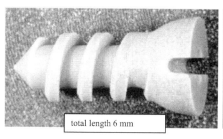
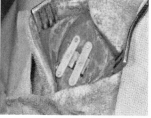

total length 6 mm

Figure 7 surgical screw and dummy implant with GPSN screw fixation, implanted into a pigs head

2.7 THERMAL ENGINEERING

In high temperature and thermal engineering, the demand for highly strong and long term corrosion resistant kiln furniture for technical ceramics as well as for sanitary and tableware has grown in the last years. This was strongly pushed by the use and implementation of robotic set up and unloading of high volume ceramic parts. Here the SSiC as a dense and high strength material with good thermal shock behaviour and excellent corrosion resistance has gained niche markets. Using SSiC setters as shown in Figure 9 in a specific application, the capacity of the furnace was tripled and investment in additional equipment could be avoided.

 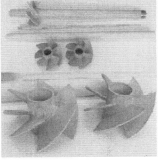

Figure 8 SSiC setter plates for high volume firing technical ceramics, RBSN high power gas burner components

2.8 TESTING EQUIPMENT

Due to it's low density, high Youngs modulus, long term mechanical stability and the high strength and fracture toughness as well as it's electrical insulating behaviour, silicon nitride is a candidate material to be used as a swinger head in high frequency vibration test equipment. A further advantage is it's very low CTE, leading to no dimensional change within a certain temperature range and so keeping a stable test frequency. This enables silicon nitride to be used as calibration vibrator for frequencies up to 25 kHz.

The complex shaped and finely structure component with very narrow tolerances and minimal wall thickness of 0.25 mm was developed in tight cooperation with TIRA, a world leading company for special testing equipment.

Figure 9 swinger head, external and internal structure and calibration shaker unit for 25 kHz

3. IMPROVED PROPERTIES

As mentioned above, silicon nitride shows improved application benefits compared to other ceramics or metallic solutions. Mainly in wear and foundry industry, solid silicon nitride components gives a much longer service time compared to solutions with coatings. Even more improved properties can be reached by silicon nitride based composite materials.

Here, a wide variety of different additives help, to tailor materials for specific use in new advanced technologies.

With the addition of TiN for example, the hardness and the electrical conductivity can be adjusted to a wide range of technical requirements. Oxide and non oxide sintering additives can help to change and improve the corrosion resistance by the reduction of wetability for metal melts. Also the thermal conductivity can be influenced by special rare earth oxides. A wide range of possible compositions is still under investigation.

4. CONCLUSION

In the paper it is shown, that a wide range of applications of silicon nitride ceramics have been achieved due to the availability of components with different sizes, shapes and complexity. Small and very precise as well as rather large and complex components can be reproducibly and economically produced with a high standard of quality. Materials and components are highly reliable, also in very though applications. New technical solutions are though possible for highly advanced processing equipment and routes.

α/β SiAlON BASED COMPOSITES INCORPORATED WITH MoSi$_2$ FOR ELECTRICAL APPLICATIONS

Erhan Ayas, Alpagut Kara, Ferhat Kara

Department of Materials Science & Engineering
Anadolu University, Eskişehir, TURKEY

ABSTRACT

Electrically conductive α-β SiAlON composites containing 30 vol. % MoSi$_2$ was prepared by gas pressure sintering (GPS). The effect of sintering atmosphere on phase formation, grain growth, mechanical and electrical properties of the composites was discussed. In order to understand the effect of sintering atmosphere, composites were sintered under N$_2$ and Ar atmospheres during gas pressure sintering. It was observed that MoSi$_2$ particles transformed to Mo$_5$Si$_3$ under N$_2$ but not under Ar. The resistivity of the produced composites varied from 10^8 Ω.m to 10^{-3} Ω.m depending on the sintering atmosphere and the phase composition.

INTRODUCTION

Particulate reinforced composite approach is an effective route to produce materials with different properties in terms of the specific features of matrix and reinforcement phases. Depending on the property to influence, it can be adjusted by means of optimum reinforcement content. According to the relevant literature, in order to achieve electrical conductivity in an insulator ceramic matrix with particle reinforcement approach, at least 30 vol. % of a conductive phase should be introduced. In the case of Si$_3$N$_4$ as a matrix material, addition of TiN[1-2], SiC[3-5], TaN[6], MoSi$_2$[7-8], Ti(C,N)[9] and TiB$_2$[10] with composite approach were extensively investigated. On the other hand, in-situ formations of conductive phases were also studied by means of chemical routes that require more complex processing[11-12]. Another effective reinforcement to achieve conductivity is the incorporation of low amounts of (% 1-3) carbon nanotubes (CNT)[13].

Regarding to SiAlONs, there is a dearth of literature on these materials as matrix in composite systems, which are also important derivatives of Si$_3$N$_4$. The most important advantage of SiAlONs over Si$_3$N$_4$ is the fact that they can be tailored for specific applications due to the flexibility of their phase composition and microstructure[14]. Similar to Si$_3$N$_4$, there are two SiAlON phases that are of interest as engineering ceramics, α-SiAlON and β-SiAlON, which are solid solutions based on α and β-Si$_3$N$_4$. In α-β SiAlON ceramics, changing the α-SiAlON:β-SiAlON phase ratio open many possibilities to prepare SiAlON ceramics with desired properties. Hardness increases markedly with increasing α-SiAlON phase content, whereas the fracture toughness increases with the increasing amount of β-SiAlON. SiAlON ceramics can be easily sintered using conventional sintering techniques such as gas pressure sintering and even pressureless sintering, which reduces the production cost and allows mass production of pre-shaped materials.

Molybdenum disilicide (MoSi$_2$) has taken much more interest than other reinforcement phases in ceramic-matrix composites because of their excellent oxidation resistance due to the occurrence of a surface glassy layer of silica that forms at high temperature under oxidizing conditions[15]. Another important advantage for using MoSi$_2$ is the use of its thermal expansion mismatch with the ceramic matrix and ductile behavior to provide toughening[16]. In addition to improved fracture toughness and oxidation resistance, the high electrical conductivity of MoSi$_2$ makes the composite machinable (into a complex shape) by an electrical discharge machining (EDM) technique similar as other conductive reinforcements[17].

In this particular work, fabrication of α-β SiAlON composites containing 30 vol. % MoSi$_2$ by using gas pressure sintering under different sintering atmospheres (N$_2$ and Ar) for possible electrical applications was studied. The densification, phase formations, mechanical and electrical properties of the composites were investigated.

EXPERIMENTAL

α-β SiAlON composition (designated as SN) with the ratio of 25 % α and 75 % β were designed by using multi cation dopants including CaO, Y_2O_3, and Sm_2O_3. As starting powders, Si_3N_4 (Ube SN E10, Japan), AlN (H.C. Starck, Grade C), Al_2O_3 (Sumitomo-Japan) and $MoSi_2$ (H.C. Starck, Grade A) were used. Mixing of the composite powder mixtures was prepared in a planetary ball mill (Fritsch -Pulverisette 5) in water-free 2-propanol for 1.5 hour using Si_3N_4 milling media. The slurry was then dried in an evaporator (Heidolph WB2000) at 60°C. Pellets were pre-pressed in a stainless steel die at a pressure of 15 kg/cm². Following the uniaxial pressing, the pellets were subjected to cold isostatic press under 300 MPa. Sintering of the pellets was carried out in a BN crucible using a GPS furnace (FCT Systeme GmbH, Germany), capable of operating at temperatures of up to 2000°C in N_2 and Ar atmosphere of up to 10 MPa pressure. During sintering, a two-stage sintering schedule was employed, which included a first stage at a sintering temperature of 1940°C for 60 min. under a low gas pressure of 0.2–0.5 MPa and a second stage at a sintering temperature of 1990°C for 60 min. under a gas pressure of 10 MPa. The heating and cooling rates were kept at 10°C/min. Relative densities of the samples were determined by the Archimedes method. Phase identification was performed using XRD (Rigaku Rint 2200-Japan) with Ni-filtered Cu Kα radiation of wavelength 1.5418 Å. Intensities of the (102) and (210) peaks of the α-SiAlON phase and the (101) and (210) peaks of the β-SiAlON phase were used for quantitative estimation of the α:β SiAlON ratio. The microstructure studies were conducted composites using SEM (Supra 50 VP, Zeiss, Germany). Vickers hardness (Hv_{10}) of the sintered samples was measured by using an indenter (Emco-Test, M1C, Germany) with a load of 10 kg. The fracture toughness (K_{IC}) of the samples was evaluated from the radial cracks formed during the indentation test. The K_{IC} values are calculated according to the Anstis formula[18]. Electrical resistivity measurements were carried out by using two probe DC method at room temperature on the disc shape samples. Au electrodes were deposited on the both sides of the samples. The volume resistivity of the composites was measured by using a Keithley 6517A Electrometer/High resistance meter. The resistivity values were calculated taking into account the thickness and cross sectional area of the samples.

RESULTS AND DISCUSSION

The properties of the produced composites and monolithic SiAlON matrix materials were given in Table 1.

Table I. General properties of the produced materials.

Composition / Sintering Atmosphere	Measured Density (gr/cm³)*	Relative Density (%)**	α/β Phase Ratio	Hv_{10} (GPa)	K_{IC} (MPa.m$^{1/2}$)	Resistivity (Ω.m)
SN-N₂	3.25	100	0.22	16.63 ± 0.16	5.22 ± 0.07	1x10¹³
SN-Ar	3.21	96	0.26	16.95 ± 0.13	5.01 ± 0.04	1x10¹³
SN30M-N₂	3.90	99.7	0	14.06 ± 0.40	6.59 ± 0.18	7.6x10⁸
SN30M-Ar	3.88	98	0	13.81 ± 0.21	6.31 ± 0.09	6.2x10⁻³

*Measured density ** Calculated density

Densification of the Composites

Representative SEM micrographs taken from the polished surfaces of the matrix material sintered under different sintering atmosphere are given in Figure 1. In the both images, typical elongated -SiAlON grains and equaixed -SiAlON grain formation can be observed. However, the SiAlON sintered under Ar contains considerable amount of isolated pores compared to that sintered under N₂ despite the fact that no decomposition of Si₃N₄ and weight loss was detected under Ar atmosphere. This confirms the positive effect that N₂ improves the densification of Si₃N₄ based ceramics. Calculated densities were also confirmed this result.

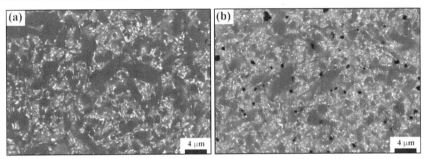

Figure 1. SEM micrographs of the matrix material sintered under a) N₂ and b) Ar.

Phase Analysis

Representative XRD spectra of the investigated composites are given in Fig. 2. Different from the matrix material, the MoSi₂ reinforced compositions sintered under both N₂ and Ar contains only -SiAlON phase. The total conversion of -SiAlON phase was attributed to the incorporation of excess SiO₂, which causes the formation of high amount of liquid phase than predicted. It is known that similar to Si₃N₄, MoSi₂ particles also have a protective SiO₂ layer. During sintering, high amount of SiO₂ incorporation into the formed liquid phase is believed to induce a shift in the predesigned matrix composition to the -SiAON region, being independent from the sintering atmosphere. Increased amount of SiO₂ enlarges the liquid phase amount, decreases its viscosity phase and accelerates α →β SiAlON transformation at the peak sintering temperature. As a result, no residual α-SiAlON phase was detected.

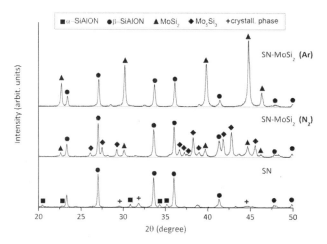

Figure 2. XRD spectra of SiAlON – MoSi₂ composites GPSed under different atmospheres

Additionally, Mo₅Si₃ phase formation occurred in the composite sintered under N₂. Guo et al. also revealed the presence of the same phase formation under N₂ and discussed the thermodynamic approach extensively[19]. The possible reaction for the conversion was proposed by the following equation:

$$15MoSi_2(s) + 14N_2(g) \rightarrow 3Mo_5Si_3(s) + 7Si_3N_4(s) \qquad (1)$$

The nitridation of MoSi₂ particles under high amount of N₂ gas used during gas pressure sintering resulted in a high amount of conversion to Mo₅Si₃. On the other hand, it was previously shown that this conversion can be prevented by employing pressure assisted techniques such as hot pressing[8]. Long soaking times during gas pressure sintering promotes this formation. Furthermore, it was stated that MoSi₂ phase is in equilibrium with Si₃N₄ at low N₂ partial pressure (around 10^{-4} bar) at 1300°C [21]. However, the formation of Mo₅Si₃ was thermodynamically favorable at N₂ partial pressure between 0.03 and 7 bar. This might explain why the transformation did not occur during the sintering of the composite under Ar atmosphere. Formation of Mo₅Si₃ phase in the composite sintered under N₂ was also observed during the microstructural analysis. In Fig. 3, comparative SEM images of the composites sintered under Ar and N₂ is presented. In the images, the gray phase represents MoSi₂ and the bright phase represents Mo₅Si₃ particles.

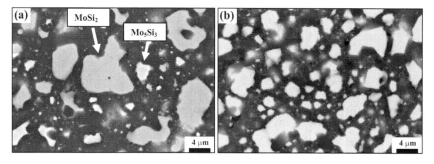

Figure 3. SEM micrographs of the SN-MoSi₂ composites sintered under a) N₂ and b) Ar.

Microstructural Analysis

Figure 4 (a) and (b) shows the polished cross sections of the composites containing 30 vol. % MoSi$_2$ sintered under N$_2$ and Ar, respectively. Although the amount of reinforcement phase is high, the dispersion of the MoSi$_2$ particles in the matrix appears to be homogenous. Composites show the same densification behavior as the matrix material. Some residual porosity remains in the composites sintered under Ar atmosphere (Fig. 4b) however not as much as matrix material. The increase in the density is attributed to the incorporation of excess SiO$_2$ originated from the surface of MoSi$_2$ particles that alters the liquid phase formation. Considering the starting particle size of MoSi$_2$ powder (d$_{50}$ = 2-3 μm) coalescence of especially MoSi$_2$ particles in composites sintered under N$_2$ can be seen. This formation prevents the formation of a chain type structure which decreases electrical conductivity.

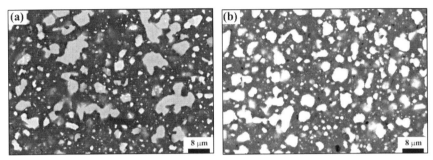

Figure 4. SEM micrographs of SN-MoSi$_2$ composites sintered under a) N$_2$ and b) Ar (low magnification).

Mechanical Properties

In general, different sintering atmosphere resulted in no significant effect on the hardness and fracture toughness of the MoSi$_2$ containing composites and the monolithic matrix material. The hardness of the matrix decreased with the incorporation of MoSi$_2$. As revealed in the XRD spectra, designed α- SiAlON composition alters to -SiAlON with the addition of MoSi$_2$ in both composites. Lack of α-SiAlON phase and incorporation of softer MoSi$_2$ are thought to be the reason in the reduced composite hardness.

Indentation fracture toughness of the composites was improved by MoSi$_2$ incorporation. The most common aspect for the toughening increment due to the introduction of MoSi$_2$ reinforcement was attributed to the residual stresses caused by the different thermal expansion coefficients of matrix (3-3.5x10^{-6} °C^{-1}) and particles (6-8x10^{-6} °C^{-1})[16]. In Fig. 5, certain MoSi$_2$ particles are fractured due to the mentioned aspect. Additionally, typical crack deflection and pull out of elongated -SiAlON grains plays an important role on the increment of fracture toughness.

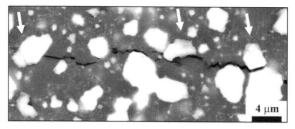

Figure 5. SEM micrograph of a typical crack profile of SN-MoSi$_2$ composite sintered under Ar.

Electrical Properties

The main mechanism of the electrical conduction in particulate reinforced composite materials is the formation of a percolating network of the conductive particles within the matrix material. The distribution of conductive particles is key factor that provides the electrical conductivity[8,19]. When the reinforcement amount reaches above 30 vol. %, a drastic decrease on the resistivity of the composites was observed due to the network of conductive particles. The resistivity of the composites changed with sintering conditions, being in the range of 10^8 Ω.m to $10^{-3}\Omega$.m. The resistivity of the composites sintered under Ar is in good agreement with that of the Si$_3$N$_4$-MoSi$_2$ composites with a resistivity value of 10^{-3} Ω.m[8]. However, composites sintered under N$_2$ atmosphere show insulating behavior with a resistivity value of 10^8 Ω.m. Mo$_5$Si$_3$ formation along with MoSi$_2$ phase is thought to be the reason for the increased resistivity of the composite. Both Mo$_5$Si$_3$ and MoSi$_2$ are good conductors. However, considering the volume effect, formation of the Mo$_5$Si$_3$ may cause to a damping on the conductivity of MoSi$_2$. If all MoSi$_2$ particles have transformed to Mo$_5$Si$_3$, these composites would also be conductive as the composites sintered under Ar.

CONCLUSION

SiAlON-MoSi$_2$ composites containing 30 vol. % MoSi$_2$ were produced by gas pressure sintering. Microstructure, phase evolution, mechanical and electrical properties of the composites were studied and compared with the matrix material in terms of different sintering atmosphere (Ar, N$_2$). Employing Ar and N$_2$ during sintering affected the relevant composited. In general densification of the composites sintered under N2 is better than composites sintered under Ar. In both sintering conditions, introduction of MoSi$_2$ phase caused the conversion of the pre-designed α-β-SiAlON matrix to β-SiAlON due to the incorporation of high amount of SiO$_2$ originating from MoSi$_2$ particles. A considerable amount of Mo$_5$Si$_3$ phase formation was observed in the composites sintered under N$_2$ atmosphere. Formation of this phase has no adverse effect on the densification and mechanical properties, but resulted in the increase in the resistivity of the composites to a value of 10^8 Ω.m. The composite sintered under Ar was a good conductor with the resistivity value of 10^{-3} Ω.m.

REFERENCES

[1]A. Bellosi, S. Guicciardi and A. Tampieri, "Development and Characterization of Electroconductive Si$_3$N$_4$-TiN Composites", J. Eur. Ceram. Soc., **9**, 83-93 (1992).

[2]S. Boskovic, F. Sigulinski and L. Zivkovic, "Liquid Phase Sintering and Properties of Si$_3$N$_4$ TiN Composites", J. Mat. Synt. and Proces., **7**, 119-126 (1999).

[3]A. Sawaguchi, K. Toda, K. Niihara, "Mechanical and Electrical Properties of Silicon Nitride-Silicon Carbide Nanocomposite Material", J. Am. Ceram. Soc., **74**, 1142-1144 (1991).

[4]W. J. Kim, M. Taya, K. Yamada, N. Kamiya, "Percolation Study on Electrical Resistivity of SiC/Si₃N₄ Composites with Segragated Distrubition", J. Appl. Phys., **83**, 2593-2598 (1998).

[5]K.Yamada, N. Kamiya, "High Temperature Mechanical Properties of Si₃N₄–MoSi₂ and Si₃N₄–SiC Composites With Network Structures of Second Phase", Mater. Sci. Eng. A, **261**, 270-277 (1999).

[6]V. Y. Petrovsky and Z. S. Rak, "Densification, Microstructure and Properties of Electroconductive Si₃N₄-TaN Composites. Part II: Electrical and Mechanical Properties", J. of Eur. Ceram. Soc., **21**, 237-244 (2001).

[7]M. Y. Kao, "Properties of Silicon Nitride-Molybdenum Disilicide Particulate Ceramic Composites", J. Am. Ceram. Soc., **76**, 2879-2875 (1993).

[8]D. Sciti, S. Guicciardi and A. Bellosi, "Microstructure and Properties of Si₃N₄-MoSi₂ Composites", J. Ceram. Process. Res., **3**, 87-95 (2002).

[9]D. T. Jiang, J. Vleugels, O. Van Der Biest, W. Liu, R. Verheyen, B. Lauwers, "Electrically Conductive and Wear Resistant Si₃N₄-based Composites with TiC₀.₅N₀.₅ Particles for Electrical Discharge Machining", Mat. Sci. Forum, **492-493**, 27-32 (2005).

[10]A. H. Jones, C. Trueman, R. S. Dobedoe, J. Huddleston, and M. H. Lewis, Production and EDM of Si₃N₄–TiB₂ Ceramic Composites, British Ceramic Transactions, **100**, 49-54 (2001).

[11]S. Kawano, J. Takahashi and S. Shimada, The Preparation and Spark Plasma Sintering of Silicon Nitride-Based Materials Coated With Nano-Sized TiN, J. Eur. Ceram. Soc., **24**, 309-312 (2004).

[12]S. Shimada and K. Kato, Coating and Spark Plasma Sintering of Nano-Sized TiN on Y-α-SiAlON, Mater. Sci. Eng. A, **443**, 47–53 (2007).

[13]J. Tatami, T. Katashima, K. Komeya, T. Meguro and T. Wakihara, Electrically Conductive CNT-Dispersed Silicon Nitride Ceramics, J Am Ceram Soc, **88**, 2889-2895 (2005).

[14]V.A. Izhevskiy, L.A. Genova, J.C. Bressiani and F. Aldinger, Progress in SiAlON Ceramics, J. Eur. Ceram. Soc., **20**, 2275-2295 (2000)

[15]K. Natesan and S. C. Deevi, Oxidation Behavior of Molybdenum Silicides and Their Composites, Intermetallics, 8, 1147-1158,

[16]J.J. Petrovic, M.I. Pena and H.H. Kung, Fabrication and Microstructures of MoSi₂ Reinforced-Si₃N₄ Matrix Composites, J. Am. Ceram. Soc. **80**,1111–1116 (1997).

[17]C. Martin, B. Cales, P. Vivier, P. Mathieu, Electrical Discharge Machinable Ceramic Composites, Mater. Sci. Eng., **A109**, 351–356 (1989).

[18]G.R. Anstis, P. Chantikul, B.R. Lawn and D.B. Marshall, A Critical Evaluation of Indentation Techniques for Measuring Fracture Toughness. I. Direct Crack Measurements, J. Am. Ceram. Soc. **64**, 533-538 (1981).

[19]Z. Guo, G. Blugan, T. Graule, M. Reece, J. Kuebler, The effect of different sintering additives on the electrical and oxidation properties of Si₃N₄–MoSi₂ composites, J. Eur. Ceram. Soc., **27**, 2153-2161 (2007).

[20]E. Heikinheimo, A. Kodentsov, J. A. Van Beek, J. T. Klomp and F. J. J. Van Loo, Reactions in the Sysytems Mo-Si₃N₄ and Ni-Si₃N₄, Acta Metall, **40**, S111-S119 (1992).

MICROSTRUCTURE CHARACTERISTICS AND HIGH-TEMPERATURE PERFORMANCE OF IN-SITU REINFORCED α-SiAlON CERAMICS

Yu Zhou, Feng Ye, Chunfeng Liu
Institute for Advanced Ceramics, Harbin Institute of Technology
Harbin, Heilongjiang Province, China

ABSTRACT

Rare-earth Lu, Y and Nd oxides were adopted to stabilize α-SiAlON by hot-pressing. The results show that the addition of excess rare-earth not only promote α-SiAlON densification, but also facilitates the development of elongated α-SiAlON grains. The obtained materials are both hard and tough. The Vickers hardness, flexural strength and fracture toughness are improved remarkably. The Lu-doped α-SiAlONs present an excellent high-temperature performance. It is attributed to the formation of the refractory intergranular phase, Lu-J'($Lu_4Si_{2-x}Al_xO_{7+x}N_{2-x}$). The flexural strength at 1400°C can reach 550MPa. A thin dense oxidation layer is formed after oxidation at 1300°C in air.

INTRODUCTION

α-Sialons, as solid solution of Si_3N_4, have been attracting more attention for engineering application because of their notably high frication, wear resistance and excellent thermal, chemical stability. The ability of α-sialon to absorb liquid constituents into its structure can potentially improve the high temperature application. Many metal cations like Ca^{2+}, Mg^{2+}, Li^+ and some rare earth metal cation, such as Y^{3+}, Yb^{3+}, Dy^{3+}, Sm^{3+}, Nd^{3+} and Er^{3+}, have been added to stabilize α-sialon [2-5]. The results show that the formation of elongated α-sialon phase contributes to the improvement of fracture toughness [1, 2]. To obtain this whisker-like microstructure, many methods, such as adjusting starting powders [1], dopant [1-5], compositions and sintering conditions [1, 5] during the fabrication.

Lutetium (Lu) possesses a high melting point (1663°C), and hence it is expected to obtain a more refractory crystallized phase using Lu_2O_3 as a sintering additive in order to improve the elevated temperature properties of α-sialon ceramics. However, the research on the Lu-α-sialon, especially high-temperature fracture and oxidation behavior was hardly reported. Moreover, the ionic radius of Lu is the smallest among the common rare-earth used in α-sialon. The microstructure difference between Lu-α-sialon and other RE-α-sialon has been seldom studied.

In this paper, three rare-earth cations with a large difference in radius were selected to stabilize α-sialon. Their microstructure characterization was investigated. The high-temperature fracture and oxidation behavior of Lu-α-sialon doped with more extra liquid phase were studied, with intention to explore the effect of and Lu_2O_3. The oxidation damage mechanisms were also discussed.

EXPERIMENTAL PROCEDURE

The composition of α-sialon investigated in this study was $Re_{0.333}Si_{10}Al_2ON_{15}$ (i.e., m=n=1 in the general formula of α-sialon, $RE_{m/3}Si_{12-(m+n)}Al_{(m+n)}O_nN_{16-n}$, RE=Lu, Y and Nd). For the selected compositions, an excess of 2wt% RE_2O_3 was added to achieve complete densification. They will be abridged as RE1010E2 in present work. An extra 4wt% Lu_2O_3 was added to the selected composition and referred to as Lu1010E4. Starting powders are α-Si_3N_4 (E10 grade, UBE Industries Ltd., Tokyo, Japan), Al_2O_3 (grade A16SG, Alcoa, Pittsburgh, PA), AlN (grade C, H.C. Starck, Berlin, Germany),

Y_2O_3 and Nd_2O_3 (Grade fine, H.C.Starck), Lu_2O_3 (99.99%, grade fine, Rare-Chem Hi-Tech Co., Ltd, Huizhou, China). When calculating the overall compositions, 2.38wt% SiO_2 and 1.83wt% Al_2O_3 on the surface of Si_3N_4 and AlN powders, respectively, were taken into account.

Table I. Starting Compositions, ionic Size of Rare-Earth and Density of the α-Sialons

Materials	Compositions (wt.%)				Ionic radius	Density
	Si_3N_4	Al_2O_3	AlN	RE_2O_3	(nm)	(g/cm^3)
Lu1010E2	74.44	0.456	12.66	12.45	0.085	3.478
Y1010E2	78.05	0.478	13.28	8.202	0.089	3.254
Nd1010E2	75.69	0.464	12.87	10.98	0.100	3.326
Lu1010E4	73.00	0.447	12.41	14.13	0.085	3.525

The starting powders were ball-milled in absolute ethanol for 20h with Si_3N_4 balls as the mixing media, and then the powder mixtures were dried at 40°C in a rotary evaporator and sieved. The dry mixtures were hot-pressed at 30MPa under 0.1MPa nitrogen atmosphere in a graphite resistance furnace. A two-step sintering proce dure was applied, i.e. sample was first heated to 1500°C at a rate of 20°C/min and held there for 1 h, and then heated to 1800°C at a rate of 30°C/min and held there again for 1 h. After sintering, the samples were cooled spontaneously in the furnace.

The microstructure was characterized in detail by transmission electron microscopy (TEM) and high resolution electron microscopy (HREM). Vickers hardness was obtained at 10kg load. Fracture toughness was evaluated by single edge notch beam (SENB) method with a span of 16mm using samples of 20mm×4mm×2mm. Flexural strength was measured in air from room temperature to 1400°C by three-point bending test using samples in size of 36mm×4mm×3mm with a span of 30mm.

The oxidation experiments were carried out at 1100–1300°C over a period of 8–32h in air. Samples of approximate dimension 3mm×4mm×15mm were cut from the as-sintered blocks and mirror polished by diamond paste lapping to 1μm. The polished specimens were ultrasonically cleaned in acetone, followed by a cleaning in ethanol. Then the specimens were placed into a high-temperature furnace preheated to 900°C and heated to the required temperature. The weight gains as a function of time were recorded every 8 hours by intermittently removing the specimens from the furnace. Three samples were used for each oxidation procedure.

The phase assemblages in the oxide scale were characterized by X-ray diffraction (XRD). The microstructure of the oxidized surfaces and their cross sections were examined by scanning electron microscopy (SEM). For the cross section analysis, the oxidized samples were immersed in polymer and polished along the cross section, and then observed in backscattered electron mode. The compositions of the individual phases were identified by energy dispersive X-ray spectrometer (EDS) fitted on the SEM.

RESULTS AND DISCUSSION
Microstructure characterization

Density measurements (listed in Table I) show that the all the samples were densified to over 99% of their theoretical densities. It indicates that the addition of 2wt% or 4wt% extra liquid phase could effectively promote densification.

XRD results of the investigated α-sialons are shown in Table II. It can be found that Lu1010E2 is composed of α-sialon phase and a small amount of secondary crystalline phase J′ ($Lu_4Si_{2-x}Al_xO_{7+x}N_{2-x}$). But only single α-sialon phase was observed in Y1010E2. For Nd-doped α-sialons, a few M′ ($RE_2Si_{3-x}Al_xO_{3+x}N_{4-x}$) phases and a small amount of β-Sialon phases were also detected, besides α-sialon as mainly crystalline phase.

Table II. Phase Assemblages, Grain Morphology and Mechanical Properties of α-Sialons

Materials	Phase assemblages	Grain morphology	Mechanical properties	
			Hardness(GPa)	Toughness(MPa·m$^{1/2}$)
Lu1010E2	α-sialon, J′	elongated	21.9±0.4	4.4±0.1
Y1010E2	α-sialon	elongated	21.1±0.2	4.4±0.3
Nd1010E2	α-sialon, β-sialon, M′	elongated	20.4±0.2	5.2±0.3

Fig.1 shows the morphologies of α-sialon grains in the materials doped with different rare-earth cations. It can be found that large amounts of elongated grains exist in Lu1010E2, Y1010E2 and Nd1010E2 ceramics. When the extra rare-earth content was increased to 4wt%, the aspect ratio of elongated α-sialon grains further increase [6]. It implies that addition of extra liquid phase contributes to the formation and anisotropic growth of elongated α-sialon grains. According to the formation mechanism of α-sialon, it is known that α-sialon forms in liquid phase by dissolution-reprecipitation. After the transformation is completed from α-Si_3N_4 to α-sialon, α-sialon grains grow following an Ostwald ripening process through a transient liquid. Small α-sialon grains dissolve into the liquid, and reprecipitate on the surfaces of large grains via liquid diffusion. The presence of additional liquid phase can accelerate the diffusion rate and assist elongated grains forming via offering a free growth room for α-sialon grains and lowing viscosity of the liquid phase [7].

Fig.1 SEM micrographs of α-sialons doped with different rare-earth
(a) Lu1010E2, (b) Y1010E2, (c) Nd1010E2

The typical TEM micrographs of α-sialons doped with different cations are shown in Fig.2, which illustrates the morphology of grains and distribution of grain boundary phases. It also indicates nearly full densified materials due to the addition of exceed 2wt% rare-earth. Elongated α-sialon grains are found embedded into the fine α-sialon matrix in all the three α-sialon ceramics. A trace amount of boundary phases present at the triple-grain regions in Y1010E2. In the investigated Lu-α-sialon, a few grain boundary phases were found as marked by white allows in Fig.2(a). They commonly exist at three- and four grain pockets. SAD and EDS results prove the intergranular phases are Lu-J' [6]. In Nd1010E2, there are also some intergranular phases M' distributed at grain junctions. In comparison, the volume and content of intergranular phases are much more than those in Y- and Lu-α-sialons.

Fig.2 TEM micrographs of α-sialons, (a) Lu1010E, (b) Y1010E2, (c) Nd1010E2

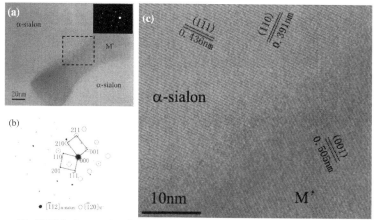

Fig.3 TEM micrographs of intergranular phase M' and HREM image of /M' interface in Nd1010E2 (a) TEM image of M' and SAD pattern inset, (b) schematic diagram showing two sets of spots of α-sialon and M', (c) HREM image of α-sialon/M' interface

The variety of phase assemblages in RE-α-sialons is affected by the type and ionic size of the doped rare-earth. For the majority of RE-Si-Al-O-N systems (Nd, Sm, Dy, Gd, Yb and Y) used in sialon production, the solid solubility range of M′ phase in respect of aluminum are determined by ionic radius of the RE element. The solubility limits in melilite solid solution decrease with decreasing ionic radius [8]. Y-M' has the lowest aluminum solubility, but Nd-M' with larger ionic radius has a higher solubility. Therefore, more M' phase can be observed in Nd-α-sialon. The subsolidus relationships in Lu-Si-Al-O-N system have not been derived, but J' phase does exist in Lu-α-sialon ceramic and other RE-Si-Al-O-N system, and its stability increases with a decrease in ionic radius of the rare-earth element [9].Fig.3 show the TEM image of M' and HREM micrograph of α-sialon/M' interface. The photograph at the upper right corner is the selected area electron diffraction pattern. Its schematic diagram is shown in Fig.3b. Two sets of diffraction spots can be separated, which belong to in the reflection direction of $[\bar{1}12]$ and M' phase with a zone axis of $[\bar{1}20]$, respectively. It means that no orientation relationship between these two phases exists. HREM (Fig.3c) was performed at the surrounded region by the imaginary line in Fig.3a. It reveals the directly bonded grain boundary without amorphous film between the adjacent α-sialon and M' phase.

Mechanical properties and high-temperature performance

The hardness and fracture toughness of the three RE-α-sialon ceramics were listed in Table II. All these ceramics are very hard. In particular, Lu- and Y-α-sialon possess a high hardness of over 21GPa. The hardness of Nd1010E2 is slight lower, due to the existence of β-sialon and a larger volume in intergranular phase M'. They also possess a high toughness of 4.4~5.2MPa·m$^{1/2}$ owing to the formation of elongated α-sialon grains.

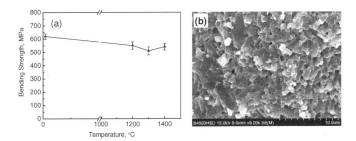

Fig.4 Flexure strength as a function of test temperature (a) and the fracture surface of Lu1010E4 after bending strength tests at 1400°

The flexural strength of Lu1010E4 as a function of test temperature was shown in Fig.4. The flexural strength decreased slowly with rising the temperature up to 1400°C [6]. It retained ~87% of its room-temperature strength at 1400°C. The fractography (Fig.4b) shows that the fracture surface was scarcely subjected to oxidation after the bending test at 1400°C. Both the grain facets remained by transgranular fracture and the traces left by the pulling-out of the elongated grains are very clear [6]. The study on phase relationship in the Si$_3$N$_4$-SiO$_2$-Lu$_2$O$_3$ system indicates that the melting temperature of the Lu-J-phase is more than 1850 °C [10]. So it is reasonable to believe that J' phase, as aluminum-containing solid solution of J phase, is a better high-temperature stable phase. Therefore, a very weak deterioration of the interface occurs when Lu-α-sialon was tested at high temperature. The high-temperature flexural strength of 611MPa was also reached for Sc^{3+}/Lu^{3+} co-doped α-sialon tested at 1400°C, and it was ~94% of the room temperature [11]. Similarly, the sintered Y-α-sialon using $BaAl_2Si_2O_8$ (BAS) as an additive could maintain the room-temperature strength (634MPa) up to 1400°C due to the extensive crystallization of BAS [12].

Fig.5a shows the plots of weight gain per unit surface area as a function of oxidation time at different temperature for Lu1010E4. The specific weight gain is only on the order of 0.08–0.65mg/cm^2 after 32h oxidation exposure at 1100–1300°C. It indicates that Lu-α-sialon has a good resistance against oxidation. Increasing the oxidation temperature, the weight gain increases. The squared weight gain $(\Delta W/A_0)^2$ versus time, t, graphs are shown in Fig.5b. The straight lines represent adherence to parabolic oxidation kinetics during the entire oxidation experiment. In the ideal case, a diffusion-controlled oxidation process will obey a parabolic rate law as following [13, 14],

$$W^2 = kt \tag{1}$$

where W is the weight gain per unit surface area, t is the oxidation time, and k is the parabolic rate constant. The results in Fig.3b suggest that the inward diffusion of oxygen through the oxide layer and the outward diffusion of metal cations are main rate-controlling oxidation mechanism for Lu-α-sialon. The oxidation rate constants, k, obtained from the slope of the lines in Fig.5b are 0.051×10^{-6}, 1.928×10^{-6} and 3.591×10^{-6} mg^2/(cm$^4 \cdot$s) for 1100, 1200 and 1300°C oxidation, respectively. The oxidation resistance is strongly affected by the thermodynamic stability of the grain boundary phase. The excellent oxidation performance of the investigated Lu-α-sialon can be attributed to the extensive crystallization of the refractory intergranular phase J', which limits the migration of rare-earth cations towards the oxidation surface during oxidation.

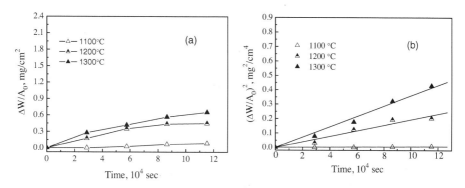

Fig. 5 Plot of specific weight gains as a function of time (a) and $(\Delta W/A_0)^2$ vs t curves (b) for Lu1010E4 at different temperature

Fig.6 indicates the phase assemblages on the oxidized surface of Lu1010E4 after oxidation at 1300°C for 32h. Large numbers of diffraction peaks of $Lu_2Si_2O_7$ appeared, together with a few peaks of mullite. Meanwhile, some peaks of α-sialon was still detected, suggesting a thin oxidation layer.

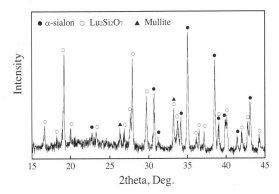

Fig.6 XRD patterns of Lu1010E4 after oxidation at 1300 °C for 32h in air

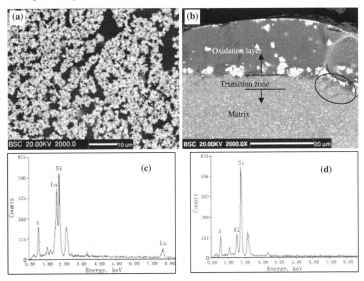

Fig.7 SEM micrographs of the oxidized surface and cross sections of Lu1010E4 and EDS patterns of the oxidized products after oxidation at 1300 °C for 32h in air
(a) SEM image of the surface oxidized, (b) SEM morphologies of the cross sections, (c) EDS pattern of the white crystalline phase in (a), (d) EDS pattern of the oxidation scale in (b)

Fig.7 shows the appearance of the oxidized surface and cross section of Lu1010E4 after oxidation at 1300°C for 32h. Large amounts of fine white phases formed on the crack-freee oxidized surface. EDS analysis to the white crystallized phase indicates that it consists of Si, Lu and O, as shown in Fig.7c. Combining with the XRD result, the crystallized phase is considered to be $Lu_2Si_2O_7$. A dense oxidation layer with a thickness of 15~17μm formed after air exposure at 1300°C for 32h (Fig.7b). It is noticed that a few gas bubbles and holes appear in the oxidation scale due to the gas expulsion during the prolonged oxidation. The oxidation diffusion is aggravated and some small loose textures can be found in the transition zone, as indicated by circle. The contrast is darker in the transition zone than that in the matrix for the depletion of Lu, as a result of their diffusion outwards to the oxidation surface. The Lu^{3+} mainly immigrates to the oxidation surface to form $Lu_2Si_2O_7$.

CONCLUSIONS

(1) Dense RE-α-sialon ceramics doped with Lu^{3+}, Y^{3+} and Nd^{3+} were obtained by a two-step hot-pressing precedure. Elongated α-sialon grains formed in all these ceramics, facilitating the self-toughening mechanism without expense their high hardness.

(2) Large cation (Nd^{3+}) doped α-sialon is composed of α-sialon, β-sialon and Nd-M' phase. The ceramic doped with smaller cation (Y^{3+}) is fully composed of α-sialon. But in the smallest cation (Lu^{3+}) doped ceramics, Lu-J' phase appears besides α-sialon.

(3) Lu-α-sialon demonstrates excellent high-temperature performance. The flexural strength of Lu1010E4 could reach 550MPa and it retained ~87% of its room-temperature strength at 1400°C.

(4) A thin, dense and crack-free oxidation layer formed on the surface of Lu1010E4 after 32h oxidation at 1300°C. Oxidation kinetics of Lu-α-sialon obeys a parabolic law with a rate constant of 0.051~3.591×10^{-6} $mg^2/(cm^4 \cdot s)$ after 32h oxidation at 1100~1300°C.

ACKNOWLEDGEMENTS

This work was supported by National Natural Science Foundation of China (Grant No.50632020), the Science Fund for Distinguished Young Scholars of Heilongjiang Province (Grant No.JC200603) and the Postdoctoral Fund of Heilongjiang Province (AUGA41100098).

REFERENCES

[1] I.-W.Chen, and A.Rosenflanz, A Tough Sialon Ceramic Based on α-Si₃N₄ with a Whisker-like Microstructure, *Nature* (London), **389**, 701-704(1997).

[2] R.Shuba and I.-W. Chen. Effect of Seeding on the Microstructure and Mechanical Properties of α-Sialon: II, Ca-α-Sialon, *J.Am.ceram.Soc.*, **85(5)**,1260-1267 (2002).

[3] F.Ye, M.J.Hoffmann, S.Holzer, Y.Zhou and M.Iwasa. Effect of the Amount of Additives and Post-Heat Treatment on the Microstructure and Mechanical Properties of Yttrium-α-Sialon Ceramics, *J.Am.ceram.Soc.*, **86(12)**, 2136-2142 (2003).

[4] S.Bandyopadhyay, M.J.Hoffmann and G..Petzow. Effect of Different Rare Earth Cations on the Densification Behavior of Oxygen Rich α-Sialon Composition, *Ceram.Int.*, **25**, 207-213 (1999).

[5] C.Zhang, K.Komeya, J.Tatami and T.Meguro. Inhomogeneous Grain Growth and Elongation of Dy-α-Sialon Ceramics at temperature above 1800°C, *J.Eur.Ceram.Soc..* **20**, 939-944 (2000).

[6] F.Ye, C.Liu, Y.Zhou, and Q.Chang, Microstructure and Properties of Self-Reinforced α-Sialon Ceramics Doped with Lu₂O₃, *Mater. Sci. Eng. A*, **469**, 143-149 (2008).

[7] H.Peng, Z.Shen, and M.Nygren, Formation of *in situ* Reinforced Microstructure in α-Sialon Ceramics: II, In the Presence of a Liquid Phase, *J.Mater.Res.*, **17(5)**, 1136-42 (2002).

[8] V.A.Izhevskiy, L.A.Genova, J.C.Bressiani, and F.Aldinger, Progress in SiAlON Ceramics. *J. Eur. Ceram. Soc.*, **20(13)** 2275-2295 (2000)

[9] Z.K.Huang, and I.-W.Chen, Rare-Earth Melilite Solid Solution and its Phase Relations with Neighboring Phases, *J.Am.ceram.Soc.*, **79(8)**, 2091-2097(1996).

[10] N.Hiroski, Y.Yamamoto, T.Nishimura, and M.Mitomo, Phase Relationships in the Si_3N_4-SiO_2-Lu_2O_3 system, *J.Am.ceram.Soc.* **85(11)**, 2861-2863 (2002).

[11] C.Liu, F.Ye, L.Liu, and Y.Zhou. High-Temperature Strength and Oxidation Behavior of Sc^{3+}/Lu^{3+} Co-doped α-Sialon, *Scr. Mater.*, **60**, 929-932 (2009).

[12] F.Ye, L.Liu, H.Zhang, and Y.Zhou. Refractory Self-Reinforced Y-α-SiAlON with Barium Aluminosilicate Glass Ceramic Addition, *Mater. Sci. Eng. A*, **488**, 352-357 (2008).

[13] C.Singhal, Thermodynamic and Kinetics of Oxidation of Hot-Pressed Silicon Nitride, *J.Mater.Sci.*, **11** 500-509 (1976).

[14] J.Persson, P.-O.Käll, and M.Nygren, Interpretation of the Parabolic and Nonparabolic Oxidation Behavior of Silicon Oxynitride, *J. Am.Ceram.Soc.*, **75**, 3377-84 (1992).

SYNTHESIS PROCESS AND MICROSTRUCTURE FOR Al₂O₃/TiC/Ti FUNCTIONALLY GRADIENT MATERIALS

Enrique Rocha-Rangel, David Hernández-Silva
Departamento de Ingeniería Metalúrgica, ESIQIE-IPN
UPALM, Av. IPN s/n, San Pedro Zacatenco, México, D. F. 07738

Elizabeth Refugio-García, José G. Miranda-Hernández
Departamento de Materiales, Universidad Autónoma Metropolitana
Av. San Pablo # 180, Reynosa-Tamaulipas, México, D. F., 02200,

Eduardo Terrés-Rojas
Laboratorio de Microscopía Electrónica de Ultra Alta Resolución, IMP,
Eje Central Lazara Cárdenas # 152, San Bartolo Atepehuacan, México, D. F., 07730

ABSTRACT

The production of Al₂O₃-based functionally gradient materials, with different amounts of fine reinforcement particles of titanium and titanium-dispersed carbide was explored. The functionally gradient materials synthesis has been induced by means of both; solid sintering of Al₂O₃-titanium powders mixed under high energy milling via pressureless sintering and formation of titanium carbide at 500 °C by a cementation pack process. SEM analyses of the microstructures obtained in cemented bodies were performed in order to know the effect of the cementant (graphite) on the titanium in each composite. In this way, it was found that different depths of titanium carbide layer were obtained as a function of the titanium content in the functionally graded materials. On the other hand, the use of reinforcing titanium enhanced significantly density and fracture toughness of the functionally materials.

INTRODUCTION

Functionally Gradient Materials have received considerable attention in recent years, primarily as potential structures for elevated temperature and high heat flux aerospace applications[1]. In contrast to the discrete interface typically formed between two dissimilar materials e.g. ceramics and metals, functionally materials are characterized by a well defined graded structure with continuous variation in composition and microstructure along the thickness. It is well know that abrupt transitions in materials composition and properties within a component often result in sharp local stress concentrations. Graded structures in compositions are expected to alleviate this problem. In this sense functionally materials offer a significant degree of durability of components exposed to applications where the operating conditions are rigorous, for example, wear-resistant materials, rocket heat shields, heat exchanger tubes, thermoelectric generators, heat-engine components and electrically insulating metal/ceramic joints. Functionally gradient materials can be constituted of carbide ceramics, oxide ceramics and metals, and they can be used in application at high temperatures such as in the construction of gas turbine engines in order to increase their thermal cycle efficiency[2-3]. In this sense, mechanical and physical properties of such sorts of functionally materials have been studied as well as their production process[4-5]. However, the high temperature cementation of metal- dispersed carbide composites has not been investigated with detail and there are no reports about the high temperature cementation of thermal barrier functionally materials coatings[6]. Since at high temperatures carbon can diffuse through an oxide matrix, metal particle dispersion will cement in the matrix. The metallic dispersion expands due to cementation thus inducing stresses in the surrounding matrix. As a result, fissures can nucleate in the neighboring matrix when the stresses generated by volume of oxide that resulted from metal particles cementation, particularly if the associated compression stresses reach the fracture strength.

Eventually, after multiple cracks form, the composite is fractured. Thus, to design functionally materials for high temperature applications, high temperature cementation is very important.

EXPERIMENTAL PROCEDURE

Functionally gradient materials were prepared in two stages; in the first one, titanium-dispersed oxide aluminum composites were produced using as a raw materials powders of Al_2O_3 (99.9 %, 1 µm, Sigma, USA) and Ti (99.9 %, 1-2 µm, Aldrich, USA). The amounts of powder employed were fixed in order to obtain Al_2O_3-based composites with 0.5, 1, 2 and 3 vol. % of Ti. The powder mixture was milled in a high energy mill (Simolyer) with ZrO_2 media, the rotation speed of the mill was of 400 rpm during 8 h. The ball-to-powder volume ratio was 20:1. With the milled powder mixture, cylindrical samples of 2 cm in diameter and 0.3 cm in thickness were fabricated by uniaxial pressing using 250 MPa pressure. Then the pressed samples were pressureless sintered in an electric furnace with an inert argon atmosphere. Heating rate was fixed at 5 °Cmin^{-1}, sintering temperature was 1500 °C and holding time 1 h, after sintering the furnace was turned off and the samples were left to cool inside the furnace. In the second stage the fabrication of the functionally materials was as follows; the composites produced in the first stage were subsequently placed inside a packed with graphite and heated at 500°C during 1 h. in vacuum and then cooled. Characterization was made as follows: the density of fired specimens was determined using the Archimedes' method. Cemented samples were analyzed using SEM and energy dispersive spectroscopy (EDS) to observe their microstructure and chemical composition respectively, as well as the thickness of carbide layer as a function of titanium carbide content in each sample. The hardness of samples was evaluated as microhardness using Vickers indentation, toughness was estimated by the fracture indentation method[7].

RESULTS AND DISCUSSION

Microstructure

Figure 1 shows SEM pictures of the surface cross section of the different samples after cementation at 500 °C during 1 h. These micrographs depict general features of the microstructure in the studied system. It is possible to view the formation of homogeneous products, because the Ti particles (white points), which retained their very fine sizes were well distributed in the alumina matrix (gray phase). In general the resulting microstructures displayed few pores present in the matrix. Figures (a), (b), (c) and (d) correspond to samples with (0.5), (1), (2) and (3) wt % of titanium respectively. In these pictures, also it can be observed that there are no Ti metal particles in the surface region to a depth of about 50, 83, 107 and 119 µm for samples with (0.5), (1), (2) and (3) wt % of titanium respectively. For the sake of argumentation, the region in which any Ti metal particles have been completely cemented is defined as the cemented layer. An important observation in all figures is that Al_2O_3-based composite is not fractured by the cementation of titanium. There is a surface layer displaying a different color with respect to the Al_2O_3-based matrix that is similar to the cemented dispersion in the cemented region. Between the cemented layer and the non-cemented region, there is an intermediate region which consists of partially-cemented Ti particles, as the case may be. From the surface, there are three regions with: (1) completely cemented metal particles, (2) partially cemented particles and (3) metal particles without cementation. In all samples, the Al_2O_3-based matrix, titanium reinforcing metal particles and titanium carbide layer were identified with the help of EDS analysis performed during SEM observations. Figure 2 reports the EDS analysis in the edge and in the core on the sample with 2 % wt Ti, confirming the presence of high carbon contents near the surface. In contrast the carbon concentration in the core of the sample is nonexistent.

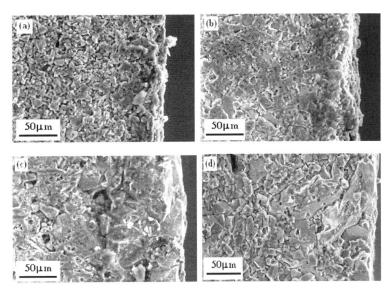

Figure 1. SEM secondary electron images of the resulting microstructures
of functionally materials produced. (a) 0.5 % Ti, (b) 1 % Ti, (c) 2 % Ti and (d) 3 % Ti.

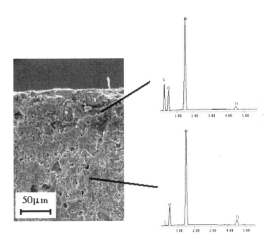

Figure 2. EDS analysis in the edge and in the core of sample with 2 % wt Ti.

Figure 3 shows the depth of the cemented layer, as a function of titanium content in the sample. Growth of the cemented regions appears to follow a parabolic behavior, because in spite of the increment on the depth of the cemented layer with the increments of titanium in the samples, this growing is not lineal, and the tendency of the curve indicates that for high titanium concentrations the depth of the cement layer will grow just until a certain limit.

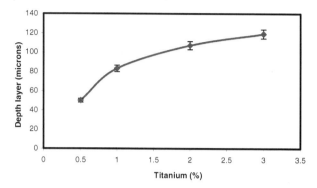

Figure 3. Depth of cemented layer as a function of titanium content in the sample.

Table 1 shows the values of the relative density and different mechanical properties measured in the final functionally materials. It can be observed that final density is rising with the increments of titanium in the composites. Such situation appears to be verified by a strong contraction of the samples after sintering. Microhardness was evaluated in two different parts of the samples, the first one at the edge of the sample and the other near the core of it. Result of microhardness are reported also in table 1, it can be observed that in all samples with titanium additions, the hardness in the edge is larger than the hardness in the core of each sample. This behavior is due to the formation of a hard material near the edge of the sample, in agreements with the next chemical reaction:

$$Ti + C \rightarrow TiC \tag{1}$$

This chemical reaction is thermodynamically possible because its free formation energy is −43.2 Kcal/mol[8]. In this sense if the TiC was formed at the edge of each studied sample it is confirmed that with the cementation process followed in this experimental sequence it is possible to fabricate Al$_2$O$_3$/TiC/Ti - functionally materials.

The values of the fracture toughness measured in each one of the materials studied are also reported in Table I. The fracture toughness of all the samples is superior to that of the pure alumina, from which it is possible to conclude that metallic particle incorporation in a ceramic matrix brings in increments to its toughness. Different authors have reported that the reinforcing mechanism here is due to the crack bridging by ductile metallic ligaments[9-10]. On the other hand, the good degree of densification reached by each of the composites is another factor that also influences considerably the toughness improvement observed.

Table I. Relative densities and mechanical properties measured in functionally materials.

System	Relative density (%)	Edge hardness (GPa)	Core hardness (GPa)	K_{IC} $(MPa \cdot m^{-1/2})$
0 % Ti	94.95	11.97 +/- 0.5	11.94 +/- 0.5	3.2 +/- 0.2
0.5 % Ti	97.64	9.76 +/- 0.3	6.80 +/- 0.3	4.1 +/- 0.2
1.0 % Ti	97.75	10.01 +/- 0.4	9.13 +/- 0.3	4.8 +/- 0.1
2.0 % Ti	97.93	10.34 +/- 0.4	9.69 +/- 0.4	5.0 +/- 0.1
3.0 % Ti	99.76	10.17 +/- 0.5	7.09 +/- 0.3	5.2 +/- 0.1

CONCLUSIONS

- Al_2O_3-based functionally materials with different titanium contents were produced successfully by the combination of techniques such as; mechanical milling, pressureless sintering and cementation.

- The functionally materials microstructures are constituted by an Al_2O_3 ceramic matrix reinforced with a carbide titanium layer and fine and homogeneous distribution of metallic particles in the core.

- The incorporation of ductile titanium inside hard ceramic matrix incremented its toughness. The probable toughening mechanism is crack bridging due to the presence of a homogeneous ductile metal in the composite's microstructure.

ACKNOWLEDGMENT

Authors would thank Universidad Autónoma Metropolitana for technical and the financial support from 2260235 project, Microscopy laboratory from IMP and ESIQIE-IPN for technical support.

REFERENCES

[1]S. Bhaduri and S. B. Bhaduri, Reactive Synthesis of Dense FGMS in the Ti-B Binary System, *Ceram. Eng. Sci. Proc.*, **20**, 235-242 (1999).
[2]J. K. Wessel, The Handbook of Advanced Materials, John Wiley & Sons, 1 (2004).
[3]Y. Miyamoto, W. A. Kaysser, B. H. Rabin, A. Kawasaki and R. G. Ford (eds.), *Functionally Graded Materials; Design, Processing and Applications*, Kluwer Academic, 1 (1999).
[4]S. Sampath, H. Herman, N. Shimoda and T. Saito, Thermal Spray Processing of FGM's, *MRS Bull.*, **20**, 27-31 (1995).
[5]B. H. Rabin and I. Shiota, Functionally Gradient Materials, *MRS Bull.*, **20**, 14-18 (1995).
[6]M. Koizumi and M. Niino, Overview of FGM Research in Japan, *MRS Bull.*, **20**, 19-26 (1995).
[7] A. G. Evans and E. A. Charles, Fracture Toughness Determination by Indentation, *J. Am. Ceram. Soc.*, **59**, 371-372 (1976).
[8]Handbook of Chemistry and Physics, 66th edition, CRS Press, R.C. Weast editor, D90 (1986).
[9]J. Rodel, M. Sindel, M. Dransmann, R. W. Steinbrech and N. Claussen, R-curve Behaviour in Ceramic Composites Produced by Directed Metal Oxidation, *J. Eur. Ceram. Soc.*, **14**, 153-161 (1994).
[10]N. Claussen, M. Knechtel, H. Prielipp and J. Rodel, Meteers- A Strong Variant of Cermets, *Ber. Dtsch. Keram. Ges.*, **71**, 301-303 (1994).

BRAZING OF MIEC CERAMICS TO HIGH TEMPERATURE METALS

S. Dabbarh, E. Pfaff, A. Ziombra, A. Bezold

Institute for Materials Applications in Mechanical Engineering (IWM)
RWTH Aachen University, Aachen, Germany

ABSTRACT

Mixed ionic/electronic conductors (MIEC) are in the center of interest as oxygen transport membranes (OTM). For OTM-reactors the perovskite ceramic Ba0.5Sr0.5Co0.8Fe0.2O3-x (BSCF) is considered as the best solution due to its high oxygen permeation flux. For an industrial application the gastight joining of BSCF components to metal partners is a special challenge in particular because of its high thermal expansion and low strength. The so far best joining solution for BSCF-tubes, which was used in an oxygen separation module at RWTH Aachen University, is the butt-to-butt joint with Reactiv Air Brazing (RAB) brazing, but the joining stresses are very high. Aim of this study is to develop a brazing technique for BSCF tubes in order to minimize joining stresses. These are mainly caused by the mismatch of the thermal expansions of the partners but can be reduced by optimization the joining design. Also the RAB braze was modified to adapt the thermal expansion to the brazing partners. The design of the joint was changed from butt-to-butt-joint to sleeve joint. Different variations of sleeve joints were simulated, tested and compared to the butt-to-butt-joint.

INTRODUCTION

As part of worldwide activities a research project at Aachen University (Germany) is aiming a CO_2-emmission-free coal combustion process [1]. The coal will be combusted with a mixture of oxygen and re-circulated exhaust gas, which consist mainly of CO_2. The non circulated part of the CO_2 will be compressed and stored underground. An essential requirement for this process is the support of pure oxygen. The core component of this process is an oxygen membrane module [2]. The tubular ceramic membranes are manufactured from $Ba_{0.5}Sr_{0.5}Co_{0.8}Fe_{0.2}O_{3-x}$ (BSCF), which is known from literature, e.g. Wang, H. et al [3], Shao, Z. et al. [4], Wang et al.[5] to have highest oxygen permeation properties and comparably good phase stability.

The materials described in literature were prepared in lab scale only so a first task is an optimization of the ceramic process to obtain best mechanical and functional properties of real components. Zwick, M. et al[6] investigated sintering temperature, atmosphere and time to reach different microstructures. The strength and Weibull-modulus could be improved. Front is material tubes can be manufactured by isostatic pressing and has then to be joint to metal parts. This is an important step from a material to an application. Compared to SOFC materials joining is a notably challenge because of the very high thermal expansion of BSCF and high temperature steels. Thermal expansion coefficients of 15-20 $10^{-6}K^{-1}$ have to be handled.

There are several approaches for gas tight joining ceramics to metals for high temperature applications. Chou, Y.-S. et al[7] used mica seals for solid oxide fuel cells. When the mica is heated up it swells and fills the gap between the parent materials. But a sufficiently hermetic joint could not be reached and thermal cycling is impossible. For joining BSCF to Alloy 310Si glass sealants were investigated by Yang et al.[8] and Emelianova et al.[9]. The most difficulty in this case is the crystallization of glass that involves the change of the properties especially the thermal expansion of the sealant. Wang[3] and Shao[4] used glass ceramic sealants to join BSCF in their oxygen permeation testing device. This worked for the measurement time but after cooling down to room temperature the mismatch of thermal expansions damaged sealant and membrane.

Another group of brazing materials are metal fillers. OTM and SOFC work in oxidizing atmosphere that means filler material also should be stable in oxygen. Only precious metals can meet this demand. Active metal brazing can be used for a wide range of material combinations but needs high vacuum for processing which decompose OTM-materials. For brazing in air the so called Reactive Air Brazing (RAB) with a Silver-Copper oxide-braze was introduced by Weil, K.S. et al.[10-13]. To overcome the insufficient wetting of the pure silver on ceramics and oxidized metal surfaces CuO was added as reactive component. Beside other ceramic parent materials Weil [10-13] tested the $La_{0.5}Sr_{0.5}Co_{0.8}Fe_{0.2}O_{3-x}$ perovskite that is similar to BSCF. Rice et al. [14] investigated the commercial braze Nioro® for reactive air brazing of Zirconia for SOFCs. It contains gold as precious metal component and nickel as reactive component. The Brazing temperature of this braze is higher due to the replacement of silver by the higher melting gold. But the thermal expansion of this braze is lower than Ag CuO braze.

To combine the good wettability of glasses on oxide surfaces with the ductility of metals Deng, X[15] developed a silver-glass composite as sealant for SOFC and sensors. But the leakage test was performed on alumina cap-to-tube joints. Aim of the present study is to optimize a cap to tube joint using RAB.

EXPERIMENTAL

Starting from a butt-to-butt joint with an inner braze meniscus the brazing process was optimized step by step. The first step was a modified design of the butt-to-butt-joint that can reduce joining stresses. The cap geometry was changed so that the meniscus is situated on the outside of the ceramic tube. A second step was the development of a cap geometry for sleeve-joints, simulated and tested practically in combination with the AgCuO braze.

Then also the chemical constitution of filler was changed. Silver was substituted by silver-palladium-alloys, two commercial precious metal filler materials Palcusil15® and Palcusil20® containing Pd and Ag as filler alloy and Copper as reactive component. Under assumption that the CTEs of the alloys show a linear relationship to the atomic ratio of the elements the thermal expansions of these materials should be better adapted to the parent materials $Ba_{0.5}Sr_{0.5}Co_{0.8}Fe_{0.2}O_{3-x}$ (BSCF) and X15CrNiSi25-20 (alloy 310Si). Combined with a new design for cap to tube joints a low stress joint should be possible. These brazes are normally used for vacuum or inert-gas brazing and were tested now for Reactive-Air-Brazing. But the desired oxidation of copper changes the properties of filler material. That's why the values from the datasheets could not be used and had to be determined again for usage in air.

The BSCF ceramic samples for the different tests, which are described below, were manufactured from BSCF powder delivered by Treibacher Industrie AG, Austria. The powder was granulated and pressed to preforms by isostatic pressing. After green shaping the samples were sintered at 1100 °C for 5 hours.

The coefficient of thermal expansion (CTE) is of essential importance for the selection of joining partners and brazes. The CTEs of BSCF, alloy 310 and the fillers were measured with a dilatometer type Netzsch 402C in the range between 200 and 900°C. A Netzsch DSC 404 was used to determine the melting range of brazes and the heat capacity of the joining partners. The thermal transport properties that are necessary for the FEM-simulation of the joining stresses were determined by a Netzsch Laser Flash LFA 427. To complete the materials' models for simulation the Young-modulus and the Shear modulus were determined by the resonance frequency method in the temperature range from RT to 1000°C. Wetting tests of different brazes on BSCF and alloy 310 substrates were performed by heating microscopy.

Brazed samples were characterized by SEM and EDS to see bonding layers, different phases in the joint and the geometry of the joint for the additional simulation. To evaluate the mechanical behaviour of the cap-to-tube-joints, they were put under a growing inner pressure until rupture. The test was performed in water to see leakages. The measured fracture loads were compared to the simulations that were performed by Bezold, A. and Chung[16].

RESULTS AND DISCUSSION

At first brazing with AgCuO using different joint geometries will be discussed. In fig. 1 is shown that the coefficients of thermal expansion of the joining partners are adapted relatively good. The thermal expansion of the AgCuO braze however, that ideally should be situated between the parent materials, is above them and results in high joining stresses. As shown below these stresses can be reduced by an optimized geometry.

The improvement of wetting by the addition of copper oxide to pure silver for BSCF as well as alloy 310 is presented in fig. 2. The wetting angle Θ of pure silver in figures 2 a) and 2 c) is wide above 90°: 115° for steel and 125° for BSCF ceramic. The wetting angle of AgCuO is 45° for steel (fig. 2 b) and 57° for BSCF (fig. 2 d).

The results of FEM simulations of the two butt-to-butt joint variations (figs. 3 and 4) reveal that a meniscus on the inside of the BSCF tube causes joining stresses in the range between 120 and 155 MPa. A meniscus on the outside of the BSCF tube in Fig. 3 causes much lower stresses, between 40 and 80 MPa. The stress singularity in fig. 3 is an artifact of the iteration in the FEM simulation process. Mechanical tests yielded strengths of about 125 MPa for a compression test of BSCF tube segments with outer diameter of 10 mm, inner diameter of 8 mm and a length of 10 mm. That means that the joining stresses in case one are in the same magnitude as the strength of the material.

For testing inner meniscus 41 joints were performed. Brazing was optical successful but the leak tightness test showed that only 9 of the tubes were gastight at an inner pressure of 1 bar. The leakages were always located in the BSCF close to the joint like it was predicted by the FEM simulations. The results of the burst tests are presented in table I. Here also 4 of 5 samples had leakages near the brazing at 1 bar inner pressure. Rupture occurred always at pressures lower than 10 bars, which corresponds to a stress of 6 to 7 MPa.

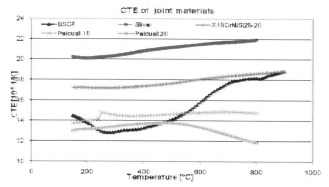

Fig.1 Coefficients of thermal expansion of parent and braze materials

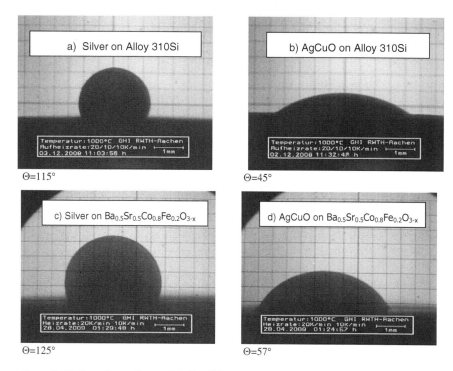

Figure 2: Wetting of pure silver and AgCuo filler

For testing the outer meniscus (fig. 4) 34 brazing joints were made, from which 7 had leakages directly in the filler but no in BSCF. These were the first joints made with this new cap geometry. The reason for these areas of missing filler material is the more difficult alignment with different outer diameters. Therefore a foam polymer was used as inner alignment support. The afterwards performed joints caused no problems and were gastight at 4 bars.

The results of the burst tests are in table I. Here the results of the tests were not as uniform as for case with inner meniscus. The 9 tested samples were all gas tight. The rupture occurred in the BSCF tube mostly.

Table I: Rupture pressure of different joint geometries

joint geometry	rupture pressure in bar		
	min	average	max
butt-to-butt, inner meniscus	3	8	10
butt-to-butt, outer meniscus	6	22	39

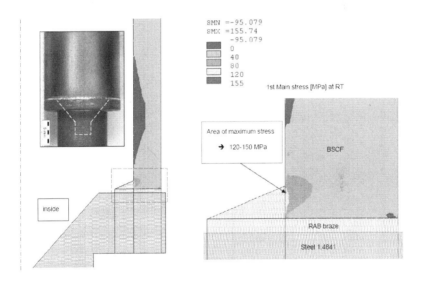

Fig. 3 FEM-Simulation of joining stresses with inner brazing meniscus

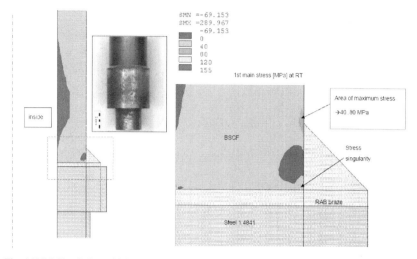

Fig. 4 FEM-Simulation of joining stresses with outer braze meniscus

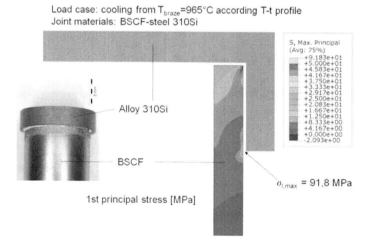

Fig. 5 Simulation of cap-to-tube sleeve joint

The joining stresses of lap joints, see fig.5 are higher than in the butt-to-butt joint. So joined samples were not gas tight. The reason for this insufficient result is still the mismatch of the thermal expansion of the AgCuO filler material.

In SEM the joints were investigated closer to see the intermediate phases between the parent materials and the filler (fig. 6 a-d). EDX Analysis showed that the intermediate layer in fig. 6 b contains Cr_2O_3 as well as Cu and Ag. In fig. 6 c) a infiltration of the AgCuO filler into BSCF can be seen. The infiltration on the chamfered edge of the ceramic is much higher than on the as fired surface, fig.6a. In fig. 6 d) the reaction phase was analyzed by EDX. It contains all the elements of the joint: Ba, Sr, Co, Fe from the ceramic, Ag and Cu from the filler and Cr, Ni and Fe from the steel. It shows that during the brazing process an intensive materials transport takes place.

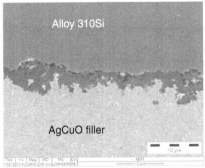

6a) overview of a lap joint

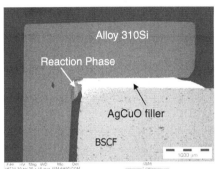

6b) detail of the filler-Alloy 310Si interface

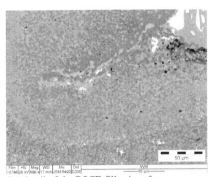

6c) detail of the BSCF-filler-interface

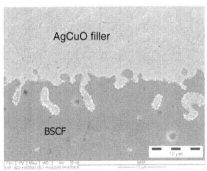

6d) detail reaction phase

Fig. 6: SEM investigation

In Palcusil15® silver and palladium have an atomic ratio of about 81:19; for Palcusil20® about 70:30. For Palcusil15® a CTE of about $15.2 *10^{-6} K^{-1}$ was calculated, for Palcusil20® about $14.7 *10^{-6} K^{-1}$ for room temperature. The measured values in fig. 1 are below the calculated values but they still match the CTEs of the parent materials very well, especially the very sensitive BSCF ceramic. The influence of the copper oxide is greater than expected. Without oxidation (that means brazing in protective atmosphere or vacuum) both brazes have thermal expansions of $18 * 10^{-6}$ K^{-1} (Information of Wesgo Ceramics GmbH, Erlangen). The melting points determined for the Palcusil® filler materials in air are shown in fig. 7 and 8. Melting of Palcusil15® according to fig. 7 starts at about 850°C, which corresponds well with the solidus temperature from the WESGO datasheet. But the liquidus temperature is about 1080°C which is far from the liquidus value of 900°C from the datasheet. The difference is related to the partly oxidation of the copper when measuring in air. The flux effect of the copper is partly lost. The brazing temperature was set to 1090°C for the brazing tests to guarantee that the filler material will be molten completely. The results for the Palcusil20® are similar to the Palcusil15®. The liquidus temperature in air is higher than in vacuum and reaches a value of 1128°C. The brazing temperature was set to 1130°C.

But these brazing temperatures of both Palcusil®-brazes in the sintering range of the BSCF and above the recommended limit of application of the steel (1000°C). Nevertheless some joining tests were made to investigate if a short term exceeding of the application limits for brazing is possible. A SEM investigation of a lap joint BSCF-Palcusil® - Alloy 310Si is presented in fig. 9. It shows that higher temperature accelerated the material transport and reaction of the joining partners. The upper reaction layer contains 4 different phases. The black phase contains Ba, Cr, Co, Fe oxides. The white spots are metallic Ag-Pd-alloy. The light gray phase contains Fe, Co, Ni, Cu, and Ba oxide. In the middle gray phase Sr, Ba and Fe oxides were found. The joint with Palcusil20® did not survive the sample preparation for SEM. The reaction between the different materials damaged the ceramic too much.

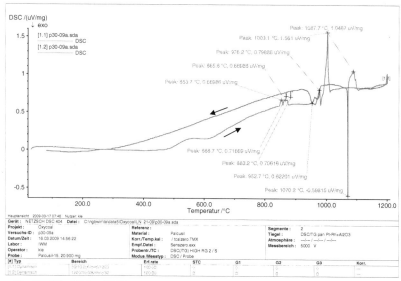

Fig. 7 DSC of Palcusil15®

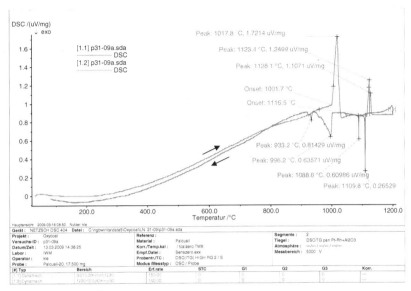

Fig.8 DSC of Palcusil20[®]

Fig. 9 a) overview Palcusil15[®] lap joint

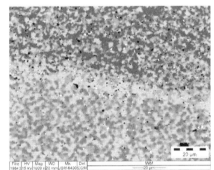

Fig. 9 b) reaction layers in the joining gap

CONCLUSION

Three different geometry variations of a BSCF-AgCuO-Alloy310Si joint were investigated by FEM and joining tests. The best solution is the butt-to-butt joint with an outer filler meniscus. The butt-to-butt joint with inner meniscus as well as the sleeve joint caused higher stresses. The importance of the joint design on the joining stresses was proved.

The commercial filler materials Palcusil15® and Palcusil20® were characterized as reactive air brazes. The liquidus temperatures proved to be too high for BSCF. A heavy reaction dissolves the ceramic. The two fillers are not suitable for joining these to parent materials.

The braze alloy has to be changed further to reach optimum joining conditions. The Pd content could be reduced and/or replaced by other precious metals in order to reduce liquidus temperature and increase CTE. An alloy with lower Copper content could reduce the reactivity of the filler, but will increases the liquidus temperature.

ACKNOWLEDGEMENT

The authors gratefully acknowledge financial support by German Ministry BMWi and MIWFT as well as companies RWE Power AG, E.On AG, Hitachi Power Europe, Linde AG, MAN Turbo and WS-Wärmeprozesstechnik.

REFERENCES

[1] E.M. Pfaff; M. Zwick: "Oxyfuel Combustion Using Perovskite Membranes" in Mechanical Properties and Performance of Engineering Ceramic and Composites III, Vol 28, **2**, 23-31 (2007)

[2] E.M. Pfaff; M.Zwick; S. Dabbarh; C. Broeckmann: „Module Design for MIEC Membranes in OXYCOAL-AC" in Proc. CD-ROM CCT 2009 Clean Coal Technologies 19.-21.05.2009, Dresden (Germany)

[3] Wang, Haihui; Cong, You; Yang, Weishen: "Oxygen permeation study in a tubular $Ba_{0.5}Sr_{0.5}Co_{0.8}Fe_{0.2}O_{3-\delta}$ oxygen permeable membrane", Journal of membrane science 210 (2002) 259-271

[4] Shao, Zongping; Yang, Weishen; Cong, You; Dong, Hui; Tong, Jianhua; Xiong, Guoxing: "Investigation of the permeation behavior of a BSCF oxygen membrane", Journal of Membrane Science 172 (2000) 177-188

[5] Wang, Haihui; Wang, Rong; Liang, David Tee; Yang, Weishen: "Experimental and modeling studies on BSCF tubular membranes for air separation", Journal of Membrane Science 243 (2004) 405-415

[6] Zwick,M., Pfaff, E.M.: "Microstructure Optimization of $Ba_{0.5} Sr_{0.5} Co_{0.8} Fe_{0.2} O_{3-\delta}$ Perovskite for OTM Reactors"; 32nd Int. Conf. on Advanced Ceramics and Composites, Daytona Beach FL, USA, 28.1 - 1.02.2008

[7] Chou, J.S.; Stevenson, J.W.; Chick, L.A.: "Ultra-Low Leak Rate of Hybrid Compressive Mica Seals for Solid Oxide Fuel Cells", Journal of power sources 112 (2002) 130-136

[8] Zhenguo Yang, Guanguang Xia, Kerry D. Meinhardt, K. Scott Weil and Jeff W. Stevenson: "Chemical stability of glass seal interfaces in intermediate temperature solid oxide fuel cells", Journal of materials Engineering and Performance, Vol. 13 (2006) 327-334

[9] Emilianova, S., Pfaff, E.M., Ziombra, A.: "Development of Glass Ceramic Sealants for Ceramic - Metal - Joints under Consideration of High Thermal Expansion; XXI. Int. Congress on Glass, Straßbourg (F) 01.-06.06.2007

[10] US-Patent No. US 7,055,733 B2, Weil et al.:"Oxidation Ceramic to Metal Braze Seals For Applications in High Temperature Electrochemical Devices and Method of Making", 2006

[11] K.S. Weil, J.S. Hardy, J.P. Rice, J.Y. Kim: "Brazing as a means of sealing ceramic membranes for use in advanced coal gasification processes", Fuel 85 (2006) 156-162

[12] Jin Yong Kim, John S. Hardy, K. Scott Weil:"Effects of CuO Content on the Wetting Behavior and Mechanical Properties of a Ag–CuO Braze for Ceramic Joining" J.Am.Ceram.Soc. 88 (2005) 2521-2527

[13] K. Scott Weil, Christopher A. Coyle, Jens T. Darsell, Gordon G. Xia, John S. Hardy: "Effects of thermal cycling and thermal aging on the hermeticity and strength of silver–copper oxide air-brazed seals", Journal of Power Sources (2005)

[14] J.P. Rice, D.M. Paxton, K.S. Weil: "Oxidation Behavior of a commercial gold-based braze alloy for ceramic-to-metal joining"; Ceramic engineering and science proceedings, 26th annual conference on composites, advanced ceramics, materials, and structures, Cocoa Beach (2002)

[15] Xiaohua Deng, Jean Duquette, Anthony Petric: "Silver-Glass Composite for High Temperature Sealing", International Journal of Applied Ceramic Technology 4 [2] 145-151 (2007)

[16] Bezold, A.; Nguyen, V.C.; Broeckmann, C.:" Structural optimization of ceramic/metal braze joints for OTM applications", 2009

Advanced
Ceramic Coatings

THERMODYNAMIC DATA FOR Y-O-H(g) FROM VOLATILIZATION STUDIES

E. Courcot[1], F. Rebillat[1], F. Teyssandier[1]

[1] Université de Bordeaux, Laboratoire des Composites Thermo-Structuraux (LCTS) UMR 5801
3 Allée de la Boétie 33600 Pessac, France

ABSTRACT

A methodology based on a weight loss measurement was used to quantify the volatility of yttria in high temperature water vapor. This method was first validated on silica. Pellets were exposed at temperatures between 1000°C and 1400°C in air with 50 kPa of water at atmospheric pressure, under a flowing gas velocity of 5 cm.s^{-1}. Besides the volatilization rate, the nature of the volatile gaseous species was determined. Knowing the nature of flows in the furnace, partial pressures of yttrium (oxy-)hydroxide in equilibrium above Y_2O_3 were calculated, and the enthalpies of formation of YO(OH) and Y(OH)$_3$ were assessed.

INTRODUCTION

Bulk silicon-based ceramics are among the main candidates for high temperature structural components. One key drawback of these materials in corrosive environment is the volatilization of the protective silica scale formed in moisture and the resulting ceramic recession[1, 2]. Therefore, the ability to use these ceramics as components depends on the development of external protection against water vapor corrosion. Some of the most promising materials seem to be rare earth silicates since they exhibit a high chemical resistance in extreme environments (high temperature, high water pressure). However, corrosion tests performed at 1400°C in moist atmosphere (P_{H_2O} = 50 kPa, P_{tot} = 100 kPa) on these materials have revealed significant vaporization by reaction with H_2O to form rare earth hydroxides (supposed to be REOOH and RE(OH)$_3$)[3]. In order to quantify the stability of these materials, a thermodynamic study has to be carried out. However very few thermodynamic data are available for yttrium hydroxides. Consequently, the aim of this work is to assess thermodynamic parameters of these compounds through the study of yttrium sesquioxide (Y_2O_3) corrosion kinetics.

EXPERIMENTAL APPROACH

In order to determine the Gibbs free energy of formation of gaseous hydroxides, their partial pressures in equilibrium over the solid oxide in a controlled atmosphere had to be quantified. For that, the two most common techniques are the Knudsen cell or the transpiration method. In order to avoid problems linked to the condensation or the detection limit of gaseous species, a methodology was developed to assess thermodynamic data from a quantitative study of oxide volatilization rates. This method is based on weight loss measurements during corrosion at high temperatures (between 1000 and 1400°C), under high water vapor pressure (from 20 kPa to 70 kPa) at atmospheric pressure. The assessment of thermodynamic data is allowed only when the partial pressures are able to be determined. This requires that the thermodynamic equilibrium is reached: the hydroxides formation rate is much higher than the evacuation one. The volatilization tests were carried out on oxide pellets ($\varnothing$ = 10 mm, thickness ~ 2 mm, S ~ 150 mm^2, porosity ~25 %) desorbed at 1300°C during 5 hours (SiO$_2$, 99.8 %, Chempur, and Y$_2$O$_3$, 99.9 %, Chempur). The conditions of corrosion are the following: temperatures between 1000 and 1400°C, a moist environment (water vapor partial pressure up to 65 kPa, 100 kPa total pressure), with different gas velocities, during times between 1.5 hours and 25 hours. A schematic of the high temperature furnace is presented in Figure 1. In order to supply air containing a controlled amount of steam at room temperature (P_{H2O} ~ 2-3 kPa), dry air goes first through the presaturator, then through the heating column to obtain the wanted water vapor partial pressure. This latter is checked by measuring the amount of liquid water used during the test. The gas flow velocity, v_{gas}, is calculated by taking into account both air and water content.

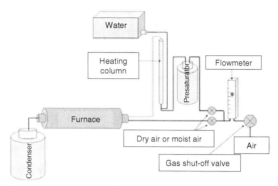

Figure 1. High temperature corrosion furnace working at atmospheric pressure.

During corrosion tests, the weight loss corresponding to reactive volatilization (1) is measured (generally it was more than 0,20 mg). In the case of a porous pellet, allowing the production of large amount of hydroxides, it can be considered that their evacuation over the sample occurred by convection, through the geometric surface (the diffusion of hydroxides inside the porosities of this pellet is a fast phenomenon in regard to their evacuation by convection). Knowing the sample surface area in contact with the moist environment, the hydroxide volatilization rate k_l, is deduced directly from the slope of the straight line representing the weight loss / surface ratio as a function of corrosion time. Three independent experiments were carried out and if similar results were obtained, data were considered as accurate. The errors bars in the different figures indicated the discrepancy between these three independent experiments.

$$\text{Oxide (s)} + H_2O \text{ (g)} \rightarrow \text{Hydroxide (g) or Oxy-Hydroxide (g)} \qquad (1)$$

In our working conditions, the moisture supply was very high compared to the consumed quantity, and further, the reaction kinetics being very slow, the low quantity of gaseous products was assumed to be easily evacuated far from the sample surface. Thus the volatilisation process was only limited by the surface reaction. From values of Reynolds numbers (equal to 45 at 1673 K), it has been first checked that the volatilization process is mainly controlled by a convective flow through the furnace. Moreover, calculations of the involved gaseous flows show that the convection would limit the volatilisation process (in regard to the diffusion one).The hydroxide partial pressure over the oxide is thus calculated from the volatilization rate (2).

$$k_l = v_{gas} \times 3{,}6 \times M_{M_aO_b} \times \frac{1}{a} \times \frac{P_{M-OH}}{R \times T_{amb}} \qquad (2)$$

Where k_l is the volatilization rate (mg.cm^{-2}.h^{-1});

v_{gas} is the gas flow velocity at room temperature (cm.s^{-1});

$M_{RE_2O_3}$ is the molar weight of RE_2O_3 (g.mol^{-1});

a is the stoichiometry ratio before $MO_x(OH)_y$ in (3);

P_{RE-OH} is the sum of the volatile (oxy-)hydroxide partial pressures (Pa);

R is the ideal gas constant (= 8,314 J.mol^{-1}.K^{-1});

T_{amb} is the ambient temperature (296 K).

In order to validate the experimental procedure, experiments were first realized on silica. Actually, silica is chosen because its behavior under a moist environment is well known[4] and all the thermodynamic parameters of silicon hydroxides are already assessed[5]. Silica pellets were prepared as the other materials in this study, to always consider the same exchange surface area.

The volatility of silica in a moist environment (P_{H2O} = 50 kPa, P_{tot} = P_{atm}) induced a linear weight loss that increases with temperature. This thermal activated process has an activation energy value around 56 kJ.mol^{-1}, that is in agreement with those in the literature (61 ± 8 kJ.mol^{-1})[6, 7]. The partial pressure of $Si(OH)_4$ (predominant species under moisture) was deduced from the volatilization rates obtained at 1100 and 1400°C (2) and then compared with the calculated partial pressures by running minimization of the Gibbs free energy of the chemical system with parameters corresponding to the experimental parameters (from thermodynamic data given by [5]) (Table I). A good agreement between experimental and calculated equilibrium hydroxide partial pressures above SiO_2 is obtained. The deviation between these two values is less than 10 %. The experimental approach can be reasonably considered as accurate and the following main assumptions are validated: the flow rate used in the furnace allows thermodynamic equilibrium to be reached along the sample surface area, and the entrainment of gaseous hydroxide species out of the solid surface is mainly controlled by gas convective flows.

Table I. Volatilization rates of silica (P_{H2O} = 50 kPa, P_{atm}, v_{gas} = 5 cm.s^{-1}) and partial pressures of $Si(OH)_4$

T (K)	k_1 exp (mg.cm^{-2}.h^{-1})	$P_{Si(OH)4}$ exp (Pa)	$P_{Si(OH)4}$ thermo (Pa)	deviation
1373	0,0145 ± 0,0002	0,0326 ± 0,0005	0,0344	9 %
1673	0,0349 ± 0,0003	0,0784 ± 0,0007	0,0837	10 %

RESULTS

In order to apply this methodology to an unknown compound, like Y_2O_3, it has first to be checked that the thermodynamic equilibrium was reached along the sample surface area. According to (2), if the convection remains the limiting step in the volatilization process, the volatilization should stay proportional to the gas flow velocity. For this purpose, volatilization tests were carried out at 1200°C in air with 50 kPa of water at atmospheric pressure, under different flowing gas velocities (Figure 2). For a gas velocity superior or equal to 4 cm.s^{-1}, the proportionality relation is confirmed: a straight line through the origin is obtained, as expected (2). The thermodynamic equilibrium appears to be reached whatever the gas velocity and the same value of partial pressure of the gaseous volatile species above Y_2O_3 is deduced from the slope. For a gas velocity much lower than 5 cm.s^{-1}, the thermodynamic equilibrium is reached but the evacuation rate of gases is lower: the limiting step becomes the diffusion across the gaseous phase around the sample surface. Consequently, in order to be at the thermodynamic equilibrium, with convective flows, a working gas velocity of 5 cm.s^{-1} is chosen.

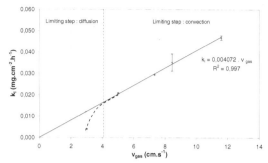

Figure 2. Volatilization rate of Y_2O_3 in function of gas velocity
(under corrosion at 1200°C, in air with 50 kPa H_2O, at P_{atm}).

Further, in order to determine the kinetics of Y_2O_3 volatility, volatilization tests are done at different temperatures in air with a water partial pressure of 50 kPa at atmospheric pressure under a gas velocity of 5 cm.s⁻¹. The variations of the ratio between the weight change and the surface of the pellet as a function of time are shown in Figure 3. At a given temperature, the interfacial reaction process induces a linear weight loss: the slope of the linear curves corresponds to the volatilization rate k_l.

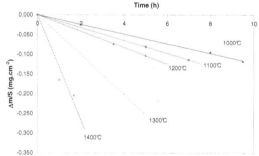

Figure 3. Weight changes during volatilization of Y_2O_3 at different temperatures
(under corrosion at 50 kPa H_2O, at P_{atm} with $v_{gas} = 5$ cm.s⁻¹).

Using an Arrhenius representation, the thermoactivation of the reaction of Y_2O_3 with the moisture showed a discontinuity around 1200°C (Figure 4). Below 1200°C, the activation energy is 41 ± 1 kJ.mol⁻¹ whereas a much larger activation energy is calculated above 1200°C: 175 ± 10 kJ.mol⁻¹. This behavior is characteristic of a change in the volatilization mechanism. Two domains of reactivity were thus assumed to take place according to temperature. They were associated with a change of the hydroxide species formed. Actually, two gaseous species, could be formed when yttria is in contact with waper vapor at high temperature (3-4).

$$Y_2O_3 + 3\ H_2O \longrightarrow 2\ Y(OH)_3$$

(3)

$$Y_2O_3 + H_2O \longrightarrow 2\ YOOH$$

(4)

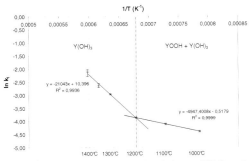

Figure 4. Determination of the apparent kinetic laws of Y_2O_3 volatilization by reaction with water (under corrosion at 50 kPa H_2O, at P_{atm} with v_{gas} = 5 cm.s^{-1}).

Now, in order to identify which species is formed in the two temperature domains, volatilization tests were carried out at different temperatures in air with different water partial pressures ranging from 17 kPa to 65 kPa at atmospheric pressure, with a gas flow velocity of 5 cm.s^{-1}. In fact, the (oxy-)hydroxides partial pressures depend on the water partial pressures according to equations 3 and 4 (5). The (oxy-) hydroxides pressures are calculated from (2) and take into account the experimental gas flow velocity.

$$K = \frac{P_{M-OH}}{P_{H_2O}^y} \qquad \ln P_{M-OH} = \ln K + y \ln P_{H_2O} \tag{5}$$

Consequently, the plot of ln P(Y-OH) vs ln P(H_2O) would yield a slope of 1,5 for the formation of $Y(OH)_3$ (2), whereas the slope would be equal to 0,5 for the formation of YOOH (3).

As expected, the yttria volatility rate increases with the water vapor partial pressures. The results are shown in figure 5 for temperatures equal to 1000, 1200 and 1400°C and indicate slopes of 0.87 ± 0.01, 1.19 ± 0.02 and 1.41 ± 0.03 respectively. From these results, formation of YOOH is expected to take place at low temperatures, whereas $Y(OH)_3$ should be the high temperature hydroxide species. The values of slopes increase with temperature, that means that the proportion of $Y(OH)_3$ raises with temperature at the expense of YOOH. As shown in Arrhenius graph (Figure 4), the change in mechanism at 1200°C corresponds to a change of gaseous species. For temperatures below 1200°C, the predominant species is YOOH, whereas for temperatures above, $Y(OH)_3$ becomes the major species. From 1300°C, $Y(OH)_3$ can be considered to be the only volatile gaseous species since the partial pressures of YOOH should be negligible.

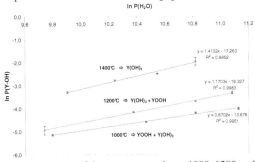

Figure 5. Determination of the nature of the gaseous species at 1000, 1200 and 1400°C through the extraction of the reaction order associated to H_2O (P_{atm}, v_{gas} = 5 cm.s^{-1}).

FREE ENERGY OF FORMATION OF YTTRIUM (OXY)-HYDROXIDES

The assessment of the thermodynamic data of each gaseous species is based on the calculation of each gaseous species partial pressure. These partial pressures are estimated from the average volatilization rate (2) from Arrhenius Law (Figure 4) obtained with a water partial pressure of 50 kPa. In this study, two cases can be distinguished: the first one at high temperature (> 1300°C), where $Y(OH)_3$ is the predominant species and the second one at low temperatures (< 1200°C), where a mixture of YOOH and $Y(OH)_3$ is present. First by using a Gibbs free energy minimization[8], the Gibbs free enthalpy of formation of $Y(OH)_3$ is assessed in the high temperature domain, where two yttrium bearing species have to be considered: $Y(OH)_3$ and YO_2. This latter species is the predominant species formed in dry air. The thermodynamic data used for these calculations are from the COACH database associated to the software, except for Y_2O_3, where its data come from Djurovic[9]. By running thermodynamic calculations using parameters corresponding to the experimental conditions (temperature, total pressure and H_2O partial pressure), the value of the Gibbs free energy of formation of the selected hydroxide is modified until to provide a partial pressure identical to the measured one. Repeating this procedure for several temperatures (1300, 1350 and 1400°C) allowed establishing the temperature dependence of the Gibbs free energy of formation of the hydroxide in the temperature range (6).

$$\Delta G_{Y(OH)_3} = 1{,}1799999999.10^{-2}\,T^2 - 5{,}756228.10^2\,T - 9{,}574324178.10^5 \tag{6}$$

Extrapolation of this equation in the low temperature domain allowed to determine the partial pressures of $Y(OH)_3$ below 1300°C. The partial pressure of YOOH was thus deduced from (7)

$$P_{\exp} = P_{Y(OH)_3} + P_{YO(OH)} + P_{YO_2} \tag{7}$$

The same procedure was applied at 1000, 1100 and 1200°C and a temperature dependent equation was obtained in the low temperature domain (8).

$$\Delta G_{YOOH} = 1{,}135\,T^2 - 3{,}13381.10^3\,T - 9{,}75679715.10^5 \tag{8}$$

Partial pressures of $Y(OH)_3$, YOOH and YO_2 in equilibrium with Y_2O_3 in a moist environment were then calculated by Gibbs free energy minimization (Figure 7). Their variations are in agreement with the experimental observations, and confirm that above 1200°C, partial pressure of YOOH are negligible in comparison with $Y(OH)_3$ one.

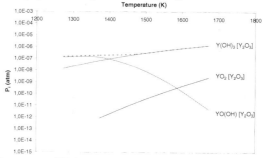

Figure 7. Partial pressures of $Y(OH)_3$, YOOH and YO_2 over Y_2O_3 at equilibrium in a moist environment (P_{Ar} = 40 kPa, P_{O2} = 10 kPa and P_{H2O} = 50 kPa) (dot lines : $P_{Y(OH)3}$ + P_{YOOH}).

The thermodynamic parameters determined in this study were checked for different water partial pressures ranging between 17 kPa and 65 kPa. The good agreement observed in figure 8 validates the methodology and the accuracy of the obtained data.

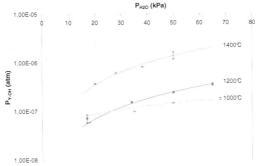

Figure 8. Variation of the partial pressures of yttrium hydroxides ($Y(OH)_3$ + YOOH) measured experimentally (points) and calculated (lines) as a function of water partial pressure
(P_{atm}, v_{gas} = 5 cm.s^{-1})

CONCLUSION

The kinetics study of yttrium sesquioxide volatilization under moist environment enabled : i) to determine the activation energy of this reactive volatilization, ii) to identify the nature of the volatile (oxy-)hydroxides, (iii) to propose a volatilization mechanism in function of temperature, and iv) to assess the Gibbs free energies of $Y(OH)_3$ and YOOH formation. Further works will be devoted to the quantification of the chemical and thermal stability of yttrium silicates.

ACKNOWLEDGMENTS

This work has been supported by the French Ministry of Education and Research through a grant given to E. Courcot. The authors are grateful to Caroline Louchet-Pouillerie and Jacques Thébault for fruitful discussions.

REFERENCES
[1] E.J. Opila, Variation of the oxidation rate of silicon carbide with water vapour pressure, *J. Am. Ceram. Soc.*, **82**, 625-636 (1999).
[2] E.J. Opila, J. L. Smialek, R.C. Robinson, D.S. Fox and J.S. Jacobson, SiC recession caused by SiO_2 scale volatility under combustion conditions: II, thermodynamics and gaseous diffusion model, *J. Am. Ceram. Soc.*, **82**, 1826-1834 (1999).
[3] E. Courcot, F. Rebillat, C. Louchet-Pouillerie, Relation between synthesis process, microstructure and corrosion resistance of two yttrium silicates, *Mat. Sci. Forum*, **595-598**, 923-931 (2008).
[4] N.S. Jacobson, E.J. Opila, D.L. Myers, E.H. Copland, Thermodynamics of gas phase species in the Si-O-H system, *J. Chem. Thermodyn.*, **37**, 1130-1137 (2005).
[5] Sandia National Laboratory
[6] A. Hashimoto, The effect of H_2O gas on volatilities of planet-forming major elements: I. Experimental determination of thermodynamic properties of Ca-, Al-, and Si-hydroxide gas molecules and its application to the solar nebula, Geochim. Cosmochim. Acta, **56**, 511-532 (1992).
[7] E. J. Opila, R. E. Hann Jr, Paralinear Oxidation of CVD SiC in Water Vapor, *J. Am. Ceram. Soc.*, **80**, 197-205 (1997).
[8] Association Thermodata GEMINI 2 Code. B.P.66, 38402 St. Martin d'Heres Cedex, France.
[9] D. Djurovic, M. Zinkevich, F. Aldinger, Thermodynamic modeling of the yttrium-oxygen system, *Comp. Coupl. Phase Diag. Thermochem.*, **31**, 560-566 (2007).

FROM THE VOLATILITY OF SIMPLE OXIDES TO THAT OF MIXED OXIDES: THERMODYNAMIC AND EXPERIMENTAL APPROACHES

E. Courcot[1], F. Rebillat[1], F. Teyssandier[1]

[1] Université de Bordeaux, Laboratoire des Composites ThermoStructuraux (LCTS) UMR 5801
3 Allée de la Boétie 33600 Pessac, France

ABSTRACT

From the partial pressures of rare earth hydroxides measured through corrosion tests on simple oxides, like RE_2O_3, the thermochemical stability of rare earth silicates was quantified. Indeed, these stabilities were first estimated by thermodynamic calculations and further confirmed with corrosion tests. Depending on the silicate rare earth composition, similar quantities of silicon and rare earth hydroxides can be formed as the result of the corrosion process. Further, the influence of the crystallization degree and of the nature of the rare earth on the corrosion resistance was discussed.

INTRODUCTION

Refractory materials which can withstand high mechanical stresses at high temperature under corrosive environment are needed in aeronautic applications, especially in combustion liners. Therefore silicon-based ceramics are among the main candidates for this kind of application. One key drawback of silicon-based ceramics is the volatilization of the protective silica scale, in steam from 1200°C and the resulting ceramic recession[1-2]. That is why the use of these ceramics components depends on the development of external protection, named Environmental Barrier Coatings (EBC), against water vapor attack. This coating requires many criteria in order to be used as an EBC[3]. First of all, it has to be resistant to aggressive environments, so it must possess a low oxygen permeability to limit the transport of oxygen. Moreover its coefficient of thermal expansion (CTE) has to be close to that of the substrate material to prevent delaminating or cracking due to CTE mismatch stress. In addition, the coating must have a stable phase in the considered temperature range. Thus, it must not have any phase transformation, which typically accompanies a volume change. Finally, the coating must be chemically compatible with the substrate to avoid detrimental chemical interaction, unless it is controlled.

Currently BSAS (Barium Strontium Alumino-Silicate) is the reference top coat[4]. However, it has been shown that at 1300°C there is an important reactivity with SiO_2, formed by oxidation of the substrate, due to an eutectic equilibrium existence[5]. Thus, a new low thermal expansion coefficient EBC generation has been developed. Some of the most promising materials seem to be rare earth silicates[6-8]. In corrosive conditions, the lifetime of these materials is strongly limited by a degradation process due to the formation of gaseous hydroxides species, mainly $Si(OH)_4$, as mentioned in the literature[9]. This work consists in quantifying the stability of these materials through thermodynamic and experimental approaches. Moreover, it demonstrates that according to their composition, the stability of rare earth silicates depends on the formation and the volatilization of gaseous rare earth hydroxides too.

THERMODYNAMIC APPROACH

Thermodynamic calculations

GEMINI2 software is used to calculate the equilibrium state between a gaseous phase and a solid material. The method consists in minimizing the total Gibbs energy of the system under constant pressure conditions. The minimization method is based on a general optimization technique which has been applied to the chemical equilibrium problems[10].

In this work, the equilibrium state is calculated in the case of a mixture of one mole of condensed phase with one mole of gaseous phase, at atmospheric pressure, in a closed system. The composition of the initial gaseous phase is: $n_{Ar} = 0,4$ mol, $n_{O2} = 0,1$ mol and $n_{H2O} = 0,5$ mol.

The chemical stability of rare earth silicates can be simulated under thermodynamic equilibrium assumption, starting from thermodynamic data of rare earth silicates. The stability of the material should be improved as soon as the partial pressures of the gaseous species in equilibrium with the material are lower than those in equilibrium with Y_2O_3 and SiO_2, formed by oxidation of the substrate.

Species taken into account in calculations

As mentioned above, the Gibbs free energy of formation of each RE-, Si-, O- and/or H-bearing species are needed to determine the stability of rare earth silicates. However, RE_2O_3-SiO_2 systems were poorly studied in thermochemistry. Only few data can be found:

- the enthalpy of formation of Yb_2SiO_5 at 298 K[11]: this is insufficient for the assessment of the Yb_2O_3-SiO_2 phase diagram ;
- all the thermodynamic data[12-13] required to plot the Y_2O_3-SiO_2 phase diagram.

So it was decided to carry out thermodynamic calculations on the Y_2O_3-SiO_2 system and to draw hypotheses on RE_2O_3-SiO_2, by analogy with Y_2O_3-SiO_2.

In the Y_2O_3-SiO_2 system, two compounds Y_2SiO_5 and $Y_2Si_2O_7$ are found, with two (A and B) and four (α, β, γ and δ) polymorphs respectively. The first has a congruent melting, whereas the second has an incongruent melting. The thermodynamic parameters of this system have already been assessed, by taking into account experimental results[12-13]. In this work, the data from Mao[7] have been considered, since they take into account the update of the thermodynamic description of yttrium (recommended by SGTE)[14]. The considered phases are the following: liquid, Y_2O_3(R), Y_2O_3(H), B-Y_2SiO_5, α-$Y_2Si_2O_7$, β-$Y_2Si_2O_7$, γ-$Y_2Si_2O_7$, δ-$Y_2Si_2O_7$, SiO_2 (cristobalite), SiO_2 (quartz), SiO_2 (tridymite). The other condensed phases taken into account in the calculation are Si, Si_2H_6, Y, YH_2, YH_3, H_2O (Liq) and H_2O_2 (Liq). Their Gibbs free energies of formation are available in the database COACH[15].

With respect to the gaseous phases, the thermodynamic data of oxide and non oxide gaseous species are from the database COACH[15]. The volatilization of materials under a moist environment is mainly due to the formation of hydroxide species. For silicon hydroxides, their free energies of formation come from the database of Sandia National Laboratory[16] and the Gibbs free energy of yttrium hydroxides has recently been assessed[17]. To sum up, the gaseous species taken into account in the calculations are the following: Ar, YO, YO_2, Y_2O, Y_2O_2, SiO, SiO_2, Si_2O_2, H_2, H, O_3, O_2, O, OH, HO_2, H_2O, H_2O_2, Si, Si_2, Si_3, SiH, SiH_2, SiH_3, SiH_4, Si_2H_6, YOOH, $Y(OH)_3$, $Si(OH)_4$ and $SiO(OH)_2$.

Stability of yttrium silicates in a moist environment

The aim of this part is to determine if the yttrium silicates are promising materials as environmental barrier coatings and to quantify their stability.

First, the stabilities of Y_2O_3 and SiO_2 in a moist environment are compared between themselves. The partial pressures of $Y(OH)_3$, YOOH and YO_2 at the equilibrium over Y_2O_3 are lower than those of $Si(OH)_4$, $SiO(OH)_2$, SiO_2 and SiO over SiO_2, for temperatures below 1320°C. Consequently, for these temperatures, Y_2O_3 is more stable than SiO_2 in moist environments. At higher temperatures, it is the contrary: SiO_2 is the most stable under a moist environment. So it was expected an increasing stability of yttrium silicates due to the effect of chemical bonds.

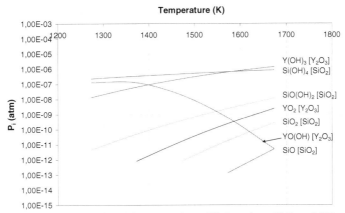

Figure 1. Evolution of partial pressures in equilibrium above Y_2O_3 and SiO_2 under 50 kPa of water at P_{atm}.

Using Gibbs free energy minimization technique, the partial pressures of gaseous species in equilibrium with Y_2SiO_5 or $Y_2Si_2O_7$ are calculated and displayed in figure 2. The partial pressures of YOOH, $Y(OH)_3$, YO_2, $Si(OH)_4$, $SiO(OH)_2$, SiO_2 and SiO at the equilibrium over these systems are lowered by at least one magnitude order with regard to those over Y_2O_3 and SiO_2. This is the result of the strong solid state bonding between yttrium and silicon oxides. Consequently, these phases are interesting since an increased thermal stability is observed compared to silica, whatever the temperature. Moreover, the ratios of the partial pressures between yttrium bearing species and silicon bearing species are similar to the stoechiometric ratio between Y and Si. Consequently, between 1000°C and 1400°C, yttrium silicates are stable compounds in moist environments (at atmospheric pressure with a water vapor pressure of 50 kPa).

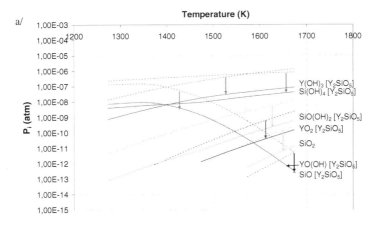

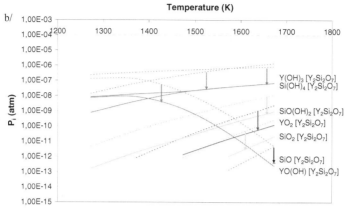

Figure 2. Partial pressures of gaseous species in equilibrium with a/ Y_2SiO_5 and b/ $Y_2Si_2O_7$ in a moist environment calculated with Gemini2 ($n_{Ar} = 0.4$ mol; $n_{O2} = 0.1$ mol and $n_{H2O} = 0.5$ mol)

Now, the stability of yttrium silicates can be quantified by calculating the recession rate k_r. This parameter takes into account the density of the material and allows to compare different materials between themselves in a given moist environment, even if they have different molar weights. These values are deduced from the calculated partial pressures in equilibrium over these materials, in agreement with experiments realized in the same corrosion furnace, whose principles were described elsewhere[17], on rare earth oxides in the same conditions[18] (1-4). Indeed, preliminary experiments were carried out in order to verify that the evacuation of the gaseous species over the sample occurred by convection, through the geometric surface (the diffusion of hydroxides inside the porosities of this pellet is a fast phenomenon in regard to their evacuation by convection). The results obtained for Y_2SiO_5 and $Y_2Si_2O_7$ are gathered in table I and compared with SiO_2 and Y_2O_3.

$$\text{For} \quad Y_2SiO_5 + 5\,H_2O \quad \rightarrow \quad 2\,Y(OH)_3 + Si(OH)_4 \tag{1}$$

$$k_r = 10^4 \times \frac{k_l}{\rho} = 3.6 \cdot 10^4 \times \frac{v_{gas} \times M_{Y_2SiO_5}}{\rho \times R \times T_{amb}} \times P_{Si(OH)_4} \tag{2}$$

$$\text{For} \quad Y_2Si_2O_7 + 7\,H_2O \quad \rightarrow \quad 2\,Y(OH)_3 + 2\,Si(OH)_4 \tag{3}$$

$$k_r = 10^4 \times \frac{k_l}{\rho} = 3.6 \cdot 10^4 \times \frac{v_{gas} \times M_{Y_2Si_2O_7}}{\rho \times R \times T_{amb}} \times \frac{P_{Si(OH)_4}}{2} \tag{4}$$

Where
k_r is the recession rate (nm.h^{-1})
k_l is the volatilization rate (mg.cm^{-2}.h^{-1});
ρ is the density of the material (g.cm^{-3})
v_{gas} is the gas flow velocity at room temperature (cm.s^{-1});
M is the molar weight of the material (g.mol^{-1});
$P_{Si(OH)4}$ is the partial pressure of $Si(OH)_4$ (Pa);
R is the ideal gas constant (= 8,314 J.mol^{-1}.K^{-1});
T_{amb} is the ambient temperature (296 K).

Y_2SiO_5 and $Y_2Si_2O_7$ have comparable recession rates in a moist environment at high temperatures. So their stability is similar and is higher than that of SiO_2. These results confirm the interest of yttrium silicates as environmental barrier coatings.

Table I. Recession rates (k_r in nm.h^{-1}) of Y_2SiO_5 and $Y_2Si_2O_7$ at different temperatures ($P_{H2O} = 50$ kPa, $P_{tot} = P_{atm}$)

k_r (nm.h^{-1})	1000°C	1100°C	1200°C	1300°C	1400°C
Y_2SiO_5	2.0	3.0	4.5	10.1	20.5
$Y_2Si_2O_7$	2.6	4.1	6.0	12.0	21.9
SiO_2	42.5	62.5	87.4	117.2	151.9
Y_2O_3	24.3	32.3	41.2	103.8	230.5

Since RE_2O_3-SiO_2 systems have similar phase diagrams, it can be supposed that rare earth silicates are stable in a moist environment at high temperatures. This assumption should be confirmed by corrosion tests, carried out for different rare earth oxides. It was expected that the classification of stability of rare earth oxides and of rare earth silicates should be similar. Consequently, as ytterbium oxide is the most stable rare earth oxides in a moist environment at high temperatures[19], ytterbium silicates should be the best candidates as environmental barrier coatings.

EXPERIMENTAL APPROACHES

First, in order to validate thermodynamic calculations done on yttrium silicates, corrosion tests were carried out between 1200 and 1400°C in air under a total pressure of 100 kPa with a water vapor partial pressure of 50 kPa under a flowing gas of 5 cm.s^{-1}, during testing times up to 25 hours. This study allows too to demonstrate the influence of the purity or of the crystallization degree of materials on corrosion resistance.

Elaboration

Y_2SiO_5 and $Y_2Si_2O_7$ were elaborated by a reactive sintering from Y_2O_3 (Chempur, 99.9 %) and SiO_2 (Chempur, 99.8 %) powders. Powders were milled, compacted in order to elaborate pellets. These last were sintered at 1300°C, 1400°C during 5 hours or 1500°C during 5 hours and 50 hours. They were characterized by X-Ray diffraction (Fig. 3). The crystallization degree and the purity increases with the sintering temperature or the duration of the heat treatment. A sintering at 1500°C for 50 h seems to be necessary to obtain quasi-pure yttrium silicates. Moreover, the polymorphic transition of $Y_2Si_2O_7$ was highlighted: once the γ-$Y_2Si_2O_7$ is formed at high temperature, it is still present at room temperature. Since γ-$Y_2Si_2O_7$ is stable from ambient temperature to high temperature (1500°C at least), γ-$Y_2Si_2O_{7\,appears}$ as the most promising material for EBC.

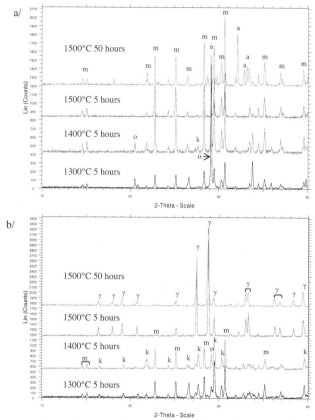

Figure 3. Influence of the sintering conditions on the crystallization degree of a/ Y_2SiO_5; b/ $Y_2Si_2O_7$ (m = Y_2SiO_5, o = Y_2O_3, a = apatite $Y_{4.67}(SiO_4)_3O$, k = Keiviite – $Y_2Si_2O_7$, γ = γ-$Y_2Si_2O_7$).

Corrosion tests on yttrium silicates

Corrosion tests were performed between 1200°C and 1400°C in a moist environment (P_{H2O} = 50 kPa, P_{tot} = P_{atm}, v_{gas} = 5 cm.s^{-1}) on pellets previously sintered. Alumina furnace tube was used for the experiments due to its corrosion resistance at high temperatures. However, in moist environment, alumina can form Al(OH)$_3$ gaseous species that may react with the surface of the pellet and lead to formation of YAG ($Y_3Al_5O_{12}$) or YAM ($Y_4Al_2O_9$) from 1200°C. To prevent as far as possible this pollution to occur, it was decided (i) to limit the experiment duration and (ii) to use a new pellet for each new experiment, and (iii) to dispose the pellet on a zirconia plate. The limited aluminum pollution was checked at the sample surface by use of various characterizations: X-Ray Diffraction, Raman spectroscopy and Energy Dispersive Spectroscopy.

During corrosion tests, a downstream white deposit is observed in the cold zone of the furnace and a weight loss corresponding to the volatilization of materials is measured. Knowing the sample surface area in contact with the moist environment, the oxide volatilization rate k_l is calculated directly from the slope of the straight line representing the weight loss – surface ratio as a function of corrosion time.

Figure 4 shows the linear weight loss recorded for $Y_2Si_2O_7$ at 1200°C and tables 3 and 4 gathered the recession rates deduced from these experimental data (2). These values are compared with the theoretical one and with those of Y_2O_3 and SiO_2.

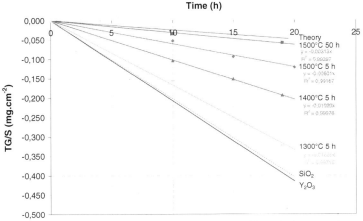

Figure 4. Weight losses of $Y_2Si_2O_7$, sintered under different conditions, during corrosion at 1200°C (P_{H2O} = 50 kPa, P_{tot} = P_{atm}, v_{gas} = 5 cm.s^{-1}).

Table 3. Recession rates (k_r in nm.h^{-1}) of Y_2SiO_5 recorded during corrosion at 1200°C, 1300°C and 1400°C (P_{H2O} = 50 kPa, P_{tot} = P_{atm}, v_{gas} = 5 cm.s^{-1}).

Sample	Corrosion	1200°C	1300°C	1400°C
Y_2SiO_5 Reactive sintering	1300°C 5 h	16,1 ± 0,4	58,3 ± 0,5	121 ± 5
	1400°C 5 h	16 ± 2	45,6 ± 0,2	99 ± 5
	1500°C 5 h	16 ± 3	42,2 ± 0,2	90 ± 5
	1500°C 50 h	9,9 ± 0,7	25 ± 1	50 ± 7
Y_2SiO_5 (Gemini2)		4,5	10,1	20,5
Y_2O_3 (Gemini2)		41.2	103.8	230.5
SiO_2 (Gemini2)		87.4	117.2	151.9

Recession rates of Y_2SiO_5 (Table 3) reveal that the crystallization degree or the purity influences the corrosion resistance. Indeed, when the crystallization degree increases or when the purity is improved, recession rates decrease and a higher stability is observed in a moist environment at high temperatures. A reactive sintering at 1500°C during 50 hours allows to have the highest thermochemical stability. This can be explained by the fact that few impurities (not expected phases) are present in these samples. Moreover, whatever the crystallization degree or the purity, the stability is higher in a moist environment than Y_2O_3 or SiO_2. That is why these materials present an interest to be used as environmental barrier coatings.

Table 4. Recession rates (k_r in nm.h^{-1}) of $Y_2Si_2O_7$ recorded during corrosion at 1200°C, 1300°C and 1400°C (P_{H2O} = 50 kPa, P_{tot} = P_{atm}, v_{gas} = 5 cm.s^{-1}).

Sample	Corrosion	1200°C	1300°C	1400°C
$Y_2Si_2O_7$ Reactive sintering	1300°C 5 h	41 ± 3	113 ± 5	267 ± 15
	1400°C 5 h	25,0 ± 0,5	81 ± 3	199 ± 8
	1500°C 5 h	15 ± 2	55 ± 3	137 ± 10
	1500°C 50 h	7,7 ± 0,3	22,8 ± 0,8	42,6 ± 0,3
$Y_2Si_2O_7$ (Gemini2)		6,0	12,0	21,9
Y_2O_3 (Gemini2)		41.2	103.8	230.5
SiO_2 (Gemini2)		87.4	117.2	151.9

With respect to $Y_2Si_2O_7$ (Table 4), the same features are observed: the recession rates decrease when the temperature of thermal treatment increases. The crystallization or the purity influences the corrosion resistance. Contrary to Y_2SiO_5, $Y_2Si_2O_7$ presents an interest to be used as environmental barrier coatings only when it is well crystallized or when few impurities are present in samples. Actually, a reactive sintering at 1500°C during 5 hours is necessary to get a corrosion resistance at least similar to silica at 1400°C.

Whatever the yttrium silicate, the experimental recession rate of the purest compound (sintered at 1500°C during 50 hours) corresponds to the double of that determined by thermodynamic calculations (by running Gibbs free energy minimization with Gemini2 software). This deviation can be explained by the purity of the samples. They possess impurities like Y_2O_3 or apatite phase ($Y_{4,67}(SiO_4)_3O$), that decreases the corrosion resistance. Moreover, even if the hydroxide partial pressures are doubled with regard to thermodynamic calculations, they remain low (1.10^{-7} atm theoretically versus 2.10^{-7} experimentally). The reliability of the theoretical partial pressures is linked to that of thermodynamic data ($Si(OH)_4$, $Y(OH)_3$ and YOOH). An error of 1 kJ.mol^{-1} on the free enthalpy leads to a deviation of 10 % on partial pressures. Moreover thermodynamics gives information about the equilibrium nature, but does not precise the mechanism of the reaction. An intermediary reaction could be formed because of the partial hydration of the compound. The stability of the compound could be decreased due to a change of the chemical bonds, and a raise of the partial pressures at the equilibrium could be induced.

Thermodynamics allows to give tendencies, which seems to be confirmed by experiments.

Furthermore, it can be mentioned that these experiments were not in disagreement with the results found in the literature. For example, the $Y_2O_3.SiO_2$ system does not show a dramatic weight loss during corrosion tests[19-20] because the furnace tube, generally in alumina or silica, volatilized itself and $Y(OH)_3$ reacts with $Si(OH)_4$ or $Al(OH)_3$ to form yttrium silicate or YAG respectively. As a consequence, a weight gain can be sometimes recorded for long exposure times. The amount of alumina from the furnace tube is in fact the major difficulty of this work that explains the particular attention for the alumina pollution.

Influence of the rare earth on corrosion resistance

Now, corrosion tests were performed on different rare earth silicates (Dy, Er, Yb, Sc) in order to compare their stability with that of yttrium silicates.

For this purpose, pellets were made from powders of SiO_2 and RE_2O_3 (where RE_2O_3 = Dy_2O_3, Er_2O_3 and Yb_2O_3 (Chempur, 99.9 %) and Sc_2O_3 (Neyco, 99.99 %). They were sintered at 1500°C during 50 hours under ambient air, since this condition leads to the best crystallization degree and purity. Pellets were characterized by X-Ray Diffraction. $Sc_2Si_2O_7$, Yb_2SiO_5 and $Yb_2Si_2O_7$ are assumed to be pure. The phase named Sc_2SiO_5 is a mixture of $Sc_2Si_2O_7$, Sc_2O_3 and

Sc_2SiO_5. The other compounds have few impurities (in small quantities) like RE_2O_3 or apatite phase.

Corrosion tests were performed on these pellets at 1200°C with a water vapor partial pressure of 50 kPa at atmospheric pressure with a flowing gas velocity of 5 cm.s^{-1}. Table 5 gathered recession rates of rare earth silicates at 1200°C, and compared them with those of RE_2O_3 and SiO_2. The recession rates of simple oxides were determined from the partial pressures calculated with Gemini2 software. The Gibbs free energy of rare earth hydroxides is taken from reference[21].

The thermochemical stability of rare earth silicates is considerably enhanced with regard to that of rare earth oxides and silica, with the exception of Sc_2SiO_5, which contains in fact Sc_2O_3, $Sc_2Si_2O_7$ and Sc_2SiO_5. Thus its recession rate is close to that of Sc_2O_3. Moreover, the recession rates of rare earth silicates and oxides seem to vary in the same way. Thus the highest stability of ytterbium silicates is confirmed in a moist environment at high temperatures and these silicates appears to be the best candidates to be used as environmental barrier coatings.

Table 5. Recession rates (k_r in nm.h^{-1}) of different rare earth silicates under corrosion at 1200°C (P_{H2O} = 50 kPa, P_{tot} = P_{atm}, v_{gas} = 5 cm.s^{-1}) (k_r (SiO_2) = 87.4 nm.h^{-1})

k_r à 1200°C (nm.h^{-1})	RE_2O_3	RE_2SiO_5	$RE_2Si_2O_7$
Sc	36.5	24 ± 2	6.8 ± 0.5
Y	41.2	9.9 ± 0.7	7.7 ± 0.3
Dy	32.9	10.2 ± 0.2	5 ± 2
Er	31.5	5.6 ± 0.8	4.0 ± 0.8
Yb	29.2	4.0 ± 0.2	2.9 ± 0.2

CONCLUSIONS

A thermodynamic study allowed to simulate the behaviour of yttrium silicates in a moist environment. The calculations of partial pressures of gaseous species in equilibrium over these materials showed that yttrium silicates are promising candidates for environmental barrier coatings. Moreover, the stability of yttrium silicates was quantified and their recession rates were deduced from the calculated partial pressures. These results were confirmed by experiments carried out between 1200°C and 1400°C on yttrium silicates elaborated by a reactive sintering. It was demonstrated that the crystallization degree or the purity influences the corrosion resistance. A thermal treatment at 1500°C during 50 hours is necessary to have the highest corrosion resistance. The influence of the nature of the rare earth on the corrosion resistance was also studied. The stability of rare earth silicates and rare earth oxides varied in the same way. Ytterbium silicates were the most stable materials in a moist environment at high temperatures.

Moreover, this work shows that as soon as an oxide is bonded in a compound, it becomes more resistant against moisture: Y_2O_3 in ZrO_2 or SiO_2 and Y_2O_3 in silicates. The gain in resistance depends on the strength of the bond between both oxides.

Future works will be devoted to the corrosion resistance of ytterbium silicates between 1200°C and 1400°C.

ACKNOWLEDGEMENTS
   This work has been supported by the French Ministry of Education and Research through a grant given to E. Courcot. The authors are grateful to Caroline Louchet-Pouillerie and Jacques Thébault for fruitful discussions.

REFERENCES
[1] E.J. Opila, Variation of the Oxidation Rate of Silicon Carbide with Water-Vapor Pressure, *J. Am. Ceram. Soc.*, **82**, 625-636 (1999).

[2] E.J. Opila, J. L. Smialek, R.C. Robinson, D.S. Fox, J.S. Jacobson, SiC Recession Caused by SiO_2 Scale Volatility under Combustion Conditions: II, Thermodynamics and Gaseous-Diffusion Model, *J. Am. Ceram. Soc.*, **82**, 1826-1834 (1999).

[3] K.N. Lee, Current status of environmental barrier coatings for Si-based ceramics, *Surf. Coat. Technol.*, **133-134**, 1-7 (2000).

[4] K.N. Lee, D.S. Fox, J.I. Eldridge, D. Zhu, R.C. Robinson, N. P. Bansal, R. A. Miller, Upper Temperature Limit of Environmental Barrier Coatings Based on Mullite and BSAS, *J. Am. Ceram. Soc.*, **86**, 1299-1306 (2003).

[5] H.E. Eaton, G.D. Linsey, Accelerated oxidation of SiC CMC's by water vapor and protection via environmental barrier coating approach. *J. Europ. Ceram. Soc.*, **22**, 2741-2747 (2005).

[6] K.N. Lee, D. S. Fox and N. P. Bansal: J. Europ. Ceram. Soc. Vol. 25 (2005), p.1705.

[7] K.N. Lee, U.S. Patent 6,759,151. (2002).

[8] Y. Ogura, M. Kondo, T. Morimoto, Y_2SiO_5 as oxidation resistant coating for C/C composites. Proceedings of ICCM-10, Whistler, Canada. August 1995, p. 767.

[9] Klemm, H., Fritsch, M. and Schenk, B., Corrosion of ceramic materials in hot gas environment, *Ceram. Eng. Sci. Proc.*, **25**, 463–468 (2004).

[10] B. Cheynet, P.-Y. Chevalier, E. Fischer, Thermosuite, *Calphad,* **26**, 167-174 (2002).

[11] J.J Liang, A. Navrotsky, T. Ludwig, H.J. Seifert, F. Aldinger, Enthalpy of formation of rare earth silicates Y_2SiO_5 and Yb_2SiO_5 and N-containing silicate $Y_{10}(SiO_4)_6N_2$. *J. Mat. Res.*, **4**, 1181-1185 (1999).

[12] O. Fabrichnaya, H.J. Seifert, R. Weiland, T. Ludwig, F. Aldinger, A. Navrotsky, Phase Equilibria and Thermodynamics in the Y_2O_3-Al_2O_3-SiO_2 system. *Zeit. Metallk.* **92**, 1083-1097 (2001).

[13] H. Mao, M. Selleby, O. Fabrichnaya, Thermodynamic reassessment of the $Y_2O_3 - Al_2O_3 - SiO_2$ and its subsystems. *Comp. Coupl. Phase Diag. Thermochem.*, **32**, 399-412 (2008).

[14] D. Djurovic, M. Zinkevich, F. Aldinger, Thermodynamic modeling of the yttrium-oxygen system, *Comp. Coupling Phase Diag. Thermochem.*, **31**, 560-566 (2007).

[15] Association Thermodata GEMINI 2 Code. B.P.66, 38402 St. Martin d'Heres Cedex, France.

[16] M.D. Allendorf, T.M Besmann, Thermodynamics Resource (database Sandia National Laboratory), http://public.ca.sandia.gov/HiTempThermo/

[17] L. Quémard, F. Rebillat, A. Guette, H. Tawil, C. Louchet-Pouillerie, Degradation mechanisms of a SiC fiber reinforced self sealing matrix composite in simulated combustor environments. *J. Europ. Ceram. Soc.*, **27**, 377-388 (2007).

[18] E. Courcot, F. Rebillat, F. Teyssandier, How to assess thermodynamic data of gaseous hydroxides from volatilization rates ? *Proceedings of PACRIM8*, Vancouver, Canada, 31 May – 5 June 2009.

[19] K. N. Lee, D. S. Fox, and N. P. Bansal, Rare earth silicate environmental barrier coatings for SiC/SiC composites and Si_3N_4 ceramics, *J. Eur. Ceram. Soc.*, **25**, 1705-1715 (2005).

[20] H. Nakayama, K. Aoyama, M. Yamamoto, H. Sumitomo, K. Okamura, T. Yamura, and M. Sato, Evaluation of Environmental Barrier Coatings for CFCC in the High-Water-Vapor-Pressurized Environments at High Temperatures, Proceedings of HTCMC-5, ed. by M. Singh, R. J. Kerans, E. Lara-Curzio, and R. Naslain, American Ceramic Society, 2004, p. 613-618.

[21] E. Courcot, F. Rebillat, F. Teyssandier, Thermochemical stability of rare earth sesquioxides under a moist environment at high temperature. *Proceedings of PACRIM 8*, Vancouver, Canada, 31 May – 5 June 2009.

NANOLAMINATED OXIDE CERAMIC COATINGS IN THE Y_2O_3-Al_2O_3 SYSTEM

Nadine K. Eils, Peter Mechnich, Martin Schmücker
German Aerospace Center (DLR)
Institute of Materials Research
51170 Cologne, Germany

Hartmut Keune
Technical University of Braunschweig
Institute of Surface Technology
38108 Braunschweig, Germany

ABSTRACT
Oxide ceramic coatings in the Y_2O_3-Al_2O_3 system were deposited by MOCVD using a screw feeding system. The pure Al_2O_3-coatings as well as the coatings containing Al_2O_3 and Y_2O_3 show a nanolaminated microstructure whereas pure Y_2O_3-coatings reveal a feathery microstructure. Further experiments with modified deposition parameters were carried out in order to control the lamellae characteristics. Therefore the stoichiometry of the starting material, the precursor feeding rate and the substrate temperature were varied systematically. Highly distinctive lamellae were found when the coating was deposited at a low substrate temperature or using an aluminum-rich starting material and a high precursor feeding rate, respectively. Annealing experiments in air and flowing water vapor of the nanolamellar structured baseline coating were carried out in order to investigate the microstructural development. Coalescence of nanopores to larger pores was observed while the lamellar coating morphology maintains.

INTRODUCTION

Oxide ceramic coatings are widely used as protective coatings for hot section components of gas turbine engines which nowadays mostly consist of nickel-based super alloys. The application of more temperature stable ceramic matrix composites (CMC) or melt grown composites (MGC) as a replacement for the metal based alloys as component material has been discussed recently in order to realize a higher operating temperature of the gas turbines.[1,2] Alumina based all-oxide CMCs like WHIPOX™ are considered suitable candidate materials for this application. In this context attention should be paid to the volatilization of alumina at higher temperatures in flowing water vapor rich atmosphere [3-5] which is expected to limit the lifetime of this material. Therefore a protection from the highly corrosive gas environment is mandatory.

Protective coatings should display a low thermal conductivity as well as a significant tolerance against mechanical load and thermal shock. Columnar structured coatings are favorable since they exhibit a high strain tolerance and thermal shock resistance. Such a columnar morphology is typical for EB-PVD coatings, but has also been observed for ceramic coatings deposited by metal-organic chemical vapor deposition (MOCVD).[6,7] Furthermore it is assumed that a nanolamellar microstructure is advantageous, since it may reduce the thermal conductivity due to the reduction of photon and phonon transports.[8] Singh and Wolfe observed this phenomenon for lamellar yttrium-stabilized zirconia (8YSZ) thermal barrier coatings produced by EB-PVD which show an up to 30% reduced thermal conductivity compared to a similar coating without lamellae.[9] Lamellar structured coatings were also produced by using MOCVD.[10-12]

Yttrium-aluminum-garnet (YAG; $Y_3Al_5O_{12}$) is a promising coating material for α-alumina based CMCs due to its similar thermal expansion and the high stability in water vapor rich atmosphere.[5] Fritsch and Klemm investigated the hot gas corrosion behavior of $Y_3Al_5O_{12}$-dip-coated alumina in the temperature range from 1200 to 1500 °C in combustion environment and observed a significantly improved hot gas stability compared to pure α-alumina.[1]

In this paper we report on single-step fabrication of nanolaminated coatings by MOCVD using flash evaporation and the development of the promising microstructure upon heat treatment and water vapor treatment at 1200 °C.

EXPERIMENTAL PROCEDURE

The laboratory-scale CVD unit consists of a flash evaporation system connected to a hot wall reactor meaning in detail a quartz glass tube (d = 4.5 cm) inside a resistor heated tube furnace. An attached vacuum pump is used to set the total pressure in the entire CVD system to 2.5 mbar. Oxygen was used as carrier and reactant gas with a volume flow rate of 30 l/h (STP). Aluminum acetylacetonate (Al(acac)$_3$, Sigma-Aldrich) and yttrium tetramethylheptandionate (Y(tmhd)$_3$, Sigma-Aldrich) were used as metal-organic precursors because of their similar vaporization behavior under the given conditions. In order to avoid agglomeration the precursor was mixed with quartz sand in a weight ratio of 2/1. The precursor sand mixture was transported by a screw feeder and dropped into a pre-heated metal crucible (T = 200 °C) where the precursor vaporizes immediately while the inert sand passes the hot zone and remains at the bottom of the crucible. The substrates were placed in the glass tube in stagnation flow arrangement. Polycrystalline α-alumina disks with a diameter of 27 mm and a thickness of about 3 mm were used as substrates. The substrate temperature was controlled by a thermocouple positioned at the outer rim of the substrate surface.

Microstructural analyses were performed by means of scanning electron microscopy (ULTRA 55, Carl Zeiss MicroImaging, Jena, Germany) and transmission electron microscopy (TECNAI F30, FEI, Eindhoven, The Netherlands). For phase analyses X-ray diffractometry (D5000, Siemens, Karlsruhe, Germany) was used. A resistor-heated furnace (Nabertherm, Lilienthal, Germany) was used for heat treatments in air at 1200°C and for a separate test at 1300°C to estimate the carbon content of the alumina coating. The test facility used for the water vapor treatments at 1200 °C consists of a steam generator (type PS 100, Stritzel Dampftechnik, Mühlheim, Germany) connected with a tube furnace (type LORA 1800-32-1000-1, HTM-Reetz, Berlin, Germany). The water vapor velocity was set to about 10 m/s.

RESULTS

In a pilot survey pure Al_2O_3 and Y_2O_3 MOCVD coatings were deposited at a substrate temperature of 950°C. In the as deposited state the alumina coating appears nearly black while the Y_2O_3-coating is white (fig. 1). XRD reveals that the alumina coating consists of a nanometer-sized very poorly

Figure 1: Top view of the dark Al_2O_3 (a) and the white Y_2O_3 (b) MOCVD coating as fabricated.

crystallized transition alumina phase, most likely γ-Al_2O_3. The Y_2O_3-coating by contrary consists of well crystallized body-centered cubic Y_2O_3 and the increased intensities of the (002), (004) and (008) reflexes reveal a texture. SEM images of the cross sections of both coatings display strongly different microstructures. The alumina coating shows a highly homogenous nanolamellar structure with an average lamellae thickness of about 350 nm whereas the structure of the Y_2O_3-coating is feathery and columnar (fig. 2).

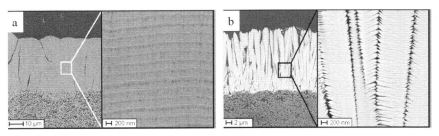

Figure 2: SEM images of an Al_2O_3-coating cross section (a) and an Y_2O_3-coating cross section (b) in low and high magnification.

Further investigations of the lamellar structured alumina coating by means of transmission electron microscopy (TEM) reveal an alternating sequence of thick porous layers of nanometer-sized grains and thin dense layers including significantly coarser grains (fig. 3). The observed high porosity could be due to the deposition of organic residues along with alumina. Weight measurements after heat treatment in air at 1300 °C for one hour of the alumina MOCVD coating reveals that the as deposited coating includes about 9 wt % organic residues.

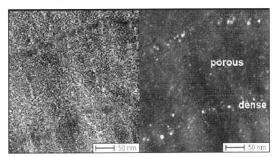

Figure 3: TEM micrographs of the as deposited Al_2O_3 MOCVD coating. Left: bright field image; right: dark field image. Lamellae are due to a sequence of highly porous and virtually dense zones.

For coating experiments in the binary system Y_2O_3-Al_2O_3 mixtures of $Al(acac)_3$ and $Y(tmhd)_3$ were used as starting materials. In a first step a parameter set was defined for a baseline experiment (tab. 1). A polycrystalline α-alumina disk was coated at a substrate temperature of 950 °C by using a precursor mixture with a Y/Al molar ratio of 3/5. The rotating speed of the screw feeder was set to 2 rpm. The as deposited baseline coating shows a dark centre and a bright outer rim (fig. 4).

Table I: Parameter set selected for the baseline experiment

Deposition parameter	Baseline experiment
Precursor stoichiometry [mol-%]	Y/Al = 3/5
Rotating speed of the screw feeder [rpm]	2
Substrate temperature [°C]	950

Figure 4: Top view of the baseline MOCVD coating in the Al_2O_3-Y_2O_3-system as fabricated.

The XRD pattern (fig. 5) reveals the coexistence of two phases and shows additionally a few peaks referred to the substrate material α-alumina. Beside the main phase yttrium-aluminum-garnet ($Y_3Al_5O_{12}$; YAG) the formation of a hexagonal phase ($YAlO_3$; YAH) with higher aluminum content is

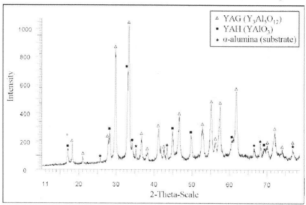

Figure 5: XRD pattern of the baseline MOCVD coating as fabricated showing the coexistence of two phases (YAG and YAH) beside the substrate material α-alumina.

observed. The stoichiometry of the coating was analyzed by means of energy dispersive spectroscopy (EDS). For that purpose 120 data points were measured along the cross section of the coating. A large-scale homogeneous stoichiometry with an average Y_2O_3-content of 32.2 mol % was observed. It must be pointed out that this Y_2O_3-content is about 5 mol-% lower than expected from the stoichiometry of the starting material. From this result it is assumed that the Y_2O_3-yield from $Y(tmhd)_3$ is lower than the

Al_2O_3-yield from $Al(acac)_3$ under the present conditions. Closer inspection of the cross section by means of SEM reveals that the columnar structured coating shows a nanolamellar microstructure (fig. 6). The lamellae with an average thickness of about 200 nm seem to be not as homogenous as observed for the pure alumina coating. The bright field TEM image at higher magnification (fig. 7) displays that these lamellae are subdivided into smaller lamellae and chains of coarse grains separated by linear porosity.

Figure 6: SEM images of the baseline coating cross section in low and high magnification showing that the columnar structured coating has a nanolamellar microstructure.

Figure 7: Bright field TEM image of the as deposited baseline coating. The thick lamellae could be subdivided into chains of coarse or fine grains separated by linear porosity.

In following runs some of the deposition parameters were systematically varied (tab. 2) and compared to the baseline experiment. First the stoichiometry of the starting material was adjusted to an

Tabelle II: Varied deposition parameters

Varied deposition parameters	Selected values
. Precursor stoichiometry [mol-%]	Y/Al = 3/7, 3/5, 1/1, 5/3
Rotating speed of the screw feeder [rpm]	1, 2, 3
Substrate temperature [°C]	900, 950, 1000

Y/Al ratio of 3/7, 1/1 and 5/3, respectively. The as fabricated coatings show significant distinctions in the coloring and the microstructure (fig. 8). Obviously a higher aluminum content of the starting material goes along with a darker color of the coating. All coatings reveal a lamellar structure but the lamellae are more distinct when the aluminum content is high.

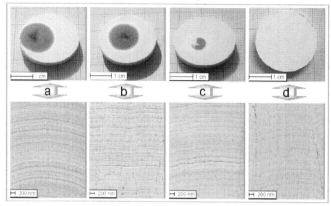

Figure 8: Top view and SEM images of the cross section of coatings deposited using a precursor stoichiometry of Y/Al of 3/7 (a), 3/5 (b), 1/1 (c) and 5/3 (d), respectively.

A similar behavior was observed for the variation of the precursor feeding rate by changing the rotating speed of the screw feeder to 1 rpm and 3 rpm, respectively. Using an increased precursor feeding rate let the deposited coatings appear darker and the lamellae become more visible (fig. 9). Moreover, it was found that the coatings deposited at higher rotating speed tend to delaminate.

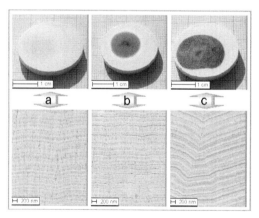

Figure 9: Top view and SEM images of the cross section of coatings deposited with a rotating speed of 1 rpm (a), 2 rpm (b) and 3 rpm (c), respectively.

In further experiments the substrate temperature was changed to 900 °C and 1000 °C, respectively. Deposition of a coating at a substrate temperature of 900 °C leads to an almost black colored coating with a distinct lamellar microstructure and a tendency to delaminate. On the other hand a coating fabricated at a substrate temperature of 1000 °C is essentially white and shows only indistinct lamellae (fig. 10).

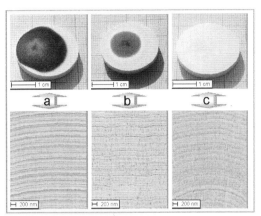

Figure 10: Top view and SEM images of the cross section of coatings deposited at a substrate temperature of 900 °C (a), 950 °C (b) and 1000 °C (c), respectively.

In order to investigate the microstructural development, heat treatments were carried out for one hour at 1200 °C in air and in flowing water vapor, respectively. Figure 11 shows the cross

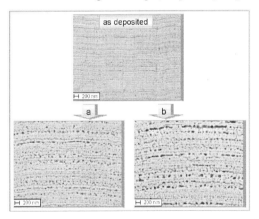

Figure 11: SEM images of the baseline coating as deposited, after heat treatment in air (a) and after heat treatment in water vapor (b), respectively. Due to heat treatment the nanopores coalesce to larger pores while the nanolamellar microstructure remains.

sectional SEM images of the baseline coating as deposited, after heat treatment in air and after heat treatment in water vapor, respectively. It is observed that the nanopores coalesce to larger pores due to heat treatment while the nanolamellar microstructure remains. Comparison of both heat treated samples reveals that the grains as well as the pores appear to be larger after treatment in water vapor. An EDS-analysis at a coarse grained area of the water vapor treated coating shows the coexistence of two phases (fig. 12). A small amount of alumina was detected beside the main component YAG. Corresponding to this result the XRD shows that the aluminum-rich phase YAH disappeared. The two heat treated coatings reveal identical XRD patterns irrespective of the atmosphere during the heating experiment (fig. 13).

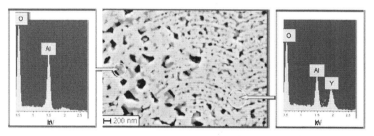

Figure 12: EDS analysis of the water vapor treated baseline coating showing the existence of a small amount of α-alumina beside the main phase YAG.

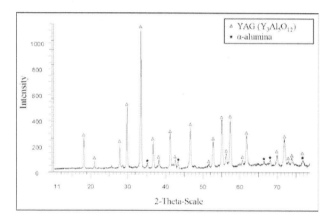

Figure 13: XRD pattern of the baseline MOCVD coating after heat treatment showing the phases YAG and α-alumina.

DISCUSSION

The cause of the formation of lamellae has not yet been clarified. Closer inspection of the alumina coating in consideration of the total coating thickness, the average lamellae thickness, the deposition time and the rotating speed of the screw feeder indicate that a single lamella grows during one screw rotation of the feeder. It is assumed that the screw induces a periodical precursor feeding leading to a fluctuating demand of oxygen. The dark color of the as deposited alumina coatings is obviously due to residual carbon[13]. That means that the precursor ($Al(acac)_3$) was not completely oxidized. Pulver[13] investigated the content of organic residues of several alumina coatings deposited by MOCVD using $Al(acac)_3$ as metal-organic precursor and observed that the coatings includes organic contents in the range between 3.8 wt % and 16.5 wt % depending on the deposition temperature. From this assumption an explanation for the formation of the lamellae is generated. It is conceivable that in periods of high precursor feeding the precursor is not completely oxidized so that organic residues were deposited along with alumina forming thick porous layers. In periods of low precursor feeding by contrast the amount of available oxygen would be high enough to oxidize the precursor completely leading to the deposition of dense coarser grained alumina layers. To confirm this it is necessary to find a reliable measuring method with a high local resolution for determining the carbon content. Glow discharge optical emission spectrometry (GD-OES) seem to be a suitable method for that purpose[14,15].

Closer inspection of the binary system shows that the distinctness of the lamellae is correlated to the color of the as deposited coatings. It is observed that the darker the coating color is the more distinct are the lamellae. A dark colored coating could be deposited by using an aluminum-rich starting material, a low substrate temperature or a high precursor feeding rate, respectively. The dependency from the aluminum content of the precursor mixture shows that $Y(tmhd)_3$ needs under the same conditions less oxygen than $Al(acac)_3$ to oxidize completely. It is assumed that fluctuations in the gas phase stoichiometry could lead to the formation of thinner lamellae superimposing the homogenous lamellae induced by the screw feeder. In this case small-scaled stoichiometry fluctuations should be found along the coating cross section. It is challenging to detect such fluctuations because the thickness of such lamellae is in the nanometer-scale and the single grains are even smaller. Preliminary tests by means of EDS were carried out to support this assumption. Figure 14 displays the Y_2O_3 content measured along lines of 2 µm in length on cross sections of an YAG single crystal, the as deposited

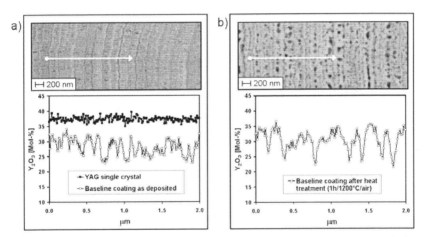

Figure 14: EDS measurements along lines on cross sections of (a) an YAG single crystal, the as deposited baseline coating and (b) the baseline coating after heat treatment in air, respectively. The baseline coating shows small-scaled stoichiometry oscillations compared to a YAG single crystal.

baseline coating and the baseline coating after heat treatment in air, respectively. Comparison of data points of the baseline coating with those of the YAG single crystal indicates the presence of nano-scaled stoichiometry oscillations in the MOCVD coating. The amplitude is about 5 mol % in the as deposited state and increases up to 7.5 mol % due to heat treatment. Additionally, the periodic time is significantly higher after heat treatment. This observation can be explained by phase transformation and grain coarsening due to heat treatment.

CONCLUSION

The MOCVD system including a screw feeder allows the fabrication of nanolaminated coatings in the Y_2O_3-Al_2O_3 system. Several experiments with systematically modified deposition parameters show that the lamellae characteristics were affected by a multitude of parameters. Due to the observation that darker coatings show more distinct lamellae, it is assumed that the inclusion of residual carbon plays a decisive role in the formation of the nanolamellar microstructure. Further investigations are necessary in order to understand this mechanism in detail. After annealing experiments in air and flowing water vapor the promising nanolamellar character of the coating maintains despite phase transformation and grain growth are observed. It is anticipated that nanometer-sized lamellae could reduce the thermal conductivity due to the reduction of photon and phonon transport. Such coatings may be highly attractive for applications as protective coatings. Measurements of the thermal conductivity are underway in order to confirm this assumption.

ACKNOWLEDGEMENT

The authors would like to thank Mr. K. Baumann and Mrs. G. Paul for the preparation of SEM and TEM samples. Financial support from the German Science Foundation (DFG) is gratefully acknowledged.

REFERENCES

[1]M. Fritsch and H. Klemm, The Water Vapor Hot Gas Corrosion Behavior of Al_2O_3-Y_2O_3 Materials, Y_2SiO_5 and $Y_3Al_5O_{12}$-coated Alumina in a Combustion Environment, *Ceram. Eng. Sci. Proc.*, **27**(3), 149-159 (2006).

[2]M. Fritsch and H. Klemm, The Water Vapor Hot Gas Corrosion of MGC Materials with Al_2O_3 as a Phase Constituent in a Combustion Atmosphere, *J. Eur. Ceram. Soc.*, **28**, 2353-2358 (2004).

[3]E. J. Opila and D. L. Myers, Alumina Volatility in Water Vapor at Elevated Temperatures, *J. Am. Ceram. Soc.*, **87** [9], 1701-1705 (2004).

[4]I. Yuri, and T. Hisamatsu, Recession Rate Prediction for Ceramic Materials in Combustion Gas Flow, *in Proceedings of ASME Turbo EXPO, June 16-19, Atlanta, Georgia*, ASME Paper GT2003-38886 (2003).

[5]H. Klemm, M. Fritsch and B. Schenk, Corrosion of Ceramic Materials in Hot Gas Environment, *Ceram. Eng. Sci. Proc.*, **25**(3-4), 463-468 (2004).

[6]G. Wahl, W. Nemetz, M. Giannozzi, S. Rushworth, D. Baxter, N. Archer, F. Cernuschi and N. Boyle, Chemical Vapor Deposition of TBC: An Alternative Process for Gas Turbine Components, *Transactions of the ASME*, **123**, 520-534 (2001).

[7]J. R. Vargas Garcia, and T. Goto, Thermal Barrier Coatings Produced by Chemical Vapor Deposition, *Sci. and Techn. of Adv. Mat.*, **4**, 397-402 (2003).

[8]J. R. Nicholls, K. L. Lawson, A. Johnstone, D. S. Rickerby, Methods to Reduce the Thermal Conductivity of EB-PVD TBCs, *Surf. and Coat. Techn.*, **151-152**, 383-391 (2002).

[9]J. Singh, D. E. Wolfe, Nano- and Macro-Structured Component Fabrication by Electron Beam-Physical Vapor Deposition (EB-PVD), *J. Mat. Sci.*, **40**, 1-26 (2005).

[10]V. G. Varanasi, T. M. Besmann, R. L. Hyde, E. A. Payzant and T. J. Anderson, MOCVD of YSZ Coatings using ß-Dikonate Precursors, *J. of Alloys and Compounds*, **470**, 354-359 (2009).

[11]V. G. Varanasi, T. M. Besmann, J. J. Henry, Jr., T. L. Starr, W. Xu and T.J. Anderson, YSZ Thermal Barrier Coatings by MOCVD, *in Proceedings of the 5th Intern. Conf. on High Temp. Ceramic Matrix Composites,* 595-601 (2004).

[12]I. Tröster, S. V. Samoilenkov, G. Wahl, W. Braue, P. Mechnich and H. Schneider, Metal-Organic Chemical Vapor Deposition of Environmental Barrier Coatings for All-Oxide Ceramic Matrix Composites, *Ceram. Eng. Sci. Proc.*, **26**, 173-179 (2005).

[13]M. Pulver, Chemische Gasphasenabscheidung von Zirkoniumdioxid, Yttriumoxid und Aluminiumoxid aus ß-Dikonaten und Alkoholaten, Ph. D. Thesis, Technical University of Braunschweig, Germany (1999).

[14]V. Lavoine, H. Chollet, J.-C. Hubinois, S. Bourgeois and B. Domenichini, Optical interfaces in GD-OES system for vacuum far ultraviolet detection, *J. Anal. At. Spectrom.*, **18**, 572-575 (2003).

[15]R. Payling, J. Michler and M. Aeberhard, Quantitative Analysis of Conductive Coatings by Radiofrequency-powered Glow Discharge Optical Emission Spectrometry: Hydrogen, d.c. Bias Voltage and Density Corrections, *Surf. Interface Anal.*, **33**, 472-477 (2002).

THERMOCHEMICAL STABILITY OF RARE EARTH SESQUIOXIDES UNDER A MOIST ENVIRONMENT AT HIGH TEMPERATURE

E. Courcot[1], F. Rebillat[1], F. Teyssandier[1]

[1] Université de Bordeaux, Laboratoire des Composites Thermo-Structuraux (LCTS) UMR 5801
3 Allée de la Boétie 33600 Pessac, France

ABSTRACT

Volatilization tests enabled to quantify the chemical and thermal stability of different rare earth sesquioxides, RE_2O_3 (where RE = Sc, Dy, Er, Yb) and to understand the corrosion process. These tests were carried out at temperatures ranging from 1000 to 1400°C in moist air with 50 kPa of water at atmospheric pressure, under a flowing gas velocity of 5 cm.s^{-1}. Besides the volatilization rate, the nature of the volatile gaseous species was determined. Moreover, the proposed experimental method allowed to assess the Gibbs free energy of formation of these gaseous volatile species. Then, the partial pressures of hydroxide gaseous species were compared in order to identify the most stable species in moist environments.

INTRODUCTION

Rare earth silicates seem to be promising materials for environmental barrier coatings, since they exhibit a high chemical resistance in extreme environments (high temperature, high water pressure)[1-2]. In order to quantify the stability of these materials, a thermodynamic study has to be carried out. However, few thermodynamic data are available for rare earth hydroxides. Consequently, the aim of this work is to assess thermodynamic parameters of these compounds through the study of the corrosion kinetics of rare earth sesquioxides.

Actually, rare earth oxides are classically divided into two groups: the ceric group and the yttric group. A few physical properties of these oxides vary with the nature of the rare earth that can be characterized by its ionic radius[3]. This paper is focused on the thermochemical stability of different rare earth oxides belonging to the yttric group, Sc_2O_3, Dy_2O_3, Er_2O_3 and Yb_2O_3 at high temperatures under a moist environment. Their volatilization rates under a moist environment are measured in order to quantify their stability, to determine the identity of the gaseous volatile hydroxides and to assess their Gibbs free energy of formation.

EXPERIMENTAL APPROACH

In order to determine the hydroxide thermodynamic data, the equilibrium hydroxide partial pressures over the solid oxide in a controlled atmosphere had to be quantified. For that, a methodology based on weight loss measurements during corrosion at high temperatures under high water vapor pressure at atmospheric pressure was applied (Figure 1)[4]. Actually, during corrosion, a reactive volatilization occurs and leads to a weight loss of oxides due to the formation of gaseous hydroxides RE(OH)$_3$ or oxy-hydroxides REOOH (1-2) :

$$RE_2O_3 + 3\ H_2O \longrightarrow 2\ RE(OH)_3 \tag{1}$$

$$RE_2O_3 + H_2O \longrightarrow 2\ REOOH \tag{2}$$

By varying either temperature or the water vapor partial pressure, experiments lead to determine the volatilization kinetics of rare earth sesquioxides in a corrosive atmosphere at high temperature and to identify the nature of the gaseous species respectively. Then, the values of the Gibbs free energies of formation of the gaseous hydroxide species were deduced from results obtained previously.

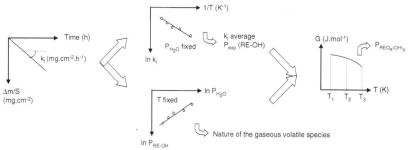

Figure 1. Experimental approach used for the assessment of the Gibbs free energy of formation of gaseous hydroxides.

The volatilization tests were carried out in a corrosion furnace, whose principle is described elsewhere [5], at temperatures between 1000 and 1400°C, in a moist air (water vapor partial pressure up to 70 kPa, 100 kPa total pressure), with a gas velocity of 5 cm.s^{-1} (in the cold zone of the furnace) during times between 1.5 hours and 25 hours on oxide pellets. These pellets were prepared as the materials described in a previous study [4], to always consider the same exchange surface area. Powders (Sc_2O_3 (Neyco, 99.99 %), Dy_2O_3, Er_2O_3 and Yb_2O_3 (Chempur, 99.9 %)) were mixed and compacted under an unidirectional pressure of 0.5 MPa during 10 minutes. Pellets ($\varnothing$ = 10 mm, thickness ~ 2 mm, S ~ 150 mm^2) were heat treated at 1300°C during 5 hours under ambient air for desorption and sintering. The porosity was around 25 %.

RESULTS AND DISCUSSION

Determination of the activation energy of the volatilization process
First, volatilization tests were carried out at temperatures ranging from 1000°C to 1400°C in air with a water partial pressure of 50 kPa at atmospheric pressure under a gas flow velocity of 5 cm.s^{-1}. The variations of the ratio between the weight change and the surface of the pellet as a function of time are shown for Sc_2O_3 in all the range of temperatures considered (Figure 2a) and for the selected rare earth oxides at 1200°C (Figure 2b). The weight loss during corrosion is linear with time and increases with temperature. The volatilization rates deduced from slopes are gathered in table I.

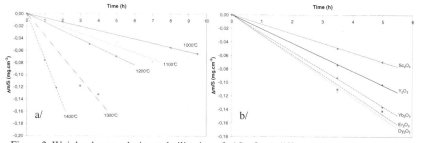

Figure 2. Weight changes during volatilization of a/ Sc_2O_3 at different temperatures ; b/ RE_2O_3 at 1200°C, under corrosion at 50 kPa H_2O, at P_{atm} with v_{gas} = 5 cm.s^{-1}.

Table I. Volatilization rates of RE_2O_3 under a moist environment at different temperatures ($P_{H2O} = 50$ kPa, $P_{tot} = P_{atm}$, $v_{gaz} = 5$ cm.s^{-1})

k_1 (mg.cm^{-2}.h^{-1})	1000°C	1100°C	1200°C	1300°C	1400°C
Y_2O_3	0,0122 ± 0,0003	0,0162 ± 0,0001	0,0208 ± 0,0003	0,051 ± 0,001	0,12 ± 0,01
Sc_2O_3	0,0069 ± 0,0001	0,0105 ± 0,0006	0,0138 ± 0,0001	0,037 ± 0,002	0,074 ± 0,001
Dy_2O_3	0,0112 ± 0,0007	0,019 ± 0,004	0,030 ± 0,002	0,058 ± 0,009	0,15 ± 0,03
Er_2O_3	0,0127 ± 0,0007	0,019 ± 0,001	0,0294 ± 0,0009	0,05 ± 0,02	0,12 ± 0,01
Yb_2O_3	0,0074 ± 0,0002	0,012 ± 0,002	0,0270 ± 0,0004	0,050 ± 0,002	0,11 ± 0,01

The thermoactivation of the reaction of RE_2O_3 with moisture is shown with an Arrhenius plot (Figure 3). As for Y_2O_3, a slope fracture was observed between 1100°C and 1300°C, corresponding certainly to a change of the volatilization mechanism. It was certainly associated to a change of the gaseous hydroxide species formed over the solid oxide.

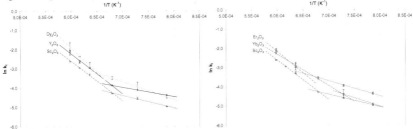

Figure 3. Arrhenius graph concerning the volatilization of the rare earth oxide under a moist environment at high temperature ($P_{H2O} = 50$ kPa, $P_{tot} = P_{atm}$, $v_{gaz} = 5$ cm.s^{-1}) (If not visible, uncertainties are hidden in the size of the dots).

The values of the activation energy and the pre-exponential term are reported in table II for each domain. At high temperature, the activation energy is higher than that of silica (61 ± 8 kJ/mol) [6, 7]. At low temperature, the activation energy is close to that of silica volatilization (with Si(OH)$_4$ formation). Even if all considered oxides belong to the yttric group, two oxide types can be distinguished: the d^1 family and the f^n one. In fact, Sc_2O_3 and Y_2O_3 have similar values of activation energy and preexponential term. Concerning Dy_2O_3, Er_2O_3 and Yb_2O_3, their activation energy and their preexponential term increase with their ionic radius.

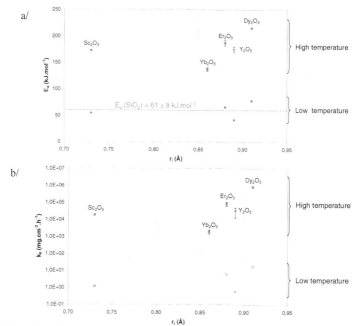

Figure 4. Variation of the values of a/ the activation energy and b/ the preexponential term determined during volatilization of rare earth oxides (P_{H2O} = 50 kPa, P_{tot} = P_{atm}, v_{gaz} = 5 cm.s^{-1}) in function of rare earth ionic radius[8] (If not visible, uncertainties are hidden in the size of the dots).

Identification of the nature of the hydroxide species

The nature of the different gaseous species can be identified by making the water partial pressure vary. Indeed, the (oxy-)hydroxide partial pressures depend on the water partial pressures (3). If the main species are RE(OH)$_3$, the plot of ln P(RE-OH) versus ln P(H$_2$O) would yield a slope of 1,5 (1), whereas a slope equal to 0,5 is expected for the formation of REOOH (2).

$$K = \frac{P_{RE-OH}}{P_{H_2O}^y} \qquad\qquad \ln P_{RE-OH} = \ln K + y \ln P_{H_2O} \qquad\qquad (3)$$

The (oxy-)hydroxide partial pressures are calculated from the values of volatilization rates (4). This equation is valuable when convective flows are predominantly present and whatever the hydroxide formed since the stoichiometric ratio before the formed gaseous species in (1) and (2) remains constant. The extraction of the partial pressures of the formed gaseous species (using (4)) allows taking accurately into account the experimental variations of gas velocity (limited with a change of gas mixture).

$$k_l = v_{gas} \times 3,6 \times M_{RE_2O_3} \times \frac{1}{2} \times \frac{P_{RE-OH}}{R \times T_{amb}} \qquad \text{where} \qquad P_{RE-OH} = P_{RE(OH)_3} + P_{REOOH} \qquad (4)$$

Where
k_l is the volatilization rate (mg.cm^{-2}.h^{-1});
v_{gas} is the gas flow velocity at room temperature (cm.s^{-1});
$M_{RE_2O_3}$ is the molar weight of RE$_2$O$_3$ (g.mol^{-1});

2 is the stoechiometry ratio before the formed gaseous species in Eq. 1 or 2;
P_{RE-OH} is the sum of the volatile gaseous species partial pressures (Pa);
R is the ideal gas constant (= 8,314 $J.mol^{-1}.K^{-1}$);
T_{amb} is the ambient temperature (296 K).

Therefore, volatilization tests were carried out at 1000, 1200 and 1400°C in air with different water partial pressures (ranging from 17 kPa to 68 kPa) under atmospheric pressure, with a gas flow velocity of 5 $cm.s^{-1}$.

The partial pressures of hydroxide species are calculated from the volatilization rates for each temperature, each water vapour partial pressure and each compound. The graph log P_{RE-OH} vs log P_{H2O} for each compound was made and their slopes are summarized in table II. The increase of the slope with temperature corresponds to an increasing proportion of $RE(OH)_3$ formed at the expense of REOOH. For temperatures below the transition, the predominant species is REOOH, whereas at higher temperatures, $RE(OH)_3$ becomes the major one. These results are in agreement with the Arrhenius graph (Figure 3). In order to validate the nature of the hydroxide species, mass spectroscopy was used, but gives no satisfactory results.

Table II. Identification of the nature of the gaseous volatile species in relation with the reaction order associated to water.

Slope of P_{RE-OH} vs P_{H2O}	Y_2O_3	Sc_2O_3	Dy_2O_3	Er_2O_3	Yb_2O_3
1000°C	0,87 ± 0,01 YOOH + $Y(OH)_3$	0,86 ± 0,01 ScOOH + $Sc(OH)_3$	0,95 ± 0,10 DyOOH + $Dy(OH)_3$	0,94 ± 0,10 ErOOH + $Er(OH)_3$	1,06 ± 0,15 $Yb(OH)_3$ + YbOOH
1200°C	1,19 ± 0,02 $Y(OH)_3$ + YOOH	0,86 ± 0,10 ScOOH + $Sc(OH)_3$	0,94 ± 0,10 DyOOH + $Dy(OH)_3$	0,97 ± 0,15 ErOOH + $Er(OH)_3$	1,24 ± 0,05 $Yb(OH)_3$ + YbOOH
1400°C	1,46 ± 0,03 $Y(OH)_3$	1,20 ± 0,14 $Sc(OH)_3$ + ScOOH	1,53 ± 0,10 $Dy(OH)_3$	1,32 ± 0,05 $Er(OH)_3$ + ErOOH	1,44 ± 0,04 $Yb(OH)_3$

Assessment of thermodynamic data

In order to assess the thermodynamic parameters of REOOH and $RE(OH)_3$, an overview of all gaseous and solid (mainly RE_2O_3) species bearing rare earth and oxygen has to be done and their associated thermodynamic data are compiled. As far as our knowledge, no data on oxide gaseous species are available for temperatures between 1000 and 1400°C. In fact, available data are valuable from 2000 K^3. The thermodynamic parameters of RE_2O_3 are well known[7] and those used in this study are compiled in table III.

The assessment of the Gibbs free energy of formation of each gaseous species is based on the calculations of their partial pressures. These partial pressures are estimated from Arrhenius law (Figure 3) obtained with a water partial pressure of 50 kPa using the expression of the average volatilization rate (4).

In this study, two cases can be distinguished: the first one at high temperature (> 1100°C - 1300°C), where $RE(OH)_3$ is the predominant species and the second one at low temperatures, where a mixture of REOOH and $RE(OH)_3$ is obtained. First the free enthalpy of formation of $RE(OH)_3$ is assessed in the high temperature domain.

The variations of Gibbs free energy of $Sc(OH)_3$ and $Yb(OH)_3$ formation between 1300 and 1400°C was determined according to the above-described procedure from measurements carried out at 1300°C, 1350°C and 1400°C. This methodology cannot be applied for $Dy(OH)_3$ and $Er(OH)_3$, since the change of gaseous species nature occurs at 1300°C. Consequently, partial pressures at 1450°C were evaluated from the Arrhenius law and (4). The values of Gibbs free energy of

Dy(OH)$_3$ and Er(OH)$_3$ formation between 1350°C and 1450°C was determined as previously from measurements at 1350°C, 1400°C and 1450°C. In the two cases, a temperature dependant equation was then fitted. Extrapolation of these equations in the low temperature domain allowed to determine the partial pressure of RE(OH)$_3$ below 1300°C. The partial pressures of REOOH were thus deduced from (5). The same procedure was applied at 1000, 1100 and 1200°C (and 1300°C for Dy$_2$O$_3$ and Er$_2$O$_3$) and a temperature dependant equation was obtained.

$$P_{exp} = P_{RE(OH)_3} + P_{REO(OH)}$$
(Eq.5)

The values of Gibbs free energy of formation of RE(OH)$_3$ and REOOH are listed in table III.

Table III. Thermodynamic data of RE$_2$O$_3$[9], RE(OH)$_3$ and REOOH
($\Delta G = a + b\,T + c\,T\ln T + d\,T^2 + e\,T^{-1} + f\,T^{-2}$)

ΔG (J.mol⁻¹)	a	b	c	d	e	f
Sc$_2$O$_3$	-1,9554720000E+06	7,5148930000E+02	-1,2843000000E+01	-4,2270000000E-03	1,8260000000E+06	-5,2000000000E+07
Sc(OH)$_3$	-9,5941361840E+05	-5,5103840000E+02	-	1,0399999999E-02	-	-
ScOOH	1,2545564653E+06	-3,5206022000E+03	-	1,2857000000E+00	-	-
Dy$_2$O$_3$	-1,9023160000E+06	6,7913130000E+02	-1,2259300000E+02	-6,9710000000E-03	5,9000000000E+01	4,0000000000E+07
Dy(OH)$_3$	-2,8540086498E+05	-1,3510733334E+03	-	2,3166666667E-01	-	-
DyOOH	1,4002661958E+05	-1,7510220500E+03	-	5,6767500000E-01	-	-
Er$_2$O$_3$	-1,9397140000E+06	6,5265620000E+02	-1,1921600000E+02	-5,8890000000E-03	4,2800000000E+02	2,0000000000E+07
Er(OH)$_3$	-1,1690949084E+06	-3,1073840000E+02	-	-7,9600000000E-02	-	-
ErOOH	3,1089014325E+05	-2,0678435000E+03	-	7,0025000000E-01	-	-
Yb$_2$O$_3$	-1,8535110000E+06	7,0275020000E+02	-1,2382100000E+02	-4,5670000000E-03	0,0000000000E+00	5,0000000000E+07
Yb(OH)$_3$	-9,3768945200E+05	-5,6595200000E+02	-	1,1999999999E-02	-	-
YbOOH	-1,2178038039E+06	2,1209860000E+02	-	-1,0910000000E-01	-	-
Y$_2$O$_3$	-1,8535110000E+06	7,0275020000E+02	-1,2382100000E+02	-4,5670000000E-03	0,0000000000E+00	5,0000000000E+07
Y(OH)$_3$	-9,3768945200E+05	-5,6595200000E+02	-	1,1999999999E-02	-	-
YOOH	-1,2178038039E+06	2,1209860000E+02	-	-1,0910000000E-01	-	-

The accuracy of assessed thermodynamic parameters was checked for different water partial pressures in the range 17 kPa to 65 kPa. The partial pressures of hydroxide species (5) experimentally obtained for P_{H2O} comprised between 17 kPa and 68 kPa are in agreement with the partial pressures calculated by Gibbs free energy minimization (Figure 5). That validates the assessed thermodynamic data.

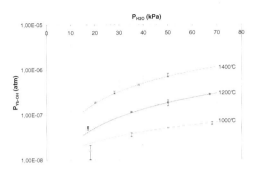

Figure 5. Comparison of experimental (symbols) and calculated (curves) hydroxides pressures using the Gibbs free energies of Yb(OH)$_3$ and YbOOH formation ($P_{tot} = P_{atm}$, v_{gaz} = 5 cm.s^{-1}) (If not visible, uncertainties are hidden in the size of the dots).

In function of temperature, partial pressures of RE(OH)$_3$ and REOOH in equilibrium over RE$_2$O$_3$ are then calculated for a water partial pressure of 50 kPa (Figure 6) . Their variations are in agreement with the hypothesis done on the predominant existence domains of each hydroxide species. If rare earth oxides are compared between themselves, lanthanide oxides appear to be more stable under a moist atmosphere at high temperatures than Y$_2$O$_3$ or Sc$_2$O$_3$. This is in agreement with the values of activation energy, which is higher for lanthanides. The most stable oxide seems to be Yb$_2$O$_3$, since the partial pressures of Yb(OH)$_3$ and YbOOH are the lowest.

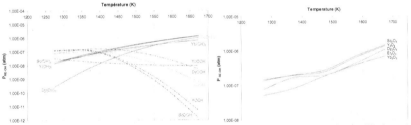

Figure 6. Comparison of partial pressures ($P_{RE-OH} = P_{RE(OH)3} + P_{REOOH}$) at the equilibrium above each oxide (P_{H2O} = 50 kPa, $P_{tot} = P_{atm}$).

CONCLUSIONS

Rare earth sesquioxide volatilization under a moist environment is characterized by two mechanisms. The first one is valuable at low temperatures, where REOOH is the major species. This reaction is characterized by a low activation energy, close to the value for silica. At high temperature, a second mechanism occurs, where RE(OH)$_3$ is the predominant species. In this case, the level of activation energy is twice or three times higher than that at low temperatures. A relation is highlighted between the ionic radius of the rare earth element and the activation energy or the preexponential term. For lanthanides, higher is the ionic radius, higher are the activation energy and the preexponential term. For Sc$_2$O$_3$ and Y$_2$O$_3$, same values of activation energy and preexponential term are obtained. Further, Gibbs free energies of REOOH and RE(OH)$_3$ formation are assessed, and their partial pressures are calculated. The relative stabilities of rare earth sesquioxides are compared and Yb$_2$O$_3$ seems to be the most stable rare earth oxide under a moist environment. Future works consist in quantifying the stability of diverse rare earth silicates by using these data.

ACKNOWLEDGMENTS

This work has been supported by the French Ministry of Education and Research through a grant given to E. Courcot. The authors are grateful to Caroline Louchet-Pouillerie and Jacques Thébault for fruitful discussions.

REFERENCES
[1] K. N. Lee, D. S. Fox, N.P. Bansal, Rare earth silicate environmental barrier coatings for SiC/SiC composites and Si_3N_4 ceramics, *J. Europ. Ceram. Soc.*, **25**, 1705-1715 (2005).

[2] M. Aparicio, A. Duran, Yttrium silicate for oxidation protection of carbon silicon carbide composites, *J. Am. Ceram. Soc.*, **83**, 1351-1355 (2000).

[3] G.-Y. Adachi, M. Imanaka, The binary rare earth oxides. *Chem. Rev.*, **98**, 1479-1514 (1998).

[4] E. Courcot, F. Rebillat, F. Teyssandier, Stability of Rare Earth Oxides in a Moist Environment at High Temperatures – Experimental and Thermodynamic Studies Part I: The way to assess thermodynamic parameters from volatilisation rates. Submitted to J. Europ. Ceram. Soc.

[5] L. Quémard, F. Rebillat, A. Guette, H. Tawil, C. Louchet-Pouillerie, Degradation mechanisms of a SiC fiber reinforced self sealin matrix composite in simulated combustor environments. *J. Europ. Ceram. Soc.*, **27**, 377-388, (2007).

[6] A. Hashimoto, The effect of H_2O gas on volatilities of planet-forming major elements: I. Experimental determination of thermodynamic properties of Ca-, Al-, and Si-hydroxide gas molecules and its application to the solar nebula, *Geochim. Cosmochim. Acta*, **56**, 511-532 (1992).

[7] E. J. Opila, R. E. Hann, Paralinear Oxidation of CVD SiC in Water Vapor. *J. Am. Ceram. Soc.*, **80**, 197-205 (1997).

[8] R.D. Shannon & C.T. Prewitt, Effective Ionic Radii in Oxides and Fluorides. *Acta Cryst.*, **B25**, 925 (1969).

[9] M. Zinkevich, Thermodynamics of rare earth sesquioxides. *Prog. Mat. Sci.*, **52**, 597-647, (2007).

MANUFACTURE OF P-TYPE ZNO THIN FILM BY CO-SPUTTERING OF ZN AND LI$_2$CO$_3$ TARGETS SIMULTANEOUSLY

Yang-Ming Lu[1], Shu-Yi Tsai[2], Shin-Yi Wu[2]

[1]Department of Electrical Engineering, National University of Tainan, Tainan, Taiwan

[2]Department of Material Science and Engineering, National Cheng Kung University, Tainan, Taiwan

ymlu@mail.nutn.edu.tw, ymlumit@yahoo.com.tw

ABSTRACT

The ZnO thin film has been the subject of much research in recent years due to its varied potential applications for optoelectronic devices. However, the p-type ZnO film conduction with low resistivity is difficult to achieve because of self-compensation effect of the opposite charge carriers in the material itself and low solubility of extrinsic doping acceptors. In this work, lithium-doped ZnO films were grown on glass substrates by RF reactive magnetron sputtering using Zn metal target with several Li$_2$CO$_3$ piece meals on it. This study examined the effect of Li content on the conduction and optical of ZnO thin films by r.f. magnetron sputtering. The structural, electrical and optical properties of the deposited films were strongly influenced by various Li content. The Li compositions of the ZnO thin films were determined by ICP-MS. Elemental depth profiles were performed by SIMS analysis. The structure characterization was used by GIAXRD measurement. Hot probes and Hall measurement were used to determine the resistivity, carrier density and mobility. The GIAXRD analysis indicated that the ZnO:Li films have (002) preferred orientation. The optical transmission spectra show a high transmittance (~80%) in the visible region. The lowest resistivity of as-grown p-type ZnO:Li film is 4.72×10^{-1} ohm-cm. A p-type conductive behavior was confirmed by the Hall effect measurement for these ZnO:Li films with a carrier concentration of 2.47×10^{19} cm^{-3} and hall mobility of 0.85 cm^2/V-sec

1. INTRODUCTION

Zinc oxide (ZnO) films have been widely studied in photoelectric devices[1,2], including a range of conductivity from metallic to insulating, high transparency, piezoelectricity, room-temperature ferromagnetism, huge magneto optic and chemical-sensing effects [3]. Due to the wide band gap (3.4 eV) and large exciton binding energy (60meV), it has promising applications in the blue and ultraviolet light emitting devices.

The p-type doping of ZnO has considerable challenges. Similar to GaN, compensation by native donors is the basic problem, since nominally the undoped ZnO is shown as n-type, due to the electrical activity of native defects, such as zinc interstitials, zinc antisites, and oxygen vacancies [4,5] as well as hydrogen impurity[6].If a p-type ZnO film with low resistivity can be made, a p-n oxide junction can be obtained and on this basis more unique semiconductor devices will develop out [7].On the other hand, the group-V elements can also serve as dopants to form p-type ZnO thin films, such as nitrogen[8] phosphorus[9] and arsenic[10] forming deep states within the energy bandgap (about 160, 130, and 100 meV above the valence band maximum, respectively),and therefore, have low ionization fractions at room temperature.

Lithium has also been considered as a potential acceptor dopant, since Li$_{Zn}$ (substitutional Li on the Zn site) was predicted to form an acceptor level much shallower than those of group V elements [11,12]. However, because of its small radius, the incorporation of Li leads to the formation of Li interstitials (Li_i), which act as shallow donors and compensates the hole concentration [13].It is generally accepted that the growth of *p*-type ZnO films is still an experimental challenge and cannot be reproducibly created due to self-compensation from oxygen vacancies or the incorporation of hydrogen as an unintentional donor. In this study, the ZnO:Li thin films were prepared by Zn target and Li_2CO_3 composite target. The effects of lithium composition and defects on electrical properties of the films were investigated.

2. EXPERIMENTAL DETAILS

ZnO films were deposited on Corning 1737 glass by RF magnetron sputtering system using Zn target with several Li_2CO_3 piece meals on it. The glass substrates were thoroughly cleaned with organic solvents and dried before loading in the sputtering system. Ar(99.995%) and O_2 (99.99%) were controlled by the electronic mass flow controller with the ratio of 10:1 (in volume) were introduced as the sputtering gases at a total pressure of 1.33 Pa. Before deposition, the chamber was pumped to an ultimate background of 2×10^{-5} torrs for half an hour. Throughout all experiments, the target was pre-sputtered for 15 mins under 145 watts RF power before the actual deposition began to delete any contamination on the target surface.

The crystalline structure of the films was confirmed by Glancing incident angle XRD (GIAXRD) using a CuK_α radiation (λ=0.15406 nm). The distribution of species in samples was investigated using a CAMECA IMS-6f secondary ion mass spectrometry (SIMS). The optical transmittance of the thin films was measured by using UV-Vis spectrophotometer measurements. The electrical properties of as-deposited films were examined by Hall-effect measurements using the Van der Pauw configuration at room temperature.

3. RESULTS AND DISCUSSION

3.1 Compositions analysis

The chemical compositions of sputtered films were characterized by ICP-MS and FE-EPMA and the results were shown in Fig.1.The results show that the lithium composition variation in the ZnO films with different piecemeal on the target surface. It can be seen that more Li_2CO_3 piecemeal on the target surface, the more lithium composition exists in the film.

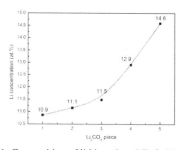

Fig.1. Composition of lithium doped ZnO thin films

The secondary ion mass spectroscopy (SIMS) was carried out to investigate the distribution of species in films. The depth profile of Li concentration in the 300nm-thick ZnO films on Si substrate analyzed by the SIMS was shown in Fig. 2. As demonstrated in Fig.2, although a higher Li content is detected near the top surface of thin film, the Li content gradually uniformly distributed in the film after 50nm depth, indicating that the Li is piled up on the surface and may be incorporated into the ZnO lattice by atomic diffusion process. The V_{Zn} defects provide empty space and facilitate the Li incorporation, hence let the Li preferentially occupy the substitutional sites behaving as an acceptor [14]. It is the presumable reason for good p-type conductivity obtained in the film.

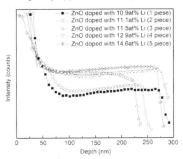

Fig. 2. SIMS depth profile of the Li-doped ZnO thin film

3.2 Structural characterization

Fig. 3 shows the X-ray diffraction spectra of ZnO:Li films as a function of Li content in the range from 10.9 to 14.6 at %.There are (100), (002), (101), (102) and (110) diffraction peaks and a strong peak of ZnO (002) observed for each sample. The positions of the XRD peaks of the samples perfectly match the standard values of ZnO (JCPDS # 800075). In addition, the diffraction angle of ZnO:Li films increased from $2\theta = 34.48°$ to $2\theta = 34.58°$ when Li concentration was increased from 10.9at.% to 11.1at.% shown in Fig.3(b). The diffraction angles of thin ZnO film were lower than the 34.47° angle observed for ZnO bulk material. Higher diffraction angles imply that the interplanar spaces of the ZnO:Li thin films were smaller than those in ZnO bulk material[15]. It is due to the Li atoms incorporates into the ZnO lattice and occupy the substitution lattice sites of it. This mechanism can be shown by equation(1) and produces a p-type ZnO film. Actually, the substitution solubility of Li in ZnO crystal can be as high as 30% [16]. From the result, initially we can observed the lattice constant of ZnO:Li films decreases as the Li doping concentration increases(up to 11.1at.%Li), which should be due to the radius of Li$^+$(0.60Å) ions is smaller than that of Zn^{2+}(0.74 Å) and the substitution mechanism is executed. When the Li concentration is more than 11.1at%, the (002) diffraction angles contrarily shift to lower values, which would result in lattice expansion owing to interstitial Li creates in the ZnO native lattice as shown in Equation(2) and deduce extra electron in the film. Hence, the conductive behavior changes from p-type to n-type. The result is similar to Dhananjay's finding[17] that the observed shifting (002) peaks to lower 2θ values suggesting the Li is not substituting Zn or O site; rather it tends to occupy interstitial positions or probably lies between the Zn–O bond.

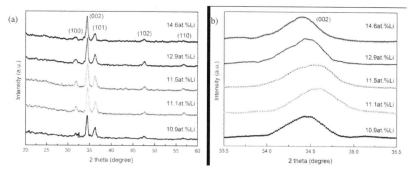

Fig. 3. GIAXRD spectra of ZnO thin films with various Li concentration deposited.

3.3 Electric performance

Fig.4 shows electron concentration, resistivity and mobility of the ZnO:Li films in this study. The p-type conductivity of Li-doped ZnO films were determined from the sign of the Hall measurement and confirmed by the hot-probe method. The p-type ZnO film with 11.1at% Li concentration has the lowest resistivity of 0.47ohm-cm, Hall mobility of 0.85cm²/V-sec and carrier concentration of 2.47×10^{19} cm⁻³. According to chemistry defect equations shown in Eq.(1)and Eq.(2), the conductive mechanism depends on the amount of Li concentration. At lower Li concentration, a substitution mechanism executes and deduces a p-type ZnO film as shown in Eq.(1). While increasing the Li concentration, the conductive mechanism changes to n-type according to Eq.(2).

$$Li_2O + 2V_{Zn} \xrightarrow{ZnO} 2Li'_{Zn} + O_o + 2h^{\bullet} \qquad (1)$$

$$Li'_{Zn} + h^{\bullet} \rightarrow Li'_i + V_{Zn} + e' \qquad (2)$$

If the Li effectively replace the Zn site in the crystal, two electron holes will deduce to maintain the electrical neutrality and a p-type behaved conduction appears. Although Li ions are good shallow acceptors and there are some experimental evidences to support the candidacy of Li for p-type doping [18,19]. However, up to now, p-type of ZnO with Li doping still remains difficult due to self-compensation by donors generated by occupying the interstitial site. When the Li concentration is more than 11.1at%, the conduction type of films changes to n-type. In high Li-content samples, may be associated with the formation of Li_{Zn}–Li_i complex[20]which is harmful to p-type doping, can be suppressed by adjusting Li content in the sputtering target.

As shown in Fig.4, the carrier concentration of ZnO:Li films increased when the Li concentration increased from 10.9at.%to 11.1at.%, which is responsible for the relatively high hole concentration of 2.47×10^{19} cm⁻³ and appears p-type behaved conduction. When Li concentration increases up to 11.5at%, more Li_{Zn}-Li_i complexes formation, which act as an ineffective acceptor and thus the n-type conduction formed. This speculation is also supported by the XRD measurement as mentioned above.

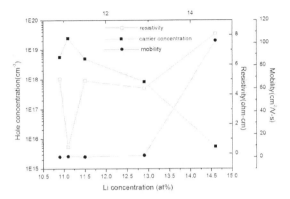

Fig. 4. Resistivity, Hall mobility, and hole concentration of ZnO:Li thin films as a function of Li concentration.

3.4 Optical properties

Fig.5 shows the transmittance spectra of ZnO films with different Li content. All the films show a high transmittance about 80% in the visible range. The films displayed a sharp fundamental absorption edge at around 380 nm. The sharp fundamental absorption edge of the high Li content film has a clear blue shift compared with the low Li content film. The blue shift is an indicator of a well-known Burstein-Moss (BM) effect [21]. For the given Li concentration range, the result implies that Li content has little effect on the transmittance of doped ZnO film and the Li atoms must have successfully incorporated into the ZnO lattice without extra phase forming.

To determine the band gap of the film, we use the theory developed for optical transitions in insulators [22, 23].The optical band gap (Eg) can be estimated by assuming a direct transition between valence and conduction bands for direct band gap semiconductors such as ZnO. The absorption coefficient (α) as a function of photon energy (hv) is expressed as [24]:

$$(\alpha h v)^2 = A\left(h v - E_g\right)$$

where A is a constant.

Fig. 6 shows the plots of $(\alpha h v)^2$ versus the incident radiation $h v$ in the fundamental absorption region. The optical band gaps were obtained by extrapolating the linear part of the curves of $(\alpha h v)^2$ as a function of $h v$ to intercept the energy axis. The optical band gap calculated from the absorption coefficient varied between 3.25 and 3.30eV. The BM effect occurs when the Fermi level enters into the conduction band. Therefore, the increase of band gap is believed to be the result of incorporation of more Li atoms in the films.

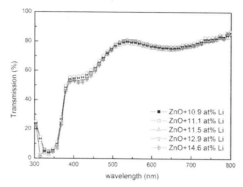

Fig.5. The UV-Visible transmittance spectra of ZnO films with different Li concentration.

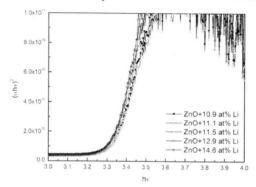

Fig.6. Plot to determine the optical band gap from $(\alpha h\nu)^2$ vs $h\nu$. The inset shows optical band gap of ZnO:Li thin films as a function of Li concentration.

4.   CONCLUSION

We have demonstrated reproducible growth of Li-doped p-type ZnO thin films by RF reactive magnetron sputtering using Zn target with several Li₂CO₃ piece meals on it. The influence of growth temperature on the structural, electrical, and optical properties of p-type ZnO:Li films were investigated in detail. All the ZnO:Li films were polycrystalline and preferentially oriented along (002). The SIMS measurements revealed that the Li concentration in the film was uniformly distributed in the film. Transmission properties of films exhibit an average transmission of 80% in visible wavelength range .The direct optical band gap of films was calculated to be 3.35~3.61eV. The resistivity could be as low as 0.47ohm-cm with the hole concentration of 2.47×10^{19} cm^{-3}, which is adequate to provide a transparent p-n junction of ZnO device.

ACKNOWLEDGEMENTS
This research was supported by the National Science Council of Republic of China, under grant 96-2221-E-024-020-MY3 and 96-2221-E-024-021-MY3.

5. REFERENCES

[1] B. Rech, S.Wieder, C. Beneking, A. Loffl, O. Kluth,W. Reetz, H.Wanger, Proceedings of the 26th IEEE Photovoltaic Specialists Conference, Anaheim,1997, p. 619.
[2] S. Bose, A.K. Barua, J. Phys. D: Appl. Phys. 32, 213 (1999).
[3] Lukas Schmidt-Mende and Judith L. MacManus-Driscoll, materialstoday volume 10, number 5 (2007).
[4] F. Oba, S. R. Nishitani, S. Isotani, H. Adachi, and I. Tanaka, J. Appl. Phys. 90, 824 (2001).
[5] G. W. Tomlins, J. L. Routbort, and T. O. Mason, J. Appl. Phys. 87, 117 (2000).
[6] C. G. Van de Walle, Phys. Rev. Lett. 85, 1012 (2000).
[7] Mathew JOSEPH, Hitoshi TABATA and Tomoji KAWAI, Jpn. J. Appl. Phys. Vol. 38 (1999).
[8] G. Xiong, K. B. Ucer, R. T. Williams, J. Lee, D. Bhattacharyya, J. Metson, and P. Evans, J. Appl. Phys. 97, 043528 (2005).
[9] D. K. Hwang, H. S. Kim, J. H. Lim, J. Y. Oh, J. H. Yang, S. J. Park, K. K. Kim, D. C. Look, and Y. S. Park, Appl. Phys. Lett. 86, 151917 (2005).
[10] Y. R. Ryu, S. Zhu, D. C. Look, J. M. Wrobel, H. M. Jeong, and H. W. White, J. Cryst. Growth 216, 330 (2000).
[11] C. H. Park, S. B. Zhang, and S. H. Wei, Phys. Rev. B 66, 073202 (2002).
[12] E. C. Lee and K. J. Chang, Phys. Rev. B 70, 115210 (2004).
[13] O. Lopatiuk, L. Chernyak, A. Osinsky, J. Q. Xie, APPLIED PHYSICS LETTERS 87, 214110 (2005).
[14] J. G. Lu, APPLIED PHYSICS LETTERS 89, 112113(2006)
[15] V. Gupta and A. Mansingh, J. Appl. Phys., 80, 1063 (1996)
[16] A. Onodera, N. Tamaki, Y. Kawamura, and H. Yamashita, Jpn. J. Appl.Phys., Part 1 35, 5160(1996).
[17] Dhananjay, J. Nagaraju, and S. B. Krupanidhi, J. Appl. Phys. 101, 104104(2007).
[18] O. F. Schirmer, J. Phys. Chem. Solids 29, 1407 (1968).
[19] D. Block, A. Herve, and R. T. Cox, Phys. Rev. B 25, 6049 (1982).
[20] Y. J. Zeng,,APPLIED PHYSICS LETTERS 89, 042106 (2006)
[21] E. Burstein, Phys. Rev., 93, 632 _1954_; T. S. Moss, Proc. Phys. Soc. London, Sect. B, 67, 775 (1954).
[22] D.L. Dexter, Proceedings of Atlantic City Photoconductivity Conference, Wiley, New York, 1954, pp. 155–183.
[23] J.E. Bardeen, F.J. Blatt, L.H. Hall, Proceedings of Atlantic City Photoconductivity Conference, Wiley, New York, 1954, pp. 146–153.
[24] D.Jiles,Introduction to the Electronic Properties of Materials, Chapman and Hall, 1994,chap.9

SYNTHESIZED ZIRCON AND ZIRCON COMPOSITE FROM LIQUID CHEMICAL PROCESS

Yanxia Lu
Corning Inc. Corning, NY 14831, USA

ABSTRACT

Zircon is an important ceramic refractory. It has been used as structure ceramics as well as refractory in many areas. Zircon generally occurs naturally. However the impurities in the natural zircon affect its performance at high temperature. Therefore, pure synthesized zircon becomes attractive. In this paper, the in-situ zircon, which is synthesized from zircon precursors including zirconium compounds and silica compounds, has been explored. The liquid form of zircon precursors is either coated on natural zircon to make zircon composite refractory or reacts to form so-gel, then mixing with natural zircon. Pure zircon with at least 95% yield has been achieved by a simple process. The process includes two major steps: 1) to select zirconia and silica precursors to make a stoichiometric solution or suspension; 2) to control the environment that is favorable for zircon formation. Zircon composite refractory compositions have been successfully developed by coating zircon precursor on natural zircon.

INTRODUCTION

Zircon is an important ceramic refractory due to its high thermal shock resistance. It exhibits a low thermal expansion coefficient [1] (about 4.1×10^{-6}/K in the range 25-1400°C), and stable mechanical strength in a broad temperature range up to 1400°C. It also possesses very low heat conductivity (5.1 W/mC at room temperature and 3.5 W/mC at 1000°C) and high chemical stability [2]. All of these properties make zircon very attractive in ceramic industry and chemical industry. It has been used as an opacifier in ceramic glazes and enamels [3], as structural ceramics [4], and as isopipe in glass fusion process [5,6].

Zircon is naturally occurring material, it is found in most igneous rocks and some metamorphic rocks as small crystals or grains. The commercial available zircon is usually from natural zircon. Depending on the zircon-crystallizing melts, the impurities in natural zircon may vary. The dominant impurity in natural zircon is hafnium oxide (HfO_2) because hafnon($HfSiO_4$) and zircon have complete solid solution [7]. Extend of this solid solution in natural zircon is typically restricted by the HfO_2 content from ~0.7 wt% - 8.3 wt% with mean of 2.0 wt%. The other impurities include the Al, Y, P, Ti, Fe, Mg, Na, etc. Most these impurities are imbedded in the zircon structure, some of them are added as sintering additives to achieve high dense zircon body. These impurities limit zircon in some applications, such as isopipe at high temperature. Therefore, pure zircon or lower impurity content is desirable.

Pure zircon has stoichiometric composition of 67.2 wt% ZrO_2 and 32.8 wt% SiO_2. The crystal structure of zircon is tetragonal, which contains two cation sites; the 4-coordinated Si-site and the distorted 8-coordinated Zr-site. Both Si and Zr are tetravalent and have ionic radii of 0.84 Å and 0.26 Å respectively [8]. Generally, pure zircon is synthesized by reacting zirconia ZrO_2 and silica SiO_2. However, the zircon yield is low if directly reacting α-silica and tetragonal ZrO_2 powder [9]. The popular method of making pure zircon is sol-gel process [10]. Depending on the sources of reactants, the process may be slightly different.

Preparation of high-purity zircon from sol-gel has been the subject of research [4,7, 11,12,13] for many years. Mori and Kanno started to make synthesized zircon from late 80's, the synthesized zircon was produced from zircon precursors, such as: silica or zirconia sols, silica or zirconia compounds. The process includes the hydrolysis of liquid precursors or making gel by controlling pH value of solution. The gel is then mixed with natural zircon seeds up to 10 wt% of total zircon. After drying, the powder is pressed; the green ware is finally calcinated up to 1800°C. Mori concluded that the high formation rates of zircon were only found for these batches with pH value less than 8, and containing natural zircon at least 0.1 wt%, and firing above 1300°C for at least 16 hours. Kanno synthesized fine spherical ZrO_2-SiO_2 by ultrasonic spray pyrolysis, anyhow they did not achieve high yield zircon, and instead they obtained the composite containing ZrO_2 (t), α-SiO_2 and zircon. Veytizou et. al. developed new sol gel process by adding ammonia solution to make gel and then calcinating above 1200°C to form zircon. The dense body (95% of theoretical density) has been achieved by isopressing. Unfortunately, the sintered bodies are always cracked, which is likely because of high shrinkage during the zircon formation from zircon precursors. Valero et. al. applied the similar method to produce porous zircon. All these pioneers provided a guiding in making pure zircon.

Basically, there are two fundamentals in making zircon: 1) the sources of reactants, zirconia precursor and silica precursor; 2) the forming conditions including the temperature, time and chemical environments of reactants. As indicated previously, the fundamental of making zircon is to make an environment that is suitable for reaction between the zircon precursors expressed as following:

$$Zr^{+4} + S^{+4} + O^{-2} \rightarrow ZrSiO_4.$$

Where the cations (Si^+ and Zr^+) are weakly bonded with anions in precursors, which makes it is possible to form Zr-O-Si bonds during hydrolysis reaction and calcinations. Based on this fundamental, the research is focused on selection of precursors and developing a process to produce pure zircon. It covers both hydrothermal reaction from liquid and sol-gel process.

SYNTHESIZED ZIRCON FROM ZIRCON PRECURSORS

The in-situ zircon (or synthesized zircon) is produced by hydrothermal synthesis of zircon precursors composed of zirconium compounds and silicon compounds. The process starts from mixing zirconium compound and silicon compound stoichiometrically with 1:1 ratio of Zr^+/Si^+. The zirconium compounds are selected in the experiment include:

      1) Zirconyl nitrate hydrate: $ZrO(NO_3)_2.6H_2O$

      2) Zirconium oxychloride: $ZrOCl_2.8H_2O$

      3) Zirconium hydrate: $Zr(OH)_4$

And silicon compounds includes

      1) Silica Sol (Ludox HS-40)

      2) Tetraethoxysilane (TEOS): $Si(OC_2H_5)_4$

      3) Silicon hydrate (fine): $Si(OH)_4$

      4) Silicon Tetrachloride: $SiCl_4$

      5) Amorphous silica fine powder

The procedure of making synthesized zircon is described in Figure 1. Zirconium compounds are in the form of solid or powder, so it needs to be dissolved first. For instance, zirconyl nitrate hydrate ($ZrO(NO_3)_2.6H_2O$) is dissolved in warm water (~50°C), after fully dissolving, a clear solution is obtained, it is a strong acidic solution with pH value about 1.0, and then adding silica sol to make colloidal suspension. The suspension is either saved for mixing with natural zircon to make zircon composite coated with synthesized zircon, or made gel by adding ammonia hydrate (NH_4OH). The addition of ammonia hydrate quickly changes the pH value of the suspension, when pH value reaches 10 or above, the gel is sudden formed. Such so-gel process produces heterogeneous gel, the gel is later dried and calcinated to form zircon.

The drying condition is set at 100-120°C for 24 hours. After drying, the gel becomes powder with phases as shown in Figure 2. During the sol-gel process and subsequent hydrolysis, the ammonium hydrate reacts with zirconium compounds, and form ammonium nitrate amorphous phase as shown in Figure 2A or ammonium chloride crystalline phase as shown in Figure 2B. Such reactions leave the Zr-O bonds in zirconium compounds. The Zr-O bonds then react with Si-O bonds in amorphous silica, which promote the formation of Zr-O-Si bonds prior to calcination at high temperature. The Zr-O-Si bond is claimed a critical component in the zircon precursor to ensure the completion of zircon formation [4].

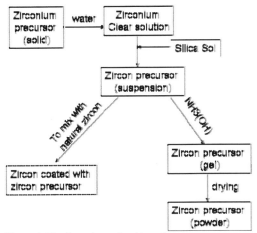

Figure 1. The flow chart of making zircon precursor

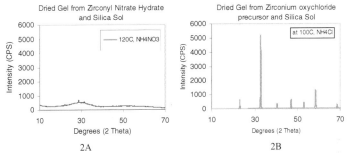

2A   2B

Figure 2. Phase analysis of dried gel. Silica sol is used in both batches. 2A. Zirconium compound is Zirconyl nitrate hydrate: $ZrO(NO_3)_2.6H_2O$; 2B. Zirconium compound is Zirconium oxychloride: $ZrOCl_2.8H_2O$.

The pure zircon is developed from dried powder which contains ammonia compounds and zirconium-silica compound during firing. A study of phase transition from zircon precursors to pure zircon was conducted using full scan x-ray diffraction (XRD) technology. The dried gel (or powder) was fired at different temperatures and then cooled to room temperature; the samples were then characterized by XRD. Figure 3 shows an example that how zircon precursors convert to zircon during firing. The precursors are Zirconium oxychloride: $ZrOCl_2.8H_2O$, and silica sol (Ludox HS40). The gel was produced by reacting zircon precursors with ammonia hydrate (NH_4OH). The dried gel is then fired at different temperatures from 120°C to 1450°C in air.

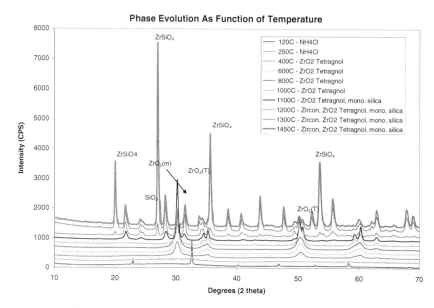

Figure 3. Phase evolution of zircon precursors (unlabeled peaks at 1450°C are also ZrSiO₄).

The gel after drying has crystalline phase of NH_4Cl, which is the product of reaction between the ammonia hydrate and zirconium oxychloride. The other compounds are in the amorphous form. The crystalline phase of NH_4Cl is gradually suppressed when zirconia phases is forming from amorphous entity. At about 600°C, NH_4Cl is completely removed from the gel due to the decomposition of NH_4Cl at 520°C. From 400°C to 1000°C, only tetragonal zirconia has been detected. At 1100°C, the tetragonal zirconia phases starts to transfer to monoclinic phase; and amorphous silica crystallizes to cristobalite silica. Both new phases have ultra-fine grain size identified by broad diffraction peaks. Simultaneously, zircon starts to crystallize and becomes detectable at 1200°C. Four phases, zirconia (tetragonal and monoclinic), cristobalite silica and zircon, are coexisted at 1200°C. As temperature further increases, zircon diffraction intensity becomes stronger and stronger while zirconia and silica diminishes. Above 1450°C, zircon phase is dominant; however, there are still some residual phases of zirconia and silica, which may continue to transfer to zircon at higher temperature.

A similar phase transition has been identified for zircon precursors from zirconyl nitrate hydrate and silica sol as shown in Figure 4. In this route, ammonia nitrate decomposes at 210°C. The tetragonal zirconia grew from amorphous gel and reached its highest around 1100°C, then gradually converted to monoclinic zirconia. Although amorphous silica started to crystallize to cristobalite silica at 1100°C, however, the transformation was suppressed by zircon formation because of reaction between the zirconia (*t* and *m*) and silica (amorphous and cristobalite). The results are identical to the observation of Kanno, where he explained that tetragonal ZrO_2 (*t*) located in the neighborhoods of SiO_2 particles contributes to the formation of zircon at low temperature. In this example, zircon formation is more favorable if ZrO_2 (*t*) reacted with amorphous silica. By 1450°C, almost all zirconia was transferred to zircon because of continuation of reaction between monoclinic ZrO_2 (*m*) and cristobalite silica at higher temperature. The residual phases of zirconia and silica are much less in this reaction path.

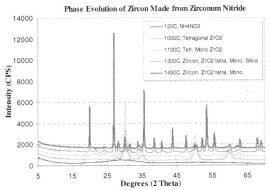

Figure 4. Phase evolution of zircon precursors from zirconium nitride and silica sol.

It has been demonstrated that pure zircon can be produced from zircon precursors using sol-gel process. The process developed in this research is much simpler than previous literatures. Actually, the pure zircon can also be formed from zircon precursor solution without sol-gel process. Figure 5 shows that similar composition has been obtained from zircon precursor solution after firing at high temperature 1450°C. For the precursor solution without seeds, where the solution is made from Zirconium oxychloride: $ZrOCl_2.8H_2O$, and silica sol (Ludox HS40), the solution was dehydrolyzed at 100°C to form a transparent precursor solid; then heated at high temperature for reactions. Zircon was fully formed at 1450°C, a residual zirconia and silica has been detected. When solution is mixed with nature zircon to make zircon composite, the residual phases of zirconia and silica is too little to be detected. The zircon composite has been used to make zircon isopipe, which will be discussed in the next section. Overall, pure zircon has been successfully produced from zircon precursors using simplified process which involves the dissolution of zirconium precursor and making stoichiometric suspension by adding silica precursor.

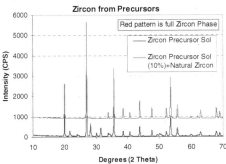

Figure 5. Zircon from zircon precursor solution with and without natural zircon seeds.

The pure zircon made from above process has very fine particle size. The SEM images reveal that nano size pure zircon has been formed from sol-gel process with size about 50 nm as shown in Figure 6. Without mixing with natural zircon, the synthesized zircon tends to form agglomerates with 1-30 μm. Since it is soft agglomerate, it can be re-dispersed by aqueous solution.

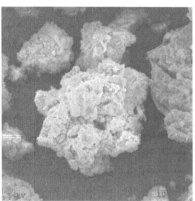

Figure 6. Morphology of synthesized zircon from Zirconyl nitrate hydrate: $ZrO(NO_3)_2.6H_2O$.

It has been experimentally demonstrated that pure zircon can be synthesized from multiple zircon precursors with nano particle size. Such nano zircon can be good sources for coating or as raw material of zircon refractory.

ZIRCON-ZIRCON COMPOSITE
One of applications of synthesized zircon is to make zircon-zircon composite. As introduced previously, the natural zircon is hard to sinter. Instead of adding sintering additives, the synthesized zircon is added in natural zircon batch to produce a dense zircon. The model of zircon-zircon composite is illustrated in Figure 7. The composite consist of commercial zircon (large grain particles) and in-situ zircon (small grains in triple points or void spaces). The commercial zircon is a major component; its particle size represents the grain size after sintering. The in-situ zircon is a minor component, and located in neck regions or voids; it acts as a sintering additive to bond the natural zircon. The creep resistance of the composite is the result of contributions from both types of zircon. The commercial zircon (natural zircon) is selected in a manner to minimize the grain boundary concentration in order to lower the Coble creep [14]. The Coble creep is a diffusional creep along the grain boundaries. The smaller grain size results in a large quantity of grain boundaries generally. Therefore, the natural zircon used in the experiment is a relative coarse zircon, called wet-milled zircon with average particle size of 7 μm from Ferro Corporation. The in-situ zircon (synthesized zircon) in the composite reduces the creep rate by making strong bond between natural zircon grains and filling the voids, which will further lower the diffusional creep caused by both Nabarro-Herring Creep and Coble creep.

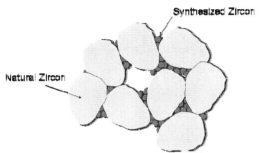

Figure 7. The model of zircon-zircon composite

The process of making zircon composite is illustrated in the flow chart as below (Figure 8). There are two approaches to make composite. Depending on the morphology of zircon precursors, the composite could be made from coating zircon liquid precursor on natural zircon particles, or mixing zircon precursor powder with natural zircon. The first method is called the coating and latter one is called the sol-gel mixing.

In the coating method, the precursor solution is prepared as described previously; the ultimate synthesized zircon in the composite is designed in a range of 0.1-10 wt%. Because of such low concentration, the zircon liquid precursor is usually diluted and then added to natural zircon powder to form slurry with about 75% solid load. The slurry is then ball milled at least 2 hours, after homogeneous mixing, the slurry is dried at 100°C over night. After drying, the powder is re-dispersed by micronizer.

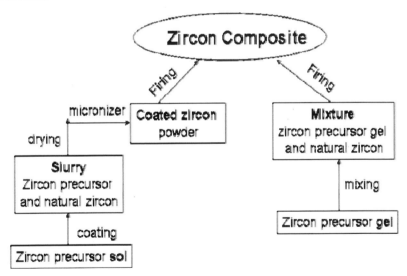

Figure 8. The flow chart of making zircon composite

For the so-gel mixing method, the zircon sol-gel precursor is re-dispersed in water prior to adding to the natural zircon. The composite contains the same concentration of synthesized zircon in a range of 0.1-10 wt%. The mixture of zircon precursor sol-gel and natural zircon went through ball milling to ensure the homogeneity, and then dried at 100°C for at least 24 hours.

After obtaining the dry powder that contains zircon precursor and natural zircon, it will be iso-pressed to form a dense solid at a condition of 18 Kpsi for 0.5-5 min at room temperature. The green body after iso-pressing has at least 65% theoretical density. After firing, the density reaches to 85-95% of theoretical density.

The firing conditions are designed based on the shrinkage study of green zircon body during heating up, measured by dilatometer. Figure 9 shows an example of shrinkage results for natural zircon. The natural zircon starts to sinter around 1240°C (minimum sintering temperature). Although the pure zircon decomposes above 1600°C, the impurities in natural zircon actually drive it sinter prior to decomposition. To make high dense refractory, such sintering is desirable because it provides strong body at ambient condition. However, for application in glass melt as isopipe, such shrinkage (or deformation) is detrimental because of the sag. Therefore, it is always looking for higher minimum sintering temperature. Above 1240°C, a linear shrinkage as function of temperature is measured. The shrinkage is continuing while holding at 1600°C. The curve suggests that natural zircon can be sintered at 1600°C without significant decomposition occurred (a sudden volume change). Based on this result, the sintering condition is selected below 1600°C for at least 12 hours for all composite zircons.

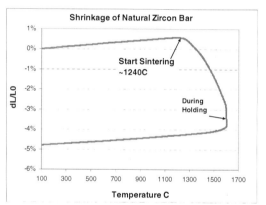

Figure 9. The shrinkage behavior of natural zircon green part.

Zircon composite has been densified through the sintering. For the composite made by coating zircon precursor on natural zircon, the density of fired part approached to theoretical density with 4.44 g/cc (theoretical density is 4.6 g/cc), which indicated the full sintering of coated zircon composite. The flexural strength is over 30000 psi. The creep rate is about 4×10^{-6}/h, which is higher than expected value ($<10^{-6}$/h). The reason of high creep is mainly because of the fine grains in the body that was produced by micronizer process after coating. The micronizer is a process to re-disperse the agglomerates after coating zircon precursor on the natural zircon. The micronizing condition set up, such as pressure, is directly related to how fine the particles will be after micronizer. It is desirable to preserve the original particle size distribution by appropriately re-dispersing the coated powder.

The composite made by coating has uniform microstructure with very fine pores as shown in Figure 10. Most pores are closed pores, which are non-permeable to mercury if measured by mercury porosimetry. The geometric density measurement indicates that the porosity is only 3.5%. The grain size is fine and in a range of 0.1μm-10μm.

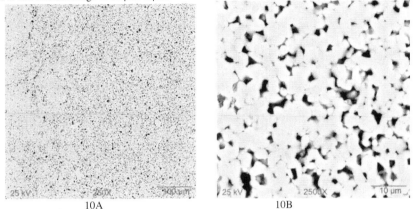

10A   10B

Figure 10. Microstructure of zircon coated with zircon liquid precursor fired at 1580°C for 48 hours. 10A shows the uniformity at low magnification; 10B shows the grain size at high magnification.

The composite made from zircon precursor gel also has relative uniform microstructure as shown in Figure 11, where the batch contains 5% synthesized zircon. The zircon precursor gel and natural zircon powder have been well mixed by ball milling prior to forming. As indicated before, the synthesized zircon is formed from nano-precursors. The gel type zircon precursors crystallize on the surface of natural zircon, and simultaneously bond two or more natural zircon grains together. Since nano-precursors are small aggregates, after reaction, the fine synthesized zircon reside between the large grains or near the large grains as shown in Figure 11B. It clearly reveals two types of zircon morphologies, fine and coarse grains. With this sample, the synthesized zircon seem excessive, lowering the synthesized zircon content can minimize the grain boundary and maximize the bulk density. The compaction of this batch is low with porosity as high as 30%. The low compaction may be due to the different forming method (extrusion) used for this batch. The extrusion has much lower forming pressure than iso-pressing, and also the batch contains organic forming agent, which may be the cause for big pores in the composite as shown in the images. The porosity can be reduced if applying the iso-pressing forming method. The optimum synthesized zircon content is less than 5%; it can be in a range of 0.5-2.0 wt%.

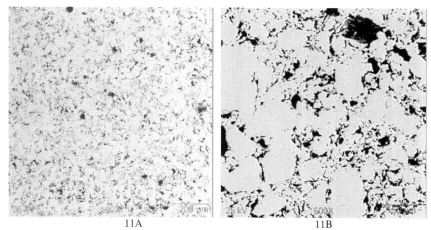

11A   11B

Figure 11. Microstructure of composite made from zircon dry precursor.

Zircon composites have been fabricated by both coating method and mixing method with uniform microstructure. The coating method produced the zircon composite with better uniform microstructure, high bulk density, and high mechanical strength. The creep at high temperature is low, but can be further reduced by controlling the powder processing after coating process. The composite made from mixing method has low density, but can be improved by using better forming method. The mixing method is easy and viable method in making zircon composite.

REFERENCES

[1] R. N. Signh, J. Am.Ceram. Soc. 73 (8) (1990) 2399

[2] T. Mori, H. Hoshino, at. al., J. Ceramic Soc. Jap. 99 (1991)227

[3] A. Weiss, Am. Ceram. Soc. Bull, 1995, 74, 162

[4] Mori, T., Yamamura, H., et. al. "Preparation of high-purity ZrSiO4 powder using sol-gel processing and mechanical properties of the sintered body", J. Am. Ceram. Soc 75(1992) 2420

[5] US 7354878 B2, John Helfinstine, D. Liebner, et. al. "Sag Control of Isopipes used in Making Sheet Glass by the Fusion Progress", 4/8/08

[6] US2005/0130830, Donald J. Ames, Ellen K. Brackman and Donald L. Guile, "Creep resistant zircon refractory material used in a glass manufacturing system", June 6, 2005

[7] John M. Hanchar & Paul Hoskin, "Reviews in Mineralogy & Geochemistry", vol 53:"Zircon", 2003

[8] Finch RJ, Hanchar JM, Hoskin PWO, Burns PC, "Rare-earth elements in synthetic zircon: Part 2. A single crystal X-ray study of xenotime substitution", Am. Mineral 86:681-689 (2001).

[9] Patent US 5032556, T. Mori, et. al, "Preparation method for zircon powder', 7/16/1991

[10] C. Veytizou, J. F. Quinson, et. al, "Preparation of Zircon bodies from amorphous precursor powder synthesized by sol-gel processing", J. of the European Ceramic Society 22 (2002) 2901-2909

[11] Y. Kanno, "Sunthesis of fine spherical ZrO2-SiO2 particles by ultrasonic spray pyrolysis", J. Mat. Sci. Lett. 7 *1988) 386-388

[12] Remi Valero, Bernard Durand, et. al. "Hydrothermal synthesis of porous zircon in basic fluorinated mechanism", Microporous and Mesoporous materials 29 (1999) 311-318

[13] Itoh, T. "Zircon ceramics prepared from hydraous zirconia and amorphous silica", J. Mater. Sci. Lett., 13, (1994) 1661-1663.

[14] Thomas H. Courtney, "Mechanical Behavior of Materials", 2nd edition, 2000, P299-303

ZrO$_2$-ENVIRONMENTAL BARRIER COATINGS FOR OXIDE/OXIDE CERAMIC MATRIX COMPOSITES FABRICATED BY ELECTRON-BEAM PHYSICAL VAPOR DEPOSITION

Peter Mechnich and Wolfgang Braue
German Aerospace Center (DLR)
Institute of Materials Research
51170 Cologne, Germany

ABSTRACT
    Oxide/oxide ceramic matrix composites were coated with Y-stabilized ZrO$_2$ (YSZ) by electron-beam physical vapor deposition (EB-PVD). The typical columnar coating morphology is strongly influenced by CMC surface characteristics. The application of thin ceramic bondcoats provides a more homogeneous microstructure and less intercolumnar separation upon heat treatment. A serious hot-corrosion problem caused by airborne inorganic particles may arise for EBC-coated CMCs in service. The interaction of EBC-coated CMCs and CMAS-type inorganic deposits was studied in laboratory-scale experiments. It turns out that melting and subsequent penetration of CMAS-type deposits is highly critical for the CMC structural integrity, in particular near the coating/CMC interface. Moreover, the presence of iron oxide gives rise to severe YSZ recession via formation of an intermediate calcia-zirconia kimzeyite-type garnet.

INTRODUCTION
    The chemical stability is a serious issue for the long-term application of mullite and alumina based ceramic matrix composites (CMC) in combustion environments, e.g. gas turbine combustors. Mullite and alumina exhibit a high chemical stability under quasi-static conditions. However, under highly dynamic flow of water-vapor rich (exhaust) gases mullite and alumina are prone to recession via formation of volatile Si- and Al-hydroxides.[1-3] Therefore, the vapor-rich gas flow and the CMC surface must be separated. Water-vapor resistant environmental barrier coatings (EBCs) are considered a solution for this problem. EBC materials must be compatible with mullite and/or alumina at least under service conditions. Since silica and alumina are major constituents of state-of-the-art CMCs, critical reactions between SiO$_2$, Al$_2$O$_3$ and EBC leading to co-melting and disintegration of the EBC/CMC system must be prevented. Due to its thermodynamic compatibility and low recession rate up to high temperatures (>1400°C), yttria stabilized zirconia (YSZ) is considered being a suitable EBC material for alumina- and mullite-based CMCs.[3] Relatively thick EBCs with additional thermal protection functionality require a strain tolerant morphology, given the significant thermal mismatch between the CMC and EBC. A favorable, columnar YSZ microstructure is produced by electron-beam physical vapor deposition (EB-PVD). Along with the water vapor corrosion problem, however, serious problems may arise in service with airborne silica- and calcia-rich dusts, impinging and sticking to hot surfaces. This phenomenon has extensively been studied for EB-PVD ZrO$_2$-coated turbine blades and is referred to as CMAS-degradation.[4] On the basis of this knowledge, a detrimental interference of CMAS-type particles with hot EBC-coated CMC combustor walls has been considered as well. A similar CMAS degradation was found for SiC/SiC-CMC with an EBC based on Ba$_{1-x}$Sr$_x$Al$_2$Si$_2$O$_8$ (BSAS).[5]
    The scope of this work was the application of the well-established electron-beam physical vapor deposition technique for the coating of oxide/oxide CMCs with YSZ. These materials were evaluated by annealing at temperatures up to 1300°C and a CMAS-type corrosion scenario.

EXPERIMENTAL

WHIPOX™ CMCs were fabricated at DLR by a filament wound process. The CMCs consist of alumina fibers (Nextel™ 610, 3M) embedded in a porous alumina matrix. From the sintered CMC plate specimens of 120 x 80 x 4 mm were cut and the surface was smoothened by wet grinding using a 320 mesh SiC paper. Approximately 200 μm thick 7 mol-% YSZ EB-PVD coatings were deposited with a 150 kW two source coater equipped with a jumping beam electron gun (von Ardenne Anlagentechnik, Dresden, Germany). The CMC plates were mounted on a rotating multi-plate sample holder. Heat treatments were performed in a resistance-heated chamber furnace (CeramAix, Agni, Aachen, Germany). Microstructural analyses were performed by scanning electron microscopy (DSM Ultra 55, Carl Zeiss MicroImaging, Jena, Germany). X-ray powder diffraction (XRD) was performed using a D5000 Diffractometer (Bruker Analytical X-Ray Systems, Karlsruhe, Germany).

RESULTS AND DISCUSSION

Bondcoat development and microstructural evolution

Since as-wound and sintered WHIPOX-type oxide/oxide CMCs typically exhibit a highly irregular surface structure, pre-grinding of the CMC surface prior to coating is mandatory. The microstructure of the YSZ coating as-deposited on a such pre-ground CMC substrate is depicted on the left-hand side of figure 1 (overview and interface region).

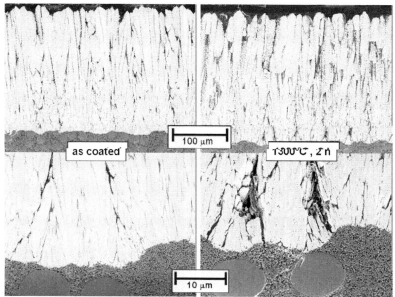

Figure 1: Overview and interface region of YSZ on WHIPOX-type oxide/oxide CMC as-deposited (left-hand side) and after annealing at 1300°C (right-hand side). The coating exhibits typical columnar growth, however with significant misalignment of the columns. Annealing leads to sintering-induced expansion of intercolumnar gaps.

The coating shows the typical microstructural features of the EB-PVD process. However, there is a significant deviation from the typical arrangement of the columns, i.e. columns grow essentially tilted with respect to the CMC surface and column alignment is not straight. A closer look at the substrate/coating interface reveals the high surface waviness which obviously induces the tilted growth of the columns. After annealing for 2 h at 1300°C, the coating exhibits significant sintering characterized by further opening of intercolumnar gaps. These flaws are dominating in areas with marked misalignment of the columns. Considering the CMC surface irregularity being the main reason for misaligned column growth, a further CMC smoothing step was considered to be helpful. However, due to a different abrasion resistance of dense fibers and porous matrix, a virtually flat surface is hardly to achieve by grinding. Moreover, a different nucleation and growth behavior on dense fibers and porous matrix is anticipated. In order to homogenize CMC surface characteristics, the concept of an additional thin ceramic layer acting as bondcoat was introduced. This bondcoat should allow bridging of CMC surface gaps and provide uniform surface characteristics. In a previous study, a reaction-bonded alumina (RBAO) coating, deriving from a dip-coated powder mixture of atomized Al (Alpate, Alcoa France) and α-Al$_2$O$_3$ (CS 400, Martinswerk, Germany) was found to be a promising approach.[6] Under identical coating conditions, this RBAO bondcoat obviously can provide a significantly higher alignment and uniformity of EB-PVD YSZ columns (Fig 2, left-hand part). This microstructural improvement is reflected also after heat treatment for 2 h at 1300°C: sintering induced gap formation is less than for CMCs without the bondcoat (fig 2, right-hand side).

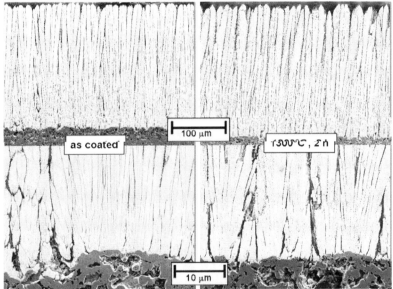

Figure 2: Overview and interface region of EB-PVD ZrO$_2$ on a CMC with RBAO-bondcoat as deposited (left-hand side) and after annealing for 2 h at 1300°C (right-hand side). Alignment of columns is improved whereas sintering-induced damaging is reduced.

Owing to the nature of the RBAO-process, the bond-coat still exhibits a significant roughness. This is mainly due to the Al starting powders having a mean diameter of about 5 microns and consequently producing a similar surface roughness during firing. In order to optimize surface characteristics an analogous bondcoat based on reaction-bonded zirconia (RBZO) was investigated. In this case, 8-YSZ (Tosoh) and zirconium nitride ZrN (H.C. Starck) powders were employed as starting materials having much lower grain size. After dip-coating and firing at 1300°C, this RBZO-bondcoat exhibits a much lower surface roughness. The effect on the EB-PVD coating is obvious: in particular at the interface YSZ columns grow with much less intercolumnar gaps while the overall alignment of the columns is much higher (fig. 3, left-hand part). Besides the lower surface roughness, also a higher nucleation density for YSZ on the chemically identical RBZO may contribute to this behavior. Upon annealing at 1300°C, 1 h , the microstructure also shows sintering phenomena, mainly intercolumnar gap formation. Especially near the coating/bond coat interface, however, the flaw density seems to be significantly lower. From this viewpoint, the RBZO-bondcoat seems to be the superior approach. However, for long term applications,the microstructural stability of the RBZO-bondcoat must also be taken into account: A closer look (fig.3) reveals significant sintering and grain coarsening, which is not observed in the case of RBAO.

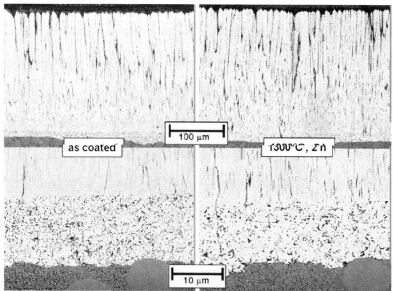

Figure 3: Overview and interface region of YSZ on WHIPOX-type CMC with RBZO-bondcoat as deposited (left-hand side) and after annealing at 1300°C, 2 h (right-hand side). In particular at the interface YSZ growth is denser, leading to lower sintering-induced widening of intercolumnar gaps. RBZO exhibits significant sintering.

Interaction of YSZ-coated oxide/oxide CMCs with CMAS-type deposits

Like EB-PVD YSZ-coated high-pressure turbine airfoils, YSZ-coated CMCs are considered to be affected by silica- and lime-rich airborne dusts. A laboratory-scale experimental approach to the CMAS corrosion was performed in analogy to the procedure reported by Krämer et al.[4]. A model CMAS (CaO 35.3, MgO 9.6, Al$_2$O$_3$ 6.9 and SiO$_2$ 48.2 mol-%) was prepared by co-precipitation of metal-nitrates and fumed silica (Merck) and subsequent co-melting in a Pt-crucible at 1300°C. The X-ray amorphous product was crushed and ground in an agate mortar. About 50 mg/cm^2 of this powder was deposited on top of YSZ coated CMC specimen with RBAO-bondcoat. After annealing at 1200°C, the artificial CMAS-deposit did not show much interaction with the YSZ, only some viscous flow and initial wetting are evident (Fig 4, left-hand side).

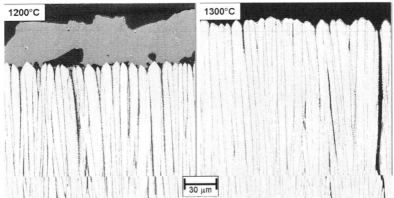

Figure 4: CMAS/YSZ diffusion pair after annealing at 1200°C (left-hand side) and 1300°C (right-hand side). Upon melting CMAS completely infiltrated the top coat as indicated by filling of intercolumnar gaps (light gray contrasted)

At 1300°C, 1 h, the CMAS deposit has been totally mobilized and infiltrated the intercolumnar gaps. Despite the infiltration, the upper region of the YSZ columns seems to be mostly unaffected. A closer look at the YSZ/RBAO-bondcoat and RBAO-bondcoat/CMC interfaces reveals a significant interaction of molten CMAS and EBC/CMC constituents. At the YSZ/RBAO-bondcoat interface, basically two phenomena occur: on the one hand, the formation of globular ZrO$_2$ grains in contact with the glassy phase (g) is observed. This may be explained by a solution/precipitation process of YSZ in contact with molten CMAS, as discussed previously.[5] At the contact between molten CMAS and RBAO newly formed facetted crystals (labeled "sp") are observed which were identified by EDS as spinel-type MgAl$_2$O$_4$ (fig. 5, left-hand side). At the RBAO-bondcoat/CMC interface (right-hand side of fig. 5) microstructural changes are obvious as well. Fibers exhibit a distinct grain growth probably induced by inward diffusion of some glassy phase, also being the reason for the contoured surface after polishing. The matrix shows drastic densification and simultaneous grain coarsening. There is no doubt that the presence of a CMAS-type deposit is responsible for the observed phenomena. Infiltration occurs spontaneously if the CMAS melting temperature is reached. Since columnar YSZ is simply penetrated, severe reactions occur at the YSZ/RBAO-bondcoat interface. However, a significant part of the molten CMAS is also able to penetrate the RBAO-bondcoat and finally reaches the porous CMC. Here, the CMAS-induced fiber grain growth and matrix sintering are considered to be detrimental for the long-term structural integrity of the CMC.

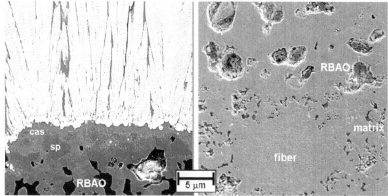

Figure 5: The YSZ/RBAO-bondcoat interface (left-hand side) shows globular ZrO$_2$ grains, Ca-Al-silicate (cas) and newly formed spinel (sp). Fibers and matrix experience massive recrystallization near the RBAO-bondcoat/CMC interface (right-hand side)

In an analogous manner the CMAS-corrosion of an EB-PVD-YSZ coated CMC with RBZO bondcoat at 1300°C was investigated. In figure 6 the interface area of an annealed specimen (left-hand side) is compared to the respective specimen with an additional CMAS contamination (right-hand side). Obviously also in this case CMAS is able to penetrate the EBC/bondcoat double layer and produces considerable damage to the CMC. Although there is no noteworthy reaction in the RBZO bondcoat, ZrO$_2$ grain growth and pore coalescence are evident.

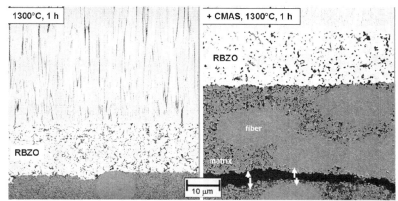

Figure 6: YSZ/RBZO-bondcoat interface annealed at 1300°C (left-hand side) as compared to a similar specimen exposed to CMAS (right-hand side). The RBZO exposed to CMAS exhibits ZrO$_2$ grain growth and pore coalescence. Fibers and CMC matrix exhibit massive recrystallization near the RBZO-bondcoat/CMC interface. The crack marks the extent of CMAS penetration (white arrows)

Again, fibers and matrix exhibit massive coarsening and sintering. A closer look reveals that the crack running through the CMC indicates the CMAS penetration front (white arrows in fig.6, right-hand side). This denotes the detrimental influence of CMAS penetration on the fracture behavior of porous all oxide CMCs. Presumably, failure will occur due to internal stresses at the penetration frontier, probably upon cooling. In any case, ZrO$_2$ with some porosity is highly penetrable for molten CMAS regardless of its microstructure. Even if the RBZO-bondcoat does not react markedly with molten CMAS, it does not provide a protective diffusion barrier for the CMC.

The CMAS approach, however, may oversimplify real conditions, i.e. usually natural dusts contain additional chemical elements. Recently, Braue found a CMAS-deposit on a high-pressure turbine airfoil containing significant amounts of Fe and Ti along with a high CaO to SiO$_2$ ratio.[7] In order to investigate the behavior of similar expanded CMAS systems, a model composition now termed FCMAS was synthesized in analogous manner with a bulk chemical composition close to the EDS analysis given by Braue (CaO 42.1, MgO 17.2, Al$_2$O$_3$ 7.0, SiO$_2$ 26.2 and Fe$_2$O$_3$ 7.5 mol-%). After synthesis at 1300°C, FCMAS is essentially crystalline consisting of a gehlenite (Ca$_3$Al$_2$SiO$_7$)/akermanite (Ca$_2$MgSi$_2$O$_7$) solid solution, merwinite (Ca$_3$MgSi$_2$O$_8$) and a ferrite phase (MgAlFeO$_4$). FCMAS deposits were fabricated similar to the CMAS experiments, specimens were annealed for 1 h at 1200°C and 1300°C, respectively. The microstructural response of an YSZ-FCMAS diffusion couple is shown in figure 7.

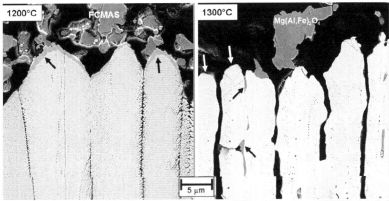

Figure 7: FCMAS-type deposit leading to reaction zone at YSZ column tips at 1200°C, 1 h (left-hand side, black arrows). After 1 h at 1300°C YSZ columns show massive recession (white arrows) and partial melt-infiltration (black arrows). FCMAS-residuum is Fe-containing MgAl$_2$O$_4$ (right-hand side). The intermediate formation of a kimzeyite-type garnet is considered to cause this phenomenon.

In contrast to CMAS, a homogeneous zone overlaying the YSZ columns tips was observed after annealing for 1 h at 1200°C (black arrows in fig. 7, left-hand side). Upon annealing for 1 h at 1300°C column tips as well as column edges are seriously affected, whereas the reaction zone completely disappeared (white arrows in fig 7, right-hand part). Some columns exhibit a pore filling which seems to be generated by melt infiltration (black arrows in fig 7, right-hand part). Also the FCMAS exhibits a significant change: only a Fe-containing MgAl-spinel remains on top of the EBC whereas gehlenite/akermanite and merwinite have disappeared, obviously via melting and subsequent infiltration. The massive degradation of YSZ in the FCMAS corrosion scenario leads to the assumption that the presence of Fe and the formation of an intermediate reaction zone are key issues. In order to study the nature of the reaction zone, a powder mixture of 8 YSZ (Tosoh) and FCMAS in a 50 to 50 weight ratio was annealed for 1 h at 1250°C. Figure 8 shows the XRD-profiles before and after annealing. It turns out that a new crystalline phase is formed at the expense of ZrO$_2$ and FCMAS constituents: Merwinite and Mg,Al-ferrite have disappeared completely whereas gehlenite is still coexisting along with YSZ. The XRD-pattern of the newly formed phase matches well the calcia-zirconia-garnet kimzeyite (Ca$_3$Zr$_2$(Al,Si,Fe)$_3$O$_{12}$). The empirical chemical formula indicates that these garnets exhibit a broad solid solution range, i.e. accommodate nearly all elements present in the YSZ/FCMAS powder mixture. Therefore, even in case of a complex FCMAS stoichiometry and phase composition a single crystalline reaction product is observed. The co-existence of minor glassy phase is possible, however is not evident regarding the flat XRD background. The results from this model experiment lead to the assumption that a Kimzeyite-type garnet forms the reaction zone between FCMAS and YSZ columns as well. Synthetic kimzeyite-type garnets with a stoichiometric Ca$_3$Zr$_2$Fe$_2$SiO$_{12}$ composition were found to have melting temperatures around 1290°C.[8] Hence, the massive recession of YSZ at 1300°C can be explained as a two-step process: First a ZrO$_2$-consuming, solid-state kimzeyite formation below the melting point takes place. After reaching the melting temperature of kimzeyite, the liquefied reaction layer penetrates the YSZ columns. The severe damaging of column edges, however, may indicate a spontaneous recession of ZrO$_2$ in contact with the kimzeyite-derived melt. Further data is required in order to confirm this assumption.

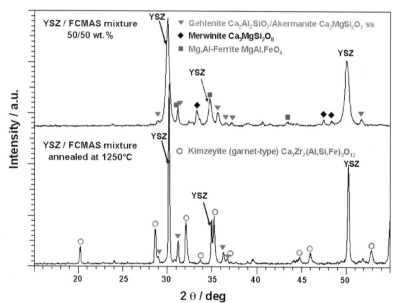

Figure 8: Phase formation in a 50/50 wt-% powder mixture of 8-YSZ and FCMAS (upper XRD-plot) after annealing at 1250°C (lower XRD-plot). A kimzeyite-type garnet Ca$_3$Zr$_2$(Al,Si,Fe)$_3$O$_{12}$ is formed at the expense of merwinite, Mg,Al-ferrite and YSZ.

Since peak wall temperatures in gas turbine combustors are intended to be at least 1300°C, this finding is considered highly critical. Once the kimzeyite reaction layer has been mobilized by melting, new YSZ surfaces are available again for a new cycle of this ZrO$_2$-consuming reaction. Still an open question is the interaction of airborne pollutants and EBC-coated CMCs in actual combustor environments. Flow conditions and presence of water-vapor may reduce or even inhibit the formation of CMAS- or FCMAS-type deposits. On the other hand, the presence of water-vapor may even lower melting temperature and viscosity of deposits. Hence, annealing experiments under water-enriched atmospheres or burner-rig studies with CMAS- or FCMAS-type EBC/CMC diffusion pairs are required.

CONCLUSION

The morphology of EB-PVD YSZ coatings deposited on all oxide CMCs can be widely influenced by application of suitable thin ceramic bondcoats. Both reaction-bonded alumina (RBAO) and reaction-bonded zirconia (RBZO) were found to improve coating homogeneity and high-temperature microstructural stability. The performance of RBZO was found superior to RBAO, however long-term sintering effects still have to be addressed. EB-PVDYSZ-coated oxide/oxide CMCs are prone to hot-corrosion by CMAS-type deposits, in particular in presence of Fe. It is anticipated that realistic combustor environments show a much higher degree of chemical complexity. Therefore, adapted corrosion scenarios including modified CMAS-type dusts, fuel-derived ashes and the presence of water-vapor are requisite.

REFERENCES

[1] K. N. Lee, Key durability issues with mullite-based environmental barrier coatings for Si-based ceramics, *J. of Eng. for Gas Turbines and Power*, **122**, 632-36 (2000).

[2] I. Yuri and T. Hisamatsu, Recession rate prediction for ceramic materials in combustion gas flow, *Proc. ASME TURBO EXPO*, paper GT 2003-38886, 1-10 (2003).

[3] M. Fritsch, H. Klemm, M. Herrmann, and B. Schenk, Corrosion of selected ceramic materials in hot gas environment, *J. Europ. Ceram. Soc.*, **26**, 3557-65 (2006).

[4] S. Krämer, J. Yang, C. G. Levi and C. A. J. Johnson, Thermochemical interaction of thermal barrier coatings with molten CaO-MgO-Al$_2$O$_3$-SiO$_2$ (CMAS) deposits, *J. Am. Ceram. Soc.*, **89** [10] 3167-75 (2006).

[5] K. M. Grant, S. Krämer, J.P.A. Löfvander and C. G. Levi, CMAS degradation of environmental barrier coatings, *Surf. and Coat. Techn.*, **202** [4-7], 653-657 (2007).

[6] P. Mechnich, W. Braue and H. Schneider, Multifunctional reaction-bonded alumina coatings for porous continuous fiber-reinforced oxide composites", *Int. J. Appl. Ceram. Techn.* **1** [4], 343-350 (2004).

[7] W. Braue, Environmental stability of the YSZ layer and the YSZ/TGO interface of an in-service EB-PVD coated high-pressure turbine blade *J. Mat. Sci.* **44**,1664-75 (2009).

[8] J. Ito and C. Frondel, Synthetic zirconium and titanium garnets, Am. Min., **52**, 773-781 (1967).

Geopolymers

DEVELOPMENT OF GEOPOLYMERS FROM PLASMA VITRIFIED AIR POLLUTION CONTROL RESIDUES FROM ENERGY FROM WASTE PLANTS

Ioanna Kourti[1], D. Amutha Rani[1,2], D.Deegan[3], C.R. Cheeseman[1], A.R. Boccaccini[2]
[1]Department of Civil and Environmental Engineering, Imperial College London, South Kensington Campus, London SW7 2AZ, United Kingdom
[2]Department of Materials, Imperial College London, South Kensington Campus, London SW7 2AZ, United Kingdom
[3]Tetronics Ltd., South Marston Business Park, Swindon, Wiltshire SN3 4DE, UK

ABSTRACT

The aim of this study was to develop low energy sustainable geopolymers using DC plasma vitrified Air Pollution Control (APC) residues as the main component. APC residues are generated from the air pollution abatement systems of Energy from Waste (EfW) plants. They are classified as hazardous waste because of their high alkalinity, heavy metal and soluble salt content. DC plasma technology combines APC residues with glass forming additives at temperatures above 1400 °C and produces a stable, non-hazardous inert glass that has been qualified as a product for potential beneficial reuse. In this research plasma vitrified glass has been used to develop geopolymers. The effect of NaOH concentration and curing time on compressive strength is reported. The microstructure of selected APC geopolymer samples was investigated by SEM, X-ray diffraction and SEM/EDS analysis. In order to evaluate the potential for using this material in a range of construction applications the density and water absorption were assessed. Optimum APC glass geopolymers were prepared with 6M NaOH in the activating solution resulting in compressive strengths of ~110 MPa after 28 days. This material had high density (2070 kg/m^3), low water absorption (11%) and low porosity. The results demonstrate that the glass derived from DC plasma vitrification of APC residues can be used to form high strength geopolymers, with potential for use in a range of applications.

INTRODUCTION

APC residues are the fine powdered waste generated from air pollution abatement systems in EfW plants processing municipal solid waste. It is a mixture of fly ash, lime and carbon and contains high concentrations of heavy metals, soluble salts (particularly leachable chlorides), and organic compounds including dioxins and furans. APC residues are classified as hazardous waste with an absolute entry in the European Waste Catalogue (EWC 19 01 07*). The incineration of 1 tonne of municipal solid waste typically produces about 30 kg of APC residues and the UK currently produces about 160,000 tonnes of APC residues per year. Management options for APC residues have recently been reviewed[1,2]. The main waste management option for APC residues has been disposal in hazardous waste landfill, but this is being challenged by both regulatory and political pressures in the UK. Alternative management options are limited, and DC thermal plasma vitrification technology has been investigated as a potential treatment method in this project[3]. This process melts the waste at temperatures above 1400 °C and incorporates the leachable components into the amorphous glassy microstructure of the final product. This reduces the volume of APC residues and produces a plasma vitrified glass that is stable and inert and has potential to be re-used[3]. Plasma vitrified APC residues in ceramic tiles[4] and glass ceramics[5] represent typical alternative reuse applications that have been investigated. However these involve the use of additional high temperature processing.

Geopolymers are synthetic aluminosilicate materials consisting of silica and alumina tetrahedra linked by sharing oxygen atoms. They are similar to cement bound materials, but due to the chemistry and production process, they incur low greenhouse gas emissions[6]. Depending on the composition of the raw materials, geopolymers can have high compressive strengths, low shrinkage, fast or slow setting, acid resistance, fire resistance and low thermal conductivity[6]. A source of silica and alumina that dissolves in alkaline solution acts as a source of geopolymer precursor species[7]. Plasma vitrified APC residue is such an aluminosilicate material and is a potentially suitable candidate for geopolymer production. It contains relatively high levels of calcium, and previous research has shown that calcium can form calcium silicate hydrate (C-S-H) gel in geopolymers[8-15]. The products that can be developed from alkali activation of aluminosilicate wastes in the presence of calcium are: M-geopolymer, where M is a non-calcium alkali element, typically Na^+ or K^+; Ca-geopolymer; C-S-H gel, calcium aluminate; calcium aluminosilicate and other precipitates[9].

This project aimed to deliver a solution that diverts APC residues away from landfill and produce geopolymer products using a process associated with low greenhouse gas emissions. Plasma vitrified APC residues have not previously been used to produce geopolymers or alkali-activated materials. The properties of geopolymers produced from plasma vitrified APC residues are investigated in this paper.

MATERIALS AND METHODS

Plasma vitrification of APC residues was carried out at Tetronics Ltd (Swindon, UK). The plasma treated glass was supplied as a granular material crushed to particles <2 mm. For the preparation of geopolymers, plasma vitrified glass was further milled to form a fine powder using a dry milling process (TEMA mill, TEMA Machinery Ltd). A SEM image of plasma vitrified glass powder is shown in Figure I. Both small and relatively large (~100 μm) glass particles can be seen, indicating a broad particle size distribution. For all experiments, the activating solution was prepared from sodium silicate solution (VWR) and sodium hydroxide pellets (Fisher Chemicals) dissolved in distilled water. The chemical composition of the amorphous plasma vitrified glass was determined by XRF analysis which showed that the major oxides present were SiO_2, CaO and Al_2O_3. The oxide composition of plasma vitrified glass is presented in Table I.

Figure I: Plasma vitrified glass powder from APC residues

Geopolymer samples were prepared by mixing plasma vitrified glass powder with the highly alkaline sodium silicate activating solution. The latter was prepared by dissolving sodium hydroxide pellets in water, allowing this to cool to room temperature before adding the required amount of sodium silicate solution. The activating solution was then mixed with the plasma vitrified glass powder for 10 minutes after which the paste was poured into rectangular moulds (80 x 25 x 25 mm^3) on a vibrating table. The samples were de-moulded after 24 hours, wrapped in cling film and cured at room temperature.

Table I: Chemical composition of plasma vitrified glass

Oxide	Composition weight %	Oxide	Composition weight %
Na_2O	2.88	CaO	32.59
MgO	2.31	TiO_2	1.19
Al_2O_3	14.78	Mn_3O_4	0.23
SiO_2	41.10	Cr_2O_3	0.06
P_2O_5	0.77	Fe_2O_3	4.07
K_2O	0.03		

The solid/liquid (S/L) and Si/Al ratios used, determined in previous experiments, were 3.4 and 2.6, respectively. Geopolymers were produced using NaOH concentrations in the activating solution ranging from 2M to 12M. These were characterised for compressive strength and the samples with the highest values were further characterised for density, using Archimedes' principle, and water absorption. The mineralogical composition of these samples was determined by X-ray diffraction (XRD, Philips PW1700 series). Scanning electron microscopy (SEM-EDS, JEOL-JSM-840A) was used to examine the microstructure of selected geopolymer mixes and to determine the chemical composition of the geopolymer samples.

RESULTS AND DISCUSSION

Figure II presents the compressive strengths of plasma vitrified glass geopolymers obtained using different NaOH concentrations in the activating solution at different curing times. This shows that the compressive strengths of APC glass derived geopolymers increased with increasing concentration of NaOH reaching a maximum value at 6M. Compressive strength was further developed after curing for 28 days reaching values of 90-110 MPa; this was expected as geopolymerisation continues after setting has occurred[16]. The dissolution of the aluminosilicate species increases with increasing NaOH concentration in the activating solution. This causes an increase in the amount of monomers available for geopolymerisation, leading to a higher degree of geopolymerisation and higher compressive strength of the final material.

A decrease in compressive strength was observed in samples prepared with NaOH concentrations greater than 10M. This is in agreement with previous research[17] which showed that excessive NaOH can adversely affect the geopolymerisation process. Very high compressive strengths and high densities have previously been attributed to the presence of Ca in geopolymer systems[9, 18].

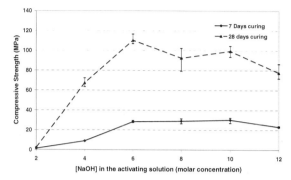

Figure II: Effect of curing time and NaOH concentration on compressive strength of APC residues plasma derived glass geopolymers

Samples prepared with 6M NaOH had the highest compressive strengths (~110 MPa) and this is considered to be the optimum mix for plasma derived glass geopolymers. These geopolymer samples had a density of 2070 kg/m^3 and a water absorption of 11%.

The XRD analysis demonstrated that the optimum geopolymer samples were completely amorphous, as indicated by the high background in the XRD data around 30° 2θ, which is in accordance with previous research[9]. The absence of crystalline peaks does not preclude the possibility of amorphous C-S-H or calcium hydroxide forming due to the high amount of calcium in the plasma vitrified glass.

All samples that were examined with SEM-EDS were polished to a 1μm surface finish before analysis. Due to the activation of plasma vitrified APC residues with sodium hydroxide and sodium silicate solution, the geopolymer network is expected to contain free sodium ions (Na$^+$) which balance the negative charge of Al^{3+} in IV-fold coordination[6]. This sodium might leach out under specific conditions but in the case of SEM-EDS polishing, the time of polishing is quite low so leaching of sodium is not expected. Moreover, debris or loose particles were not observed during the polishing process. Micrographs of selected geopolymer samples prepared after compressive strength tests are shown in Figure III (a) – (c). All images show a heterogeneous microstructure containing un-reacted APC glass particles of various sizes and shapes, surrounded by a geopolymer binder phase. This suggests that small APC glass particles react completely while the larger particles are only partially reacted during geopolymerisation.

The amount of unreacted material present is related to the NaOH concentration in the activating solution. It can be seen that the sample prepared with 4M NaOH contained larger particles than the samples prepared with 6M or 10M NaOH in the activating solution since the dissolution of the solid aluminosilicate APC residue glass powder increases with increasing NaOH concentration in the activating solution. Furthermore, the SEM images reveal that geopolymer samples prepared with 6M NaOH and above had very low porosity.

Figure III (d) shows the micrograph of the optimum geopolymer used for EDS analysis. It can be seen that the cracks developed during compressive strength tests are in the binder phase. These are deflected around residual APC glass grains and do not seem to pass through them. This type of crack deflection behaviour is believed to contribute to the high strength and toughness of the material.

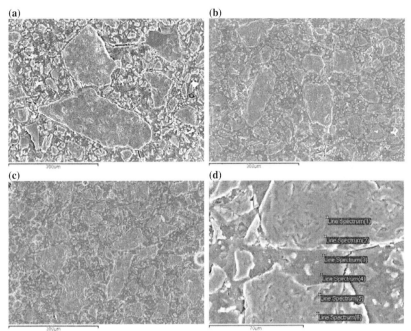

Figure III: SEM images of APC glass geopolymers with: (a) NaOH = 4, (b) NaOH = 6, (c) NaOH = 10, (d) NaOH = 6 (Micrograph used for EDS analysis)

SEM-EDS analysis of the optimum sample (Figure IV and Table I) show that the major elements present in the unreacted APC glass particles (Spectra 1, 2, 5 and 6) are O, Ca, Si, Al, Cl, Fe, Mg, and Ti, while in the binder phase (Spectra 3 and 4) sodium is also present. This sodium was provided by the activating solution which contained sodium silicate and sodium hydroxide. SEM-EDS analysis did not show a clear distinction between a geopolymer phase and a phase containing C-S-H hydration products. No evidence of calcium hydroxide precipitates was found. The high level of calcium in the plasma derived glass from APC residues indicates that the binding phase was probably a mixture of geopolymeric gel (M or Ca geopolymer) combined with the types of hydration products observed in geopolymers prepared from metakaolin and other high calcium materials such as ground granulated blast furnace slag (GGBFS) [11].

Line Spectrum (1)   Line Spectrum (2)

Line Spectrum (3)   Line Spectrum (4)

Line Spectrum (5)   Line Spectrum (6)

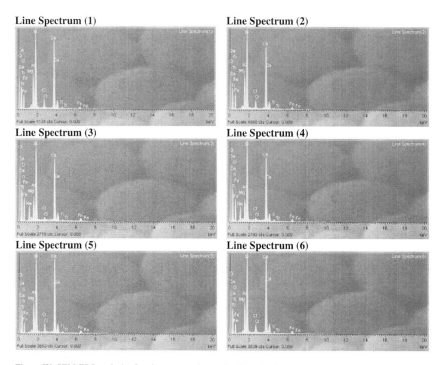

Figure IV: SEM-EDS analysis of optimum geopolymer developed in this study

Table II: Chemical composition of optimum geopolymer by SEM-EDS analysis

Spectrum	O	Na	Mg	Al	Si	Cl	Ca	Ti	Fe	Total
Spectrum(1)	34.46	ND*	0.92	8.36	21.48	3.1	28.02	1.03	2.63	100
Spectrum(2)	36.17	ND*	0.79	8.55	21.6	3.12	26.14	0.85	2.78	100
Spectrum(3)	40.71	5.92	0.86	5.47	19.65	0.83	22.82	0.82	2.92	100
Spectrum(4)	40.51	5.71	0.59	5.43	20.11	0.76	23.88	0.74	2.27	100
Spectrum(5)	30.44	ND*	0.83	8	21.85	3.23	31.38	1.09	3.18	100
Spectrum(6)	28.83	ND*	0.64	8.03	22.31	2.93	32.8	0.95	3.51	100

*ND = Not detected

CONCLUSIONS

Plasma derived glass obtained from APC residues is an inert, homogeneous, glassy material that it can be transformed into an amorphous, high-strength geopolymer that contains un-reacted APC residue glass particles and a geopolymer binder phase. Activating solutions containing sodium hydroxide concentrations up to 10M promote geopolymerisation while a higher concentration adversely affects the process. An activating solution with a NaOH concentration of 6M produces material with very high strength (~110 MPa), high density (2070 kg/m^3), low water absorption (11%) and low porosity after curing for 28 days. The research shows that plasma vitrified APC residue glass can be used to form geopolymers and the resulting materials have the potential to be used in a range of construction products.

ACKNOWLEDGMENT

This work was completed as part of the project 'Integrated solution for air pollution control residues (APC) using DC plasma technology' funded by the UK Technology Strategy Board and Defra, through the Business Resource Efficiency and Waste (BREW) programme.

REFERENCES

1. Quina, M.J., Bordado, J.C., and Quinta-Ferreira, R.M., *Treatment and use of air pollution control residues from MSW incineration: An overview*. Waste Management, 2008. 28(11): p. 2097-2121.
2. Amutha Rani, D., Boccaccini, A.R., Deegan, D., and Cheeseman, C.R., *Air pollution control residues from waste incineration: Current UK situation and assessment of alternative technologies*. Waste Management, 2008. 28(11): p. 2279-2292.
3. Amutha Rani, D., Gomez, E., Boccaccini, A.R., Hao, L., Deegan, D., and Cheeseman, C.R., *Plasma treatment of air pollution control residues*. Waste Management, 2008. 28(7): p. 1254-1262.
4. Amutha Rani, D., Boccaccini, A.R., Deegan, D., and Cheeseman, C.R., *Glass-ceramic tiles prepared by pressing and sintering DC plasma vitrified air pollution control residues*. International Journal of Applied Ceramic Technology 2009. Accepted for publication, in press.
5. Amutha Rani, D., Roether, J.A., Gomez, E., Deegan, D.E., C.R., C., and Boccaccini, A.R., *Glass-ceramics from plasma treated air pollution control (APC) residues*. Glass Technology - European Journal of Glass Science and Technology Part A, 2009. 50(1): p. 57-61.
6. Davidovits, J., *Geopolymer - Chemistry and Applications*. 2nd ed. 2008, Saint-Quentin, France: Institute Geopolymere.
7. Divya, K. and Rubina, C., *Mechanism of geopolymerization and factors influencing its development: a review*. Journal of Materials Science, 2007. 42(3): p. 729-746.
8. Yip, C.K., Provis, J.L., Lukey, G.C., and Van Deventer, J.S.J., *Carbonate mineral addition to metakaolin-based geopolymers*. Cement and Concrete Composites, 2008. 30(10): p. 979-985.
9. Yip, C.K., Lukey, G.C., and Van Deventer, J.S.J., *The coexistence of geopolymeric gel and calcium silicate hydrate at the early stage of alkaline activation*. Cement and Concrete Research, 2005. 35(9): p. 1688-1697.
10. Alonso, S. and Palomo, A., *Calorimetric study of alkaline activation of calcium hydroxide-metakaolin solid mixtures*. Cement and Concrete Research, 2001. 31(1): p. 25-30.

11. Yip, C.K. and Van Deventer, J.S.J., *Microanalysis of calcium silicate hydrate gel formed within a geopolymeric binder*. Journal of Materials Science, 2003. 38(18): p. 3851-3860.
12. Pacheco-Torgal, F., Castro-Gomes, J., and Jalali, S., *Tungsten mine waste geopolymeric binder: Preliminary hydration products investigations*. Construction and Building Materials, 2009. 23(1): p. 200-209.
13. Temuujin, J., van Riessen, A., and Williams, R., *Influence of calcium compounds on the mechanical properties of fly ash geopolymer pastes*. Journal of Hazardous Materials, 2009. In Press, Corrected proofs.
14. Yip, C.K., Lukey, G.C., Provis, J.L., and Van Deventer, J.S.J., *Effect of calcium silicate sources on geopolymerisation*. Cement and Concrete Research, 2008. 38(4): p. 554-564.
15. Alonso, S. and Palomo, A., *Alkaline activation of metakaolin and calcium hydroxide mixtures: influence of temperature, activator concentration and solids ratio*. Materials Letters, 2001. 47(1-2): p. 55-62.
16. Van Deventer, J.S.J., Provis, J.L., Duxson, P., and Lukey, G.C., *Reaction mechanisms in the geopolymeric conversion of inorganic waste to useful products*. Journal of Hazardous Materials, 2007. 139(3): p. 506-513.
17. Lampris, C., Lupo, R., and Cheeseman, C.R., *Geopolymerisation of silt generated from construction and demolition waste washing plants*. Waste Management, 2009. 29(1): p. 368-373.
18. Xu, H. and Van Deventer, J.S.J., *Microstructural characterisation of geopolymers synthesised from kaolinite/stilbite mixtures using XRD, MAS-NMR, SEM/EDX, TEM/EDX, and HREM*. Cement and Concrete Research, 2002. 32(11): p. 1705-1716.

SYNTHESIS OF ZEOLITE-X FROM WASTE PORCELAIN USING ALKALI FUSION

Takaaki Wajima[1] and Yasuyuki Ikegami[2]

[1] Faculty of Engineering and Resource Science, Akita University
1-1 Tegata-gakuen-cho, Akita, 010-8502 Japan
[2] Institute of Ocean Energy, Saga University
1-48 Kubara, Yamashiro-cho, Imari 849-4256, Japan

ABSTRACT

Unsold fired products from the ceramics industry are discharged as waste ceramics. In this study, we attempted to convert waste porcelain to crystalline zeolite-X using the alkali fusion method. By varying the experimental conditions different types of product were obtained, e.g. zeolite-X, zeolite-P, hydroxysodalite and nepheline. The siliceous minerals in the waste, quartz and mullite, were transformed into soluble phases after 24 h of agitation following alkali fusion, and then converted into zeolitic phases by heating at 80 °C. Almost all the mineral phases in porcelain were transformed into a soluble phase after fusion when the mix ratio of NaOH to porcelain was more than 1.2 and the fusion temperature was above 200 °C. Zeolite-X was synthesized from the aged material with a high content of soluble phase, while meta-stable zeolite-X was transformed into the stable hydroxysodalite phase with a long synthesis time. When the zeolite-X product was prepared with a mix ratio of 1.2, fusion temperature of 600 °C and synthesis time of 12 h, the obtained product had a high specific surface area (412 m^2/g) which was almost 700 times higher than that of the raw material, and a unique micropore diameter (13 Å), indicating a zeolite-13X structure. The product obtained with a mix ratio of 1.6, fusion temperature of 600 °C and synthesis time of 6 h had the same NH_4^+ sorption ability as commercial zeolite-X, which suggested that the product can be applied in environmental fields such as water purification.

INTRODUCTION

Zeolites are a group of more than 40 crystalline hydrated aluminosilicate minerals. Their structure is based on a three-dimensional network of aluminum and silicon tetrahedra linked by shared oxygen atoms. Due to their specific pore sizes and large surface areas, zeolites can be used in various applications, e.g., molecular sieves, adsorbents and catalysts.[1] Several researchers have reported the synthesis of zeolites from a wide variety of starting materials containing high amounts of Si and Al, e.g., kaolin, halloysite, interstratified illite-smectite, montmorillonite, bentonite and incinerated ash.[2–9]

Over 5 million metric tons of glass and ceramic wastes are discharged every year in Japan, and unsold fired products from the ceramics industry are discharged as waste ceramics. Waste from tile plants is a serious problem in many countries. In Italy and Spain, for example, approximately 20000 and 50000 metric tons per annum, respectively, of porcelain polishing wastes are discharged from tile plants. Some is used as artificial aggregates, for cement production, or for other minor applications.[10] The remaining waste ceramics are deposited in landfill sites, and because of the limited capacities of landfill sites, they cause both social and environmental problems.

In our previous study, we converted waste porcelain into zeolitic materials using a two-step alkali conversion at a low temperature (80 °C).[11] Zeolitic material with a high cation exchange capacity (ca. 170 cmol/kg) can be synthesized from waste porcelain using an alkali reaction as the first step, and has the ability to remove heavy metals from acidic solution.[12] The waste solution with a high content of Si could be converted into pure zeolite crystals (e.g. zeolite-A, zeolite-X, zeolite-P) by addition of aluminate solution in the second step. However, only the amorphous glass phase is used to synthesize the zeolite phases, and almost all Si and Al content in the crystalline phases remains in the product and is not used in zeolite synthesis. In addition, the two-step process is complex. In this work, we propose a new method that is both effective and simple.

We aimed to convert waste porcelain into crystalline zeolite-X using the alkali fusion method. Zeolite-X (faujasite type) is highly functional zeolite, which can be used not only as a cation exchanger and adsorbent but also as molecular sieves and a catalyst. It has found widespread application in the purification and separation of gases and organic compounds. Therefore, zeolite-X is an important target for the synthesis of zeolite from waste. In this study, the amorphous and crystalline phases in the waste were first converted into soluble phases by alkali fusion, and were then used to synthesize zeolite-X crystals. We determined the conditions of alkali fusion for a large amount of soluble phases to obtain high productivity of zeolite-X. The characteristics of the products as regards specific surface area and pore size distribution were examined. In addition, as one application of the obtained product, the ability to adsorb NH_4^+ was investigated. NH_4^+ is usually the leading cause of eutrophication. We focused our attention on NH_4^+ removal from an aqueous solution simulating domestic and agro-industrial wastewater.

MATERIALS AND METHODS

Raw Materials

Waste porcelain was ground with a mill. Particles less than 1 mm in size were sorted and washed with distilled water. Table 1 shows the chemical composition of the powdered waste porcelain, determined by X-ray fluorescence spectrometry (XRF, RIGAKU, ZSX 101e). The porcelain comprised SiO_2 and Al_2O_3, corresponding to 69.8 wt.% and 18.5 wt.%, respectively, with the rest being Na_2O, K_2O, MgO, CaO, Fe_2O_3 and ZnO. This is a typical composition of waste porcelain.[11]

Table 1. Chemical Composition of Powdered Waste Porcelain

	SiO_2	Al_2O_3	Na_2O	K_2O	MgO	CaO	Fe_2O_3	ZnO	Total
Powdered Waste Porcelain	69.8	18.5	0.6	6.2	0.3	3	0.7	0.7	99.8

Alkali fusion

To investigate the relationship between the properties of the fusion conditions and the fused materials produced, alkali fusion was carried out as follows. 10 g of powdered porcelain was mixed with 4, 8, 12, 16 or 20 g of NaOH and ground to obtain a homogeneous mixture. This was then heated in a nickel crucible in air at 200, 400, 600 or 800 °C for 1 h. The resultant fused mixture was cooled to room temperature and ground again. 0.5 g of each mixture was added to 2 mL distilled water in a 10 mL polymethylpentene tube. Aging was carried out with vigorous agitation by reciprocal shaker at room temperature. After 24 h of agitation, the solid product was filtered, washed with distilled water, and dried overnight in a drying oven at 60 °C.

Characterization of the solid phase was carried out using XRD. To determine the solubilities of Si and Al from the solids obtained under the different conditions, 0.1 g of solid from each treatment was added to 10 mL of 1 M HCl solution, and shaken for 6 h. This mixture was filtered, and the concentrations of Si and Al in the filtrate were determined by ICP-AES (Shimadzu, ICPS-7500).

Zeolite synthesis after fusion

To synthesize zeolitic phases from the solids obtained from all the different fusion conditions, the slurry obtained after agitation was heated at 80 °C for 0, 3, 6, 12, 24 and 48 h in a water bath. The solid product was filtered, washed with distilled water, and dried overnight in a drying oven at 60 °C. The phases of the obtained product were analyzed by XRD, and the concentrations of Si, Al and Na in the filtrate were determined by ICP-AES. The morphologies of the raw material, fused material, aged material and product were observed by scanning electron microscopy (SEM, TOPCOM, SM-200). The specific surface areas and distributions of pore sizes in the raw material, aged material and product were measured by a nitrogen adsorption method (BET) using Sorpmatic (Thermoquest) at 77 K.

Adsorption of NH_4^+

The NH_4^+ adsorption properties of the products were investigated using 3 samples: (1) raw powdered porcelain, (2) product, (3) commercial zeolite-13X (Wako). The product was synthesized at 80 °C for 6 hours from the fusion material (fusion conditions: mix ratio of NaOH to cake 1.6, fusion temperature 600 °C). An aqueous solution of 0.8 mM NH_4^+ was prepared with NH_4Cl. 0.1 g of sample was added to a 400 mL portion of aqueous 0.8 mM NH_4^+ solution. During stirring, the pH was kept neutral (6.0-8.0). 2 mL aliquots of the suspension were removed at various time and filtered. The concentration of NH_4^+ in the filtrate was determined by the method of Koyama et al.[13] The corresponding removal (R) was determined from the material balance as follows.

$$R = \frac{(C_0 - C) \bullet 100}{C_0}$$

where C_0 and C are the initial and measured concentrations of NH_4^+ in the filtrate (mmol/ L).

RESULTS AND DISCUSSION

Alkali fusion

  To prepare the best solid for zeolite synthesis with alkali fusion, the relationships between the soluble contents and mineral phases were determined. Figures 1 and 2 show the remaining phases and the amounts of soluble Si and Al, respectively, in the solid after 24 h of agitation following alkali fusion. The mineral phases in waste porcelain were mainly quartz and mullite as crystalline substances, and an amorphous glass phase. The soluble contents of Si and Al in waste porcelain were almost zero. With an increasing ratio of NaOH to porcelain, the intensities of quartz and mullite decreased (Fig. 1), and the amounts of soluble Si and Al in the solid increased (Fig. 2). The increases in soluble Si and Al were due to the decrease in siliceous minerals (quartz and mullite), which means that crystalline siliceous minerals are transformed into soluble phases by alkali fusion. The results in Figs. 1 and 2 show that the optimum conditions for transforming most of the Si and Al contents in the porcelain into soluble phases consisted of a mix ratio of NaOH to porcelain of more than 1.2 and a fusion temperature higher than 200 °C.

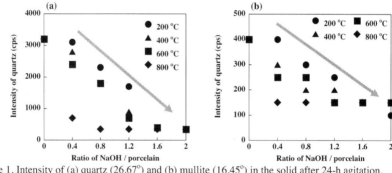

Figure 1. Intensity of (a) quartz (26.67°) and (b) mullite (16.45°) in the solid after 24-h agitation.

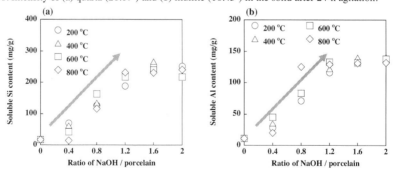

Figure 2. Amounts of soluble (a) Si and (b) Al in the solid after 24-h agitation following the alkali fusion.

Zeolite synthesis

Figure 3 shows the product phases obtained for each fusion condition. Zeolite-X, zeolite-P, hydroxysodalite and nepheline were obtained using the fusion method. In the case of a low ratio of NaOH (< 0.8) at a high temperature (800 °C), nepheline was synthesized regardless of the reaction time. Zeolite-X was synthesized when a solid with high soluble contents of Si and Al was obtained after 24 h of agitation following the alkali fusion (NaOH/porcelain > 1.2; fusion temperature > 200 °C). That is, zeolite-X was synthesized before 12 h of reaction time, and a mixture of zeolite-X and hydroxysodalite was frequently identified after 12 h of reaction time. Since zeolite-X is a metastable phase, zeolite-X was easily transformed into hydroxysodalite after 12 h of reaction. Therefore, a short synthesis time is important for obtaining zeolite-X phases.

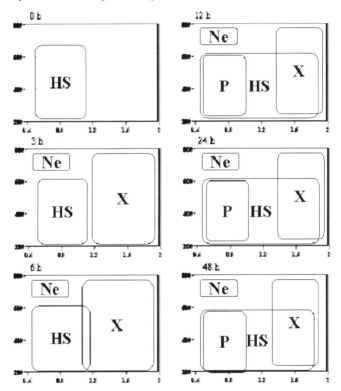

Figure 3. Product phases at each reaction time under various fusion conditions.
X: Zeolite-X; HS: Hydroxysodalite; P: Zeolite-P; Ne: Nepheline

Conversion process

We investigated the conversion process from waste porcelain into the product when the product was synthesized at 80 °C for 12 hours from the fusion material (fusion conditions: mix ratio of NaOH to cake of 1.2, fusion temperature of 600 °C).

Figure 4 shows the XRD patterns obtained for (a) raw material, (b) fused material, (c) aged material and (d) product. The raw material (waste porcelain) comprised an amorphous glass phase, and crystalline phases such as quartz and mullite (Fig. 4 (a)). After fusion, most of the phases in the raw material were initially converted into soluble phases, such as sodium silicate, sodium aluminate and sodium aluminosilicate (Fig. 4 (b)), and then to amorphous material after agitation for 24 h (Fig. 4 (c)). Zeolite-X phase was detected in the product (Fig. 4 (d)).

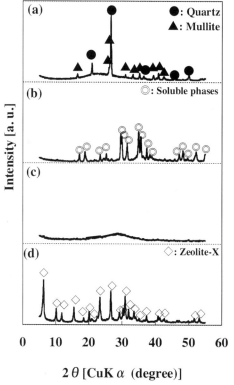

Figure 4. XRD patterns of (a) raw material, (b) fused material, (c) aged material and (d) product.

Figure 5 shows the SEM micrographs obtained for (a) raw material, (b) fused material, (c) aged material and (d) product. The raw material was fragments of porcelain (Fig. 5 (a)). The fused material was a large particle with a melted surface resulting from the formation of sodium silicate and aluminate by the alkali fusion method (Fig. 5(b)). After agitation, the aged material looked like a gel-like particle with an amorphous phase (Fig. 5 (c)), whereas the final product consisted of octahedral crystals (zeolite-X) (Fig. 5 (d)). Zeolite-X crystallizes in a typical octahedral form, and is a member of the faujasite group. Its structural framework comprises 4- and 8- membered rings.

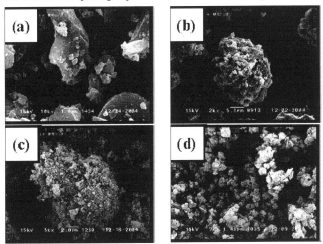

Figure 5. SEM micrographs of (a) raw material, (b) fused material, (c) aged material and (d) product.

Table 2 shows the specific surface area of raw material, aged material and product, and Figure 6 shows the distribution of pore sizes for (a) raw material, (b) aged material and (c) product. The specific surface area of raw material was only 0.6 m^2/g because waste porcelain has a small pore volume, but the specific surface area of aged material was 36.3 m^2/g due to the creation of an amorphous gel aggregate with macropores. The final product had a high specific surface area (412 m^2/g), almost 700 times higher than that of the raw material. The pores of the product had a diameter of 13 Å, due to the zeolite-13X crystals.

Table 2. Specific surface areas of raw material and product

	Specific surface area (m^2/g)
Raw material	0.6
Aged material	36.3
Product	412.0

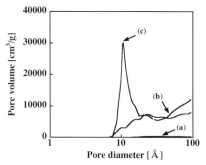

Figure 6. Distribution of pore sizes for (a) raw material, (b) aged material and (c) product.

It can be considered that waste porcelain was first converted into soluble material by alkali fusion, then transformed into an amorphous gel and finally zeolite-13X crystals can be synthesized from this gel.

Figure 7 shows the Si, Al and Na concentrations in the solution (Figure 7 (a)), and intensities of zeolite-X and hydroxysodalite in the product, during the synthesis, when the fusion conditions were a mix ratio of NaOH to cake of 1.2 and a fusion temperature of 600 °C. As can be seen in this figure, the Si concentration decreased quickly during the first stage of the synthesis, while the Al concentration decreased gradually. The Na concentration was almost constant during the synthesis. The intensity of zeolite-X increased abruptly, reaching a maximum at 12 h, followed by a slow decrease, while that of hydroxysodalite increased at 48 h. These results indicated that zeolite-X was synthesized during the first stage of the reaction, and was then transformed into hydroxysodalite. Therefore, zeolite-X can be synthesized from waste porcelain using alkali fusion, and it is important to synthesize zeolite-X phases for a shorter time than that at which transformation of zeolite-X into hydroxysodalite starts.

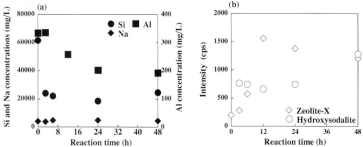

Figure 7. Si, Al and Na concentrations in the solution (a), and intensities of zeolite-X and hydroxysodalite in the product (b), during the synthesis.

Removal of NH_4^+

We investigated the ability of the product synthesized from the fusion material at 80 °C for 6 hours (fusion conditions: mix ratio of NaOH to cake of 1.6, fusion temperature of 600 °C) to remove NH_4^+, and compared this with the NH_4^+ removal abilities of the raw material and commercial zeolite-13X. Figure 8 shows the NH_4^+ removal abilities of (a) waste porcelain, (b) product and (c) commercial zeolite-13X during the reaction. Although the raw material had no NH_4^+ removal ability at all, the product was able to remove NH_4^+. The product could remove NH_4^+ completely (100 %) in 20 min, which was almost the same as the results obtained for the commercial material. These results show that current product could be used for water purification.

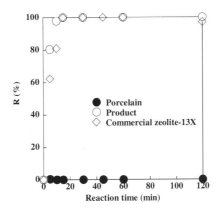

Figure 8. NH_4^+ removal by waste porcelain, the product and commercial zeolite-13X as a function of reaction time.

CONCLUSION

Zeolitic material can be synthesized from waste porcelain using the alkali fusion method. The siliceous minerals in the porcelain were converted into soluble material by alkali fusion, and zeolite-X, zeolite-P, hydroxysodalite and nepheline were obtained. The fusion conditions for the synthesis of zeolite-X are a mixing ratio of NaOH to porcelain of more than 1.2, a fusion temperature of over 200 °C, and heating time at 80 °C of less than 12 h. The product has a unique pore size (13 Å) and high specific surface area (412 m^2/g). The product has a similar NH_4^+ adsorption ability to that of commercial zeolite-13X, and one potential application is water purification. The synthesized product would be a highly valued zeolite that could be used as a molecular sieve or adsorbent.

REFERENCES

[1] R. M. Barrer, *Zeolites and Clay Minerals as Sorbents and Molecular Sieves*, Academic Press, London (1978).

[2] R. M. Barrer, R. Beaumont, C. Colella, Chemistry of soil minerals. Part XIV. Action of some basic solution on metakaolinite and kaolinite, *Journal of the Chemical Society, Dalton Transaction*, 934-941 (1974).

[3] R. Ruiz, C. Banco, C. Pesquera, F. Gonzalez, I. Benito, J. L. Lopez, Zeolitization of a bentonite and its application to the removal of ammonium ion from waste water, *Applied Clay Science*, **12**, 73-83 (1997).

[4] A. Baccouche, E. Srasra, M. E. Maaoui, Preparation of Na-P1 and sodalite octahydrate zeolites from interstratified illite-smectite, *Applied Clay Science*, **13**, 255-273 (1998).

[5] A. F. Gualtieri, Synthesis of sodium zeolites from a natural halloysite, *Physics and Chemistry of Minerals*, **28**, 719-728 (2001).

[6] D. Boukadir, N. Bettahar, Z. Derriche, Synthesis of zeolites 4A and HS from natural materials, *Annual de Chimie Science des Materiaux*, **27**, 1-13 (2002).

[7] X. Querol, N. Moreno, J. C. Umaña, A. Alastuey, E. Hernández, Synthesis of zeolites from coal fly ash: an overview, *International Journal of Coal Geology*, **50**, 413-423 (2000).

[8] G. C. C. Yang, T.-Y. Yang, Synthesis of zeolites from municipal incinerator fly ash, *Journal of Hazardous Materials*, **62**, 75-89 (1998).

[9] T. Wajima, K. Kuzawa, H. Ishimoto, O. Tamada, T. Nishiyama, The synthesis of zeolite-P, Linde type A, and hydroxysodalite zeolites from paper sludge ash at low temperature (80 °C): Optimal ash-leaching condition for zeolite synthesis, *American Mineralogist*, **89**, 1694-1700 (2004).

[10] N. Ay, M. Űnal, The use of waste ceramic tile in cement production, Cement and Concrete Research, **30**, 497-499 (2000).

[11] T. Wajima, Y. Ikegami, Synthesis of zeolitic materials from waste porcelain at low temperature via two-step alkali conversion, *Ceramics International*, **33**, 1269-1274 (2007).

[12] T. Wajima, Y. Ikegami, Zeolitic adsorbent synthesized from powdered waste porcelain, and its capacity for heavy metal removal, *ARS SEPARATORIA ACTA*, **4**, 86-95 (2006).

[13] M. Koyama, T. Hori, Y. Kitayama, *IARC Report, Kyoto University*, **2**, 11-14 (1976).

THE AGEING PROCESS OF ALKALI ACTIVATED METAKAOLIN

C. H. Rüscher[1]; E. Mielcarek[1]; W. Lutz[2]; A. Ritzmann[2]; W. M. Kriven[3]

[1.] Institute of Mineralogy, Leibniz University of Hannover, Hannover, Germany.

[2.] Süd-Chemie Zeolites GmbH Labor, Berlin, Germany.

[3.] Materials Science and Engineering, University of Illinois at Urbana-Champaign, Urbana, IL, USA.

ABSTRACT

The ageing of metakaolin-based geopolymer of special composition $(1.2K_2O \bullet Al_2O_3 \bullet 4.2SiO_2 \bullet 14H_2O)$ has been studied by flexural force measurements, a molybdate analytical method and Fourier transform infrared spectroscopy (FTIR). The results enable new insight into details of the structural development responsible for early strength development between 25 h and 100 h and the unfavorable weakening for ageing above 200 h. In particular the increase in strength is related to the slow development of an aluminosilicate network enclosing preformed polysilicate chains. A fast condensation of silicate chains occurs during the dissolution of metakaolin in the first 25 h of the ageing process. The subsequent destruction of polymeric silicate chains finally causes the weakening.

INTRODUCTION

Novel methods including FTIR and MAS-NMR spectroscopic techniques has been applied in many studies for a better understanding of the mechanism of geopolymerisation and the optimisation of complex compositions, structure and mechanical properties relationships, e.g. [1-13]. However, less attention has been given to the disadvantage of some compositions relative to their significant weakening, although a reasonable strength was gained during the first hours or days of ageing. Thus it was our main aim to address the question about what determines the properties of a most simple composition of alumosilicate gel during ageing and hardening, rather than to optimize its properties as suitable building materials. For this reason we investigated the alkali activation of metakaolin, which could be seen as a model system for the processes which are involved during ageing and hardening of a geopolymer. Using the combination of mechanical testing and infrared absorption spectroscopy, together with investigations by the molybdate measurement - a method known to be sensitive to the characterization and quantification of monomer, dimer and oligomer silicate units [14-17], a distinction of the silicate species responsible for the main variation in strength during ageing could be achieved. Our new results are presented here.

EXPERIMENTAL

The aluminosilicate (geopolymer) samples were prepared by stirring metakaolin (aluminum and silicon source) in a solution of potassium water glass (alkali and silicon source). The chemical composition of both raw materials is given in Table 1.

Table 1   Composition of raw materials for preparation of aluminosilicates

Source	Chemical Composition (mass %)				
	SiO_2	Al_2O_3	K_2O	H_2O	Density g/cm^3
Water glass	19.3	-	24.5	56.2	1.52
Metakaolin	61.5	37.8	-	0.7	

The ageing process was followed at 25° C between 0.5 h and 750 h. The mass ratio of potassium water glass to metakaolin was 15/9 according to a molar ratio of aluminium to potassium of about 1 and a Si/Al ratio of 2 as seen from Table 2.

Tab. 2   Composition of the 15/9 aluminosilicate mixture of water glass and metakaolin

Mixture	Composition (mass %)			
15/9	SiO_2	Al_2O_3	K_2O	H_2O
Total Mass	843.0	340.2	367.5	849.3
Molar	4.2	1	1.2	14.2

The cylindrical tiles of samples were 3 cm in length by 1.5 cm in diameter. In order to hinder any evaporation of water during sample ageing, the cylinders were cured in sealed containers. For characterization of the physical properties, the relative flexural bending strength of the 3 cm x 1.5 cm cylindrical aluminosilicate pieces were measured on a universal testing machine (MEGA 2-3000-100 D), by a specifically designed device for measurement of small testing pieces.

The molybdate method [14-17] was used for characterization the condensation of silicate species in the aluminosilicate solid. The yellowish silico-molybdic acid complex was due to the reaction of silicate monomers and molybdic acid. Its rate of formation was controlled by the use of photometric analysis (at a wavelength of 400 nm). For example Si-O-Al bonds rapidly decompose in 0.1 molar hydrochloric acid, giving rise to an immediate yellow color in the presence of molybdate. The decomposition of dimeric, trimeric, and oligomeric ($-Si-O-Si-$)$_n$ bonds requires a longer time, which leads to a time shift in the formation of the silico-molybdic acid complex.

IR absorption spectra were taken in a Bruker FTIR IES66v spectrometer (KBr pressed pellets were prepared from a 200 mg KBr, 1 mg sample). Acid leached samples were measured also using the KBr method. These samples were produced by using 1 percent hydrochloric acid and leaching the samples for 10 min. followed by holding the filtrate for 12 h at -18 °C before making the KBr pressed pellet.

RESULTS AND DISCUSSION

Molybdate measurements [14-17] are helpful in characterizing alumosilicate gels, providing a determination of chain length of silicate units in silicates and aluminosilicates. Whereas the monomeric, dimeric and oligomeric silicate molecules of water glass solution are molybdate active, higher polymeric SiO_2 units, which consist of more than about 30 silicate units) are molybdate inactive. Fig. 1 shows the molybdate active silicate of the 15/9 mixture aged at 25 °C. The total amount of molybdate active portion of the silicate species decreased upon ageing between 13 h and 25 h, from about 82 % to about 78%. During further ageing the molybdate active species monotonously increased to 99 % at 525 h ageing. Above 525 h the amount decreased again to about 61% at 700 h ageing. It is interesting to note that the amount of molybdate inactive silicate species corresponds to the difference of molybdate active species and the total amount of silicate species which can also directly be deduced from Fig. 1. Thus the polycondensed fraction increased between 13 h and 25 h from 18 % to 22%, which turns into a gradual decrease up to the total disappearance at 525 h. Further ageing showed a renewed increase up to about 39 % of polysilicate units at 700 h. Subdividing the molybdate active silicate further reveals a significant decrease of oligomeric silicate molecules at the ageing process from 13 h (59 %) to 25 h (57 %). During further ageing up to 525 h the oligomeric fraction increased to about 81% followed by a decrease to 50% at 700 h. The fraction of monomeric and dimeric units decreased during ageing between 13 h and 25 h from about 25 % to 19%. It might be meaningful that this decrease turned over into a slight increase at 48 h up to 21 % followed by a very smooth decrease down to 19 % during further ageing. However, for ageing between 25 h and 700 h the gain and loss of oligomeric units are directly related to the loss and gain of the polymeric fraction, respectively. The important question is, if and how is the gain and loss in strength related to the continuous reaction observed by the change in molybdate activity. For better comparison, the development of the flexural strength is shown in Fig. 1. It can be seen that the gradual decrease in the polymeric fraction (increase of oligomeric fraction) during ageing between 25 and 525 h covers both the increase and decrease in strength. Therefore, it must be concluded that the kinetics of at least two parallel reactions control the development of the mechanical properties.

The infrared absorption spectrum of the raw material metakaolin, potassium water glass solution and potassium water glass solution polycondensed by the addition of hydrochloric acid are shown in Fig. 2, together with a sequence of samples aged between 1 h and 700 h. Metakaolin shows the peak maximum of the asymmetric Si-O stretching frequency of the SiO_4 units at about 1070 cm^{-1}, which could be related to polymeric sheet fragments from the SiO_2 sheets in kaolinite. The position of the peak maximum of the asymmetric Si-O stretching are denoted in the following as DOSPM (density of states peak maximum) due to its envelope nature, including vibrations of pure silicate units as well as alumosilicate units which are hard to separate in the spectra [28]. This fact may be demonstrated by spectra of pure, more or less condensed water glass. The potassium water glass solution as used for alkali activation of metakaolin, and the water glass solution polycondensed by acid treatment show DOSPM at 1002 and 1044 cm^{-1}, respectively. For potassium water glass the DOSPM can be taken as an indication for a majority of monomeric/dimeric Q0/Q1 silicate units whereas the DOSPM of polycondensed water glass corresponds to a mixture of mainly chain-type Q2 together with some higher condensed Q3 units. Thus, it can be seen that there is an equivalence in DOSPM between Q4(4Al), Q4(3Al), Q4(2Al) and Q0, Q1, Q2 for the alumosilicates and silicates, respectively, which should be noted for the following discussion.

After stirring of the metakaolin with diluted water glass solution and taking a spectrum after 1 and 5 h the DOSPM is observed to have a maximum at about 1025 cm^{-1} and 1010 cm^{-1}, respectively (Fig. 2). This trend still continues to 13 h which is indicated by the DOSPM at about 1007 cm^{-1}. The shift to smaller wavenumbers can be related to the rapid dissolution of the metakaolin. However this process consumes hydroxide solution taken from the potassium water glass. According to this observation, the water glass solution invariably polymerizes. This effect leads to the shift in the DOSPM up to about 1025 cm^{-1} observed for ageing at 25 h. The increase in concentration of polymeric silicate units is seen in the molybdate activity (Fig. 2).

Above 25 h a systematic shift in the DOSPM occurs to 1020 cm^{-1} after 48 h, and further to 1016 cm^{-1} after 72 h. Further ageing leaves the DOSPM largely invariant (Fig. 2). However, the increasing molybdate activity indicates structural changes which decrease the amount of polycondensed silicate units continuously during ageing from 25 h to 525 h (Fig. 1). A support that the constant DOSPM during further ageing above 100 h contains changing contents of silicate and alumosilicate units is given here by further acid leaching experiments. A minimum shift in the DOSPM for acid leached samples is observed for samples aged around 500 h (Fig. 3). This observation is in line with the results obtained by the molybdate method observing the highest contribution of molybdate activity at 525 h ageing. It may be noted that acid treatment within the molybdate method is followed by complexation of all silicate forming monomers which leads to the characteristic yellow colour. On the contrary, in conducting the acid leaching for the infrared absorption experiment it is the degree of polymerisation of silicate units which lead to the shift in the DOSPM. Any meaningful changes in the DOSPM can only be seen in kinetically controlled leaching experiments interrupted at short times. In the limit, at longer time acid treatment the DOSPM always tends to 1080 cm^{-1} for all samples irrespective of ageing time. Therefore a minimum shift in the DOSPM for acid treated samples indicates here the highest stability against acid which can be assumed for a most suitable protection of Al-O-Si bonds within an alumosilicate network.

CONCLUSION

Alkali activation of metakaolin, for the mixtures used here and at room temperature, leads to the formation of an alumosilicate network in a continuous reaction during ageing between 25 h and 525 h. This conclusion can be based upon detailed experiments using the molybdate method for samples as prepared in this study. The forming alumosilicate network probably includes monomeric (Q4(4Al)) and dimeric (Q4(3Al)) alumosilicate units, which were formed within the first 25 h and which comprise about 20 % of the total silicate units. A further 20 % of polycondensed silicate units are formed during ageing between 13 h and 25 h. These units originate necessarily from the water glass. We conclude that the weakening upon ageing above 200 h is related to a significant decreasing fraction of polycondensed silicate. The increase in strength during ageing between 25 h and 100 h is related to the slowly forming alumosilicate network which overcompensates for the effect of decreasing concentration of polymeric silicate. Therefore, the coexistence of two different structural units, i.e. the alumosilicate network and silicate chains, in the right proportion and spatial distribution maximizes the mechanical strength. The destruction of the polymeric silicate chains can be understood as a consequence of network formation producing hydroxide solution.

It can be suggested that all the oligomeric units become part of the alumosilicate network during ageing up to about 525 h which is supported by details in the infrared absorption behaviour. The driving force for the further reaction can be related to the fact that the alumosilicate should tend to a Si/Al ratio of 1 thus leading to a disproportionation.

Acknowledgement: EM thanks for the financial support by "Lichtenbergstipendium" of the "Land Niedersachsen" and support during her stay abroad at the University of Illinois at Urbana-Champaign.

REFERENCES

1. MacKenzie, K. J. D., What are these things called geopolymer? A Physico-Chemical perspective. *Ceramic Transaction-Advances in Ceramic Matrix Composites IX*, 153, 175-186, **2003**.

2. Lee, W. K., W., and van Deventer, J. S. J., Use of infrared spectroscopy ro study geopolymerisation of heterogeneous amorphous aluminosilicates. *Langmuir*, 19, 8726-8734, **2003**.

3. A. Palomo, S. Alonso, A. Fernandez-Jimenez, I. Sobrados, J. Sanz, Alkaline activation of fly ashes. A NMR study of the reaction products. *J. Am. Ceram. Soc.*, 87, 1141-1145, **2004**.

4. Duxson, P., Provis, J. L., Lukey, G. C., Mallicoat, S. W., Kriven, W. M., and van Deventer, J. S. J., Understanding the relationship between geopolymer, composition, microstructure and mechanical properties. *Colloids and Surfaces A – Physicochemical and Engeneering Aspects*, 269, 47-58, **2005**.

5. Fletcher, R. A., MacKenzie, K. J. D., Nicholson, C. L., and Shimada, S., The composition range of alumosilicate geopolymers. *J. Europ. Ceramic Soc.*, 25, 1471-1477, **2005**.

6. Fernandez-Jimenez, A., and Palomo, Mid-Infrared spectroscopic studies of alkali-activated fly ash structure. *Microporous Mesoporous Mater.*, 86, 207-214, **2005**.

7. Duxson, P., Lukey, G. C., and van Deventer, J. S. J., Evolution of gel structure during thermal processing of Na-geopolymer gels. *Langmuir*, 22, 8750-8757, **2006**.

8. Fernandez-Jimenez, A., Palomo, A., Sobrados, I., and Sanz, J., The role played by the reactive alumina content in the alkaline activation of fly ashes. *Microporous Mesoporous Mater.*, 91, 111-119, **2006**.

9. Lecomte, I., Henrist, C., Liegeois, M., Maseri, F., Rulmont, A. and Cloote, R., (Micro)-structural comparison between geopolymers, alkali-activated slag cement and Portland cement. *J. Europ. Ceramic Soc.*, 26, 3789-3797, **2006**.

10. Duxson, P., Mallicoat, S. W., Lukey, G. C., Kriven, W. M., van Deventer, J. S. J., The effect of alkali and Si/Al ratio on the development of mechanical properties of metakaolin-based geopolymers. *Colloids and Surfaces A: Physicochem. Eng. Aspects*, 292, 8-20, **2007**.

11. Criado, M., Fernandez-Jimenez, A., and Palomo, A., Alkali activation of fly ash: Effect of the SiO_2/Na_2O ratio Part I: FTIR study. *Microporous Mesoporous Mater.*, 106, 180-191, **2007**.

12. Criado, M., A. Fernandez-Jimenez,A., Palomo, A., Sobrados, I., and Sanz, J., Effect of the SiO_2/Na_2O ratio on the alkali activation of fly ash. Part II: 29Si MAS-NMR survey. *Microporous Mesoporous Mater.*, 109, 525-534, **2008**.

13. MacKenzie, K. J. D., Komphanchai, S., and Vagana, R., Formation of inorganic polymers (geopolymers) from 2:1 layer lattice alumosilicates. *J. Europ. Ceramic Soc.*, 28, 177-181, **2008**.

14. Thilo, E., Wieker, W., and Stade, H., Über Beziehungen zwischen dem Polymerisationgrad silicatischer Anionen und ihrem Reaktionsvermögen mit Molybdänsäure. *Z. anorg. allg. Chem.*, 340, 261, **1965**.

15. R. K. Iler, *The Chemistry of Silica*, Wiley & Sons, New York **1979**.

16. B. Fahlke, P. Starke, V. Seefeld, W. Wieker, K.-P. Wendlandt, *Zeolites*, 209-213, **1987**.

17. Lutz, W., Rüscher, C. H., and Heidemann, D., Determination of the framework and non-framework $[SiO_2]$ and $[AlO_2]$ species of steamed and leached faujasite type zeolites: calibration of IR, NMR, and XRD data by chemical methods. *Microporous Mesoporous Mater.*, 55, 193-202, **2002**.18. Jirasit, F., Rüscher, C. H., and Lohaus, L., A study on the substantial improvement of fly ash-based geopolymeric cement with the addition of metakaolin. *Int. conf. on pozzolan, concrete and geopolymer*, Thailand, pp. 1-15, **2006**.

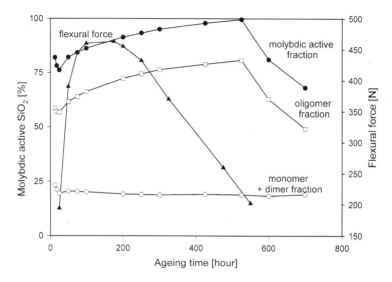

Fig. 1. Development of molybdate active species (left scale) and flexural force (right scale) of alkali activated metakaolin on ageing time (black dots: total molybdate active fraction, squares: oligomer fraction; open circles : monomer+dimmer fraction, black triangles: flexural force; lines as guide to the eyes).

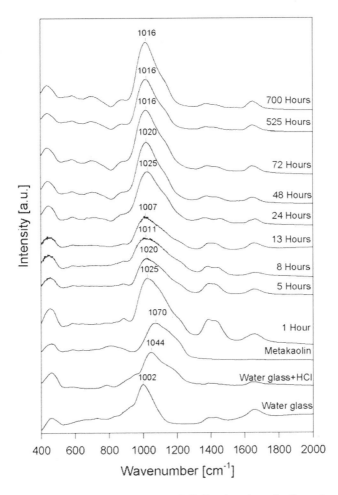

Fig. 2. Infrared absorption spectra of a series of alkali activated metakaolin ageing taken at ageing times as denoted together with spectra of metakaolin, water glass and condensates obtained from water glass by addition of hydrochloric acid. The positions of the peak maxima in the range of asymmetric SiO$_4$ stretching vibration has been specified by the wavenumber and is denoted as DOSPM (Density of States Peak Maxima) in the text.

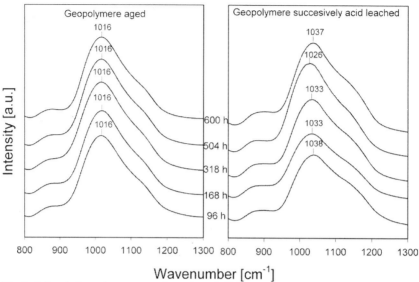

Fig. 3. Infrared absorption spectra of a series of alkali activated metakaolin (left) and acid treated (right) in the range of the asymmetric SiO4 stretching mode. The DOSPM (see Fig. 2) is denoted by wavenumber and the duration of ageing before the acid treatment is given in the middle of the figure.

TESTING OF GEOPOLYMER MORTAR PROPERTIES FOR USE AS A REPAIR MATERIAL

Wanchai Yodsudjai, Prasert Suwanvitaya, Weerapong Pikulprayong and Banjerd Taweesappaiboon
Department of Civil Engineering, Faculty of Engineering, Kasetsart University
Bangkok, Thailand

ABSTRACT

The objective of this research was to develop geopolymer mortar as a concrete repair material. The compressive strength test of goepolymer mortar was performed with different mix proportions. The tensile strength test, the slant shear test and the rapid chloride test of geopolymer mortar were performed and the results compared with those of the commercial concrete repair materials.

It was found that the geopolymer mortar had potential to develop as a concrete repair material. The compressive strength of geopolymer mortar was more than 40 MPa at the 28 days age by room temperature curing. In addition, the tensile strength and the bond strength of geopolymer mortar at 28 days age were close to those of commercial concrete repair materials. However, chloride permeability of geopolymer mortar test by the rapid chloride test ws higher than those of the commercial concrete repair materials.

INTRODUCTION

Generally, reinforced concrete structures are durable. However, if they are exposed to severe environments such as marine environment or exposed directly to dangerous substances, the reinforced concrete structures deteriorate rapidly. When reinforcement corrodes, the deterioration is always progressive. Effective action depends absolutely on having a proper understanding of how corrosion is caused and how it can be controlled. The problem associated with the corrosion of reinforcement is that the product of corrosion exerts stresses within the concrete, and the concrete therefore cracks. This leads to a weakening of the bond and anchorage between concrete reinforcement which directly affects the serviceability and ultimate strength of concrete elements within the structures. To repair reinforced concrete structures deteriorated by reinforcement corrosion, consideration must first be given to the cause of the problem. This is fundamental to the success or failure of the repair. Repair of the deteriorated structures refers to the remedial action taken to prevent or retard its further deterioration and reduces the possibility of damage. The principal objectives of repairs are (1) to stop further corrosion of rebars and (2) where chloride ions are involved, to remove or eliminate the chloride ions in concrete. The durability of repair greatly depends both on its adherence to the substrate concrete and the protection it can afford to the steel reinforcement against corrosion[1].

For effective repair, selection of appropriate materials are needed. This requires an understanding of material behavior in both cured and uncured states in the anticipated service and exposure conditions. There are two main categories of factors that need to be considered in selecting the best repair method. The first is the compatibility of the patch with the adjoining concrete and the second is service and application conditions[2]. There are various commercial repair materials for repairing corrosion related deteriorated concrete structures. These repair materials are sometimes expensive and produced in accordance with standards and limits encountered abroad. There is still no research to confirm whether these repair materials are appropriate for the Thailand climatic condition. Therefore, the development of repair materials which is appropriate for Thailand climatic condition is necessary.

Most of the commercial repair materials contains portland cement as cementitious material. It has been reported that the transportation industry and the portland cement industry happen to be the two largest producers of carbon dioxide. The latter is responsible for approximately 7% of the world's carbon dioxide emissions[3]. Many researches have attempted to develop new materials to replace the portland cement. Geopolymer, with associated positive aspects such as abundant raw resource, little

CO_2 emission, less energy consumption, low production cost, high early strength, fast setting, is an interesting alternative. These properties afford geopolymer great applications in many fields and industries such as civil engineering, automotive and aerospace industries, non-ferrous foundries and metallurgy, plastics industries, waste management, art and decoration, and retrofit of buildings. Geopolymerisation is a geosynthesis, that is, a reaction that chemically integrates minerals involving naturally occurring silico-aluminates. The silicon (Si) and aluminium (Al) atoms react to form molecules that are chemically and structurally comparable to those binding natural rock. It is assumed that geosyntheses are carried out through oligomers (dimer, trimer), which provide the actual unit structure of the three-dimensional macromolecular edifice.

To use the geopolymer mortar as the repair material; however, it is necessary to study the engineering properties of geopolymer mortar; such as, strength and durability properties which are the important properties for the repair materials. Therefore, this study aimed to develop the geopolymer mortar for using as the repair materials by testing the engineering properties under Thailand's climatic condition. It should be noted that it is beneficial for the construction industry in Thailand as the development of the new repair material may help decrease dependence of imports from abroad. In addition, the raw material for producing geopolymer mortar is a by-product which is beneficial for the environmental aspect.

EXPERIMENTS

Geopolymer Mortar Preparation
Table 1 shows the mix proportion of geopolymer mortar, varying the $Na_2O.SiO_3/NaOH$ ratios (0.7 to 1.1) and sand contents (640 to 768 kg/m^3). Constant fly ash content of 404 kg/m^3 and water content of 40 kg/m^3 were used for the total 25 mixes (GM 1 to GM 25).

Table 1. Mix Proportion of Geopolymer Mortar by Varying $Na_2O.SiO_3/NaOH$ Ratios and Sand Contents

Mix Symbol	Ratio $Na_2O.SiO_3/NaOH$	Materials (kg/m^3)				
		Sand				
GM 1 – GM 5	0.7					
GM 6 – GM 10	0.8					
GM 11 – GM 15	0.9	640	672	704	736	768
GM 16 – GM 20	1.0					
GM 21 – GM 25	1.1					

Table 2 shows the mix proportion of geopolymer mortar by varying the sand contents (672 and 704 kg/m^3) and water contents (35 and 30 kg/m^3). Constant of fly ash content of 404 kg/m^3 is used for the total 6 mixes (GM 26 to GM 31).

Table 2. Mix Proportion of Geopolymer Mortar by Varying Water Contents

Mix Symbol	Ratio Na₂O.SiO₃/NaOH	Materials (kg/m³)		
		sand	Fly ash	Water
GM 26	0.9	672	404	35
GM 27	1.0	704	404	35
GM 28	1.1	704	404	35
GM 29	0.9	672	404	30
GM 30	1.0	704	404	30
GM 31	1.1	704	404	30

Table 3 shows the mix proportion of geopolymer mortar, with various sand sizes. Constant fly ash content of 404 kg/m³ was used for the total 6 mixes (GM 32 to GM 37). The sand content used was similar to the mix proportion in Table. 2. However, the sand sizes were different. The sand used was 80% by weight of sand passing sieve No. 30, 15% by weight of sand passing sieved No.50 and 5% by weight of sand passing sieve No. 100. The three highest compressive strengths of all mix proportions (37 mixes) were investigated and selected. It was found that the mix proportion of GM 32, GM 36 and GM 27 gave the highest compressive strength among the 37 mixes. Therefore, the other properties; that is; flexural strength, bonding strength, setting time and chloride penetration were investigated comparing with those of conventional repair materials. It should be noted that the geopolymer specimens are cured in the plastic bag (Figure 1.) at room temperature (30 ° C) as the repair work was generally performed without thermal curing.

Table 3. Mix Proportion of Geopolymer Mortar by Varying Sand Sizes

Mix Symbol	Ratio Na₂O.SiO₃/NaOH	Materials (kg/m³)			
		Sand passing sieve NO.100	Sand passing sieve NO.30	Sand passing sieve NO.50	Water
GM 32	0.9	33	539	100	35
GM 33	1.0	35	564	105	35
GM 34	1.1	35	564	105	35
GM 35	0.9	33	539	100	30
GM 36	1.0	35	564	105	30
GM 37	1.1	35	564	105	30

Figure 1. Curing of Geopolymer Mortar

Test Specimens

The compressive strength of the geopolymer mortar was determined with the cubic specimens of 50 x 50 x 50 mm. in accordance with ASTM C109 as can be seen in Figure 2 and Figure 3. The flexural strength test was performed with the prism specimens of 40 x 40 x 160 mm. as can be seen in Figure 4 and Figure 5. The shear strength test was performed by the slant shear test in accordance with ASTM C882. The specimens for slant shear test were cylinder shaped with the 76.2 mm. diameter and 142.2 mm. height as shown in Figure 6 and Figure 7. The initial and final setting time were performed by the penetration resistance test in accordance with ASTM C403. The chloride penetration test was determined with cylindrical specimens with 60 mm. in diameter and 100 mm.in height. The specimens were soaked in 3% by weight chloride solution for 7, 14 and 28 days. Then, the specimens were cut and asprayed with the 0.1N silver nitrate solution. The chloride penetration depth was measured by the white precipitation of silver chloride[4] as can be seen in Figure 8. In addition, to compare the engineering properties of the geopolymer mortar with the commercial repair materials, the properties of the 2 commercials repair materials were tested.

Figure 2. Specimens Preparation for Compressive Strength Test

Figure 3. Compressive Strength Test

Figure 4. Specimens Preparation for Flexural Strength Test.

Figure 5. Flexural Strength Test.

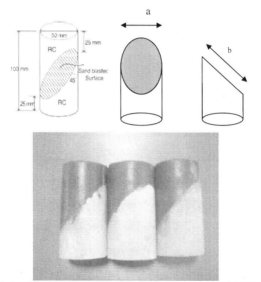

Figure 6. Specimen for Slant Shear Test in Accordance with ASTM C882.

Figure 7. Slant Shear Test

Figure 8. Chloride Penetration Depth Test.

RESULTS AND DISCUSSION

Figure 9 shows the compressive strength of the geopolymer mortars (the 3 mixes that had the highest compressive strength) compared with the compressive strength of the commercial repair mortars. As shown, the compressive strength of the geopolymer mortar (all 3 mixes) increased as age increased. Compared with commercial repair mortars, the compressive strength of the geopolymer mortar was still lower than that of the commercial repair mortar.

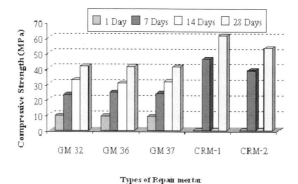

Figure 9. Compressive Strength.

Figure 10 shows the flexural strength. As can be seen, the flexural strength of the geopolymer mortar was lower than that of the commercial repair mortar at the age of 1 to 14 days. At 28 days age, the flexural strength of the geopolymer mortar was close to that of commercial repair mortar type 2; however, it was still lower than that of the commercial repair mortar type 1.

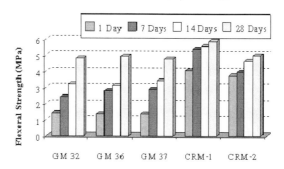

Figure 10. Flexural Strength.

Figure 11 shows the bond strength between repair material and substrate tested by slant shear test. Similar to the result of flexural strength, the bond strength between geopolymer mortar and substrate was lower than that of commercial repair mortar and substrate at the age of 1 to 14 days. At 28 days age, the bond strength of the geopolymer mortar and substrate was close to that of the commercial repair mortar type 2 and substrate; however, it was still lower than that of commercial repair mortar type 1 and substrate.

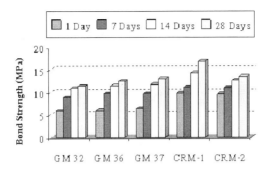

Figure 11. Bond Strength between Repair Mortar and Substrate.

Figure 12 shows the initial and final setting times of the repair mortar. As shown, both the initial and final setting times of the geopolymer mortar were shorter than that of the commercial repair mortar.

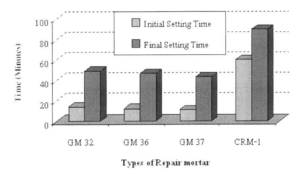

Figure 12. Setting Time.

Figure 13 shows the chloride penetration depth of the geopolymer mortar compared with the commercial repair mortar and conventional cement mortar. As shown, the chloride penetration depth of all mixes of the goepolymer mortar was higher than that of the commercial repair mortar and conventional cement mortar at all ages. This implied that the use of the geopolymer mortar as the repair material still was not effective enough in term of the durability. Further research work should be conducted to develop this property of the geopolymer.

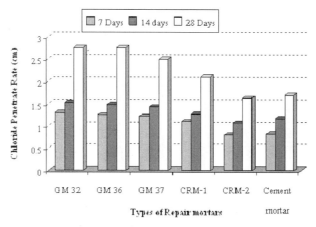

Figure 13. Chloride Penetration Depth

CONCLUSIONS

There is a potential to develop the geopolymer mortar as a repair material. However, further research is necessary to find ways to overcome low early strength and increase the chloride penetration resistance. It is possible to produce more than 40 MPa compressive strength of geopolymer mortar which is enough for using as the repair mortar without the heat curing. The flexural strength and bond strength between geopolymer mortar and substrate are close to the commercial repair mortar at the age of 28 days. However, the setting time of the geopolymer morar is still longer than that of the commercial repair mortar.

ACKNOWLEDGEMENT

The authors would like to acknowledge the Thailand Research Fund (TRF.) and the Siam City Cement Public Company Limited for financial support. The work described is a part of the research grant No. RDG 5150007.

REFERENCES

[1]G. Nounnu and Z.U.H. Chaudhary, Reinforced Concrete Repair in Beams, *Construction and Building Materials*, **13**, 195-212 (1999)

[2]N. Mailvaganam and L. Mitchell, Repairs to Restore Serviceability in Concrete Structures, *Construction Technology* Update, **59**, Institute for Research in Construction-National Research Council of Canada, Ottawa, Canada

[3]P.K. Mehta, Concrete Technology for Sustainable Development, *Concrete International*, **21**, 47-52 (1999)

[4]N. Otsuki, S. Nagataki and K Nakashita, Evaluation of AgNO₃ Solution Spray Method for Measurement of Chloride Penetration into Hardened Cementitious Matrix Materials, *ACI Materials Journal*, **89**, 587-592 (1992)

Author Index

Author Index